John B. Bachelder

Popular Resorts, and How to Reach Them

Combining a Brief Description of the Principal Summer Retreats in the United

States and the Routes of travel leading to them

John B. Bachelder

Popular Resorts, and How to Reach Them
Combining a Brief Description of the Principal Summer Retreats in the United States and the Routes of travel leading to them

ISBN/EAN: 9783337144210

Printed in Europe, USA, Canada, Australia, Japan

Cover: Foto ©Andreas Hilbeck / pixelio.de

More available books at **www.hansebooks.com**

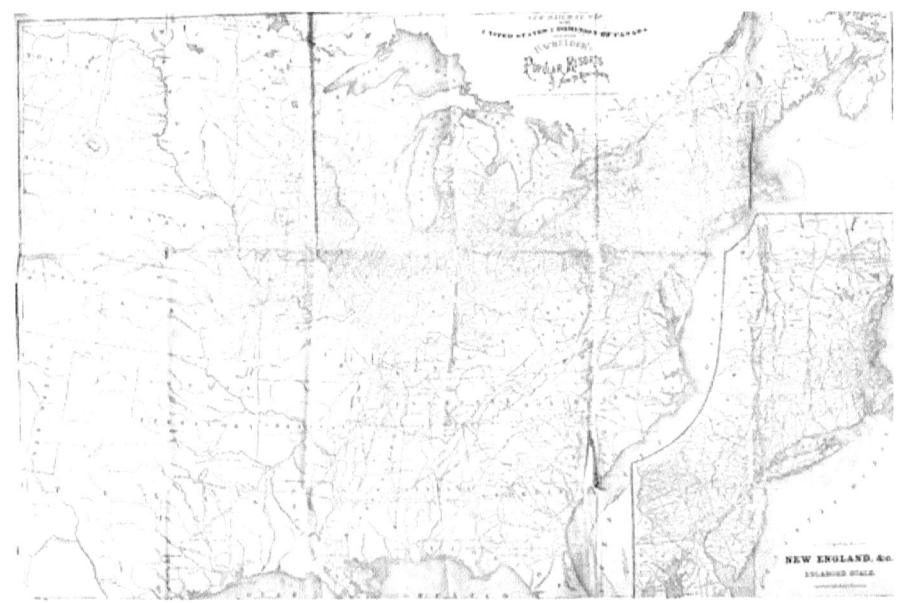

Popular Resorts,
AND HOW TO REACH THEM.

Engraved expressly for Bachelder's "Popular Resorts, and How to Reach Them."

POPULAR RESORTS

POPULAR RESORTS,

AND HOW TO REACH THEM.

COMBINING A BRIEF DESCRIPTION OF

THE PRINCIPAL SUMMER RETREATS IN THE UNITED STATES.

ROUTES OF TRAVEL LEADING TO THEM.

BY

JOHN B. BACHELDER,

Author of "*The Illustrated Tourists' Guide,*" "*Gettysburg, What to see, and How to see it,*"
"*The Isometrical Drawing of the Gettysburg Battlefield,*" "*Descriptive Key to the
Painting of Longstreet's Assault at Gettysburg,*" *Designer of the
Historical Paintings of the Battle of Gettysburg,
Last Hours of Lincoln, &c.*

Illustrated by One Hundred and Fifty=Two Wood=Cuts
BY THE BEST ENGRAVERS,
MANY OF THEM FROM ORIGINAL SKETCHES
BY THE AUTHOR.

BOSTON:
JOHN B. BACHELDER, PUBLISHER,
41-45 FRANKLIN STREET.
(At Lee & Shepard's.)
1875.

PREFACE TO THIRD EDITION.

Each year adds to the popularity of summer travel. The *vacation fever* returns annually with "the season," and custom demands that every well-to-do family prepare for it. No class of society is exempt. The mechanic and merchant, the banker and clerk, the student and professional man, are alike affected by its seductive influences, and, in the pleasure it brings, seek that respite from the cares of life which exhausted nature requires.

To know *how* to travel is a matter of great importance to the tourist. Many persons pass unheeding by the picturesque beauties of a pleasant route, expecting to find awaiting them at the end of their journey, the combined pleasures which others have described. In some instances the points to be visited are places of celebrity, but the route of approach lies through an uninteresting region. This is the case with many of our ocean watering-places. In others, as much pleasure may be derived *en route* as from the resort itself. This is particularly true of mountain travel, where every turn opens up new and interesting scenes.

Where a single excursion is to be made for the season, it is a matter of importance for the tourist to select an objective point and a route of travel, the peculiarities of which are congenial to his tastes. Public resorts which may furnish abundant sources of pleasure to one person frequently present little of interest to another. While one would be satisfied with a single day at the seashore, another would never tire of watching the waves break upon a rock-bound coast. The waving forests, the rugged grandeur of the mountains, and the deepening mystery of the glens, which to many prove sources of great delight, for others have no attractions. Hence the importance of carefully choosing desirable "Popular Resorts, and Routes to Reach Them."

Of all the celebrated watering-places on the coast, the famous springs, or the frequented mountain-houses, no two are alike, yet each locality possesses an individual interest, and finds its patrons; and the routes which lead to them have their attractions, either in the safety and comforts afforded the traveller, or in the picturesque beauty of the region through which they pass. It is to lay this subject clearly before the public that these pages with their illustrations are presented; thus enabling the tourist to choose in advance the character of scene he would visit.

The favorable reception of the two former editions of this work, and the universal request of patrons for its continuance, have determined the author to publish *annually* a volume devoted exclusively to the interests of travel, which shall give not only general information regarding the "Popular Resorts, and How to Reach Them," but furnish a standard medium through which proprietors or agents, who represent houses or routes, may describe or illustrate the merits of their respective interests. The illustrated routes have been generally prepared from sketches and notes taken on the spot by the compiler of this volume; while the "item" notices have either been written expressly for it, or compiled from the best published accounts.

It should be distinctly understood that this volume *is* not, and *does* not claim to be, a *Guide-Book*. It is rather a GAZETTEER OF PLEASURE TRAVEL; and, although it may give much useful information, it is not intended to take the place of a guide. For the details of travel, the tourist is advised to secure a current number of the TRAVELLER'S OFFICIAL GUIDE, — or "Official Time Tables," an abbreviation of it, if a more compact book is desired.

The author here takes occasion to thank those who responded to his circular invitation for data. Brief descriptions of other resorts are requested for the next edition, the compilation of which has already been commenced.

In view of the *Centennial Anniversary* at Philadelphia in 1876, special attention will be given to illustrated pleasure routes leading from every section of the country to that city, and the immediate attention of persons interested in such routes is invited.

THE AUTHOR.

CONTENTS.

	PAGE.
Summer Recreation	13
Pedestrianism	13
Equestrianism	15
Wagon Riding	17
Camping out	18
Harbor and Coastwise Excursions	21
Hints to Tourists	23
Boston	25
Mountains	26
New Hampshire Mountains	27
White Mountains	28
Routes of Approach	31
Franconia Mountains	32
PLEASURE ROUTE No. 1.	33
Boston, Concord, and Montreal Railroad	33
Stage Route from Plymouth to the Franconia Mountains	49
PLEASURE ROUTE No. 2.	63
Passumpsic and South-Eastern Railroad	63
PLEASURE ROUTE No. 3.	66
Eastern Railroad	66
To the Interior	70
PLEASURE ROUTE No. 4.	73
Boston and Maine Railroad	73
Portland and Vicinity	75
Portland, Bangor, and Machias Steamboat Company	78
Mount Desert	78
Up the Penobscot	79
PLEASURE ROUTE No. 5.	80
Portland and Ogdensburg Railway	80
North Conway	85

	PAGE.
PLEASURE ROUTE No. 6.	92
Grand Trunk Railway	92
PLEASURE ROUTE No. 7.	94
Maine Central Railway	94
Eastern Provinces	96
PLEASURE ROUTE No. 8.	97
Old Colony Railroad	97
Old Colony Steamboat Company	105
PLEASURE ROUTE No. 9.	108
Excursion to Oak Bluffs and Katama	108
PLEASURE ROUTE No. 10.	119
Central Vermont Railroad	119
PLEASURE ROUTE No. 11.	126
Shore Line	126
PLEASURE ROUTE No. 12.	131
Stonington Line	131
Middle States	133
New York City	133
PLEASURE ROUTE No. 13.	134
Up the Hudson	134
Catskills. — Approach from Kingston Station	137
" " " Catskill "	139
PLEASURE ROUTE No. 14.	142
Central Railroad of New Jersey, North Pennsylvania, Albany and Susquehanna, Rensselaer and Saratoga, and New York and Canada Railroads	142
North Pennsylvania Railroad	142
Pennsylvania Scenery	144
Central Railroad of New Jersey	145
Delaware and Hudson Canal Company's Railroads	169
Saratoga Springs	175
Lake George	177
" Champlain	178
New York to Long Branch	182

	PAGE.
PLEASURE ROUTE No. 15.	193
New York Central and Hudson River Railroad	115
Black River Railroad	195
Rome, Watertown, and Ogdensburg Railroad	196
PLEASURE ROUTE No. 16.	204
Erie Railway	204
Atlantic and Great Western Railway	210
Philadelphia	211
PLEASURE ROUTE No. 17.	214
Philadelphia and Reading Railroad	214
Long Branch	231
PLEASURE ROUTE No. 18.	232
Cape May	232
PLEASURE ROUTE No. 19.	233
Philadelphia, Wilmington, and Baltimore Railroad	233
PLEASURE ROUTE No. 20.	239
Fortress Monroe	239
PLEASURE ROUTE No. 21.	242
Pennsylvania Railway	242
Branch Roads	254
PLEASURE ROUTE No. 22.	255
Baltimore and Potomac and Northern Central Railways	256
Southern States	268
Virginia	268
Richmond	270
PLEASURE ROUTE No. 23.	272
Chesapeake and Ohio Railroad	272
Passage of the Alleghany	279
PLEASURE ROUTE No. 24.	289
Atlantic Coast Line	289
PLEASURE ROUTE No. 25.	294
Piedmont Air Line	294

	PAGE.
PLEASURE ROUTE No. 26.	301
Great Southern Mail Route	301
PLEASURE ROUTE No. 27.	308
Louisville and Great Southern Route	308
Mammoth Cave	309
Western States	314
PLEASURE ROUTE No. 28.	315
Grand Rapids and Indiana Railroad	315
Beyond Chicago	318
PLEASURE ROUTE No. 29.	320
Chicago and North-Western Railroad	320
Wisconsin	320
Chicago and St. Paul Line	332
Minnesota	334
PLEASURE ROUTE No. 30.	337
Chicago, Burlington, and Quincy Railroad . . .	337
PLEASURE ROUTE No. 31.	340
Union and Central Pacific Railroads	340
Dakota	340
Colorado	340
Montana	341
Utah	344
Idaho	344
California	345

List of Illustrations.

	PAGE.
FRONTISPIECE, Popular Resorts	2
Trout Fishing	14
Out-of-Door Life	16
Camp Stove	17
Camping Out	19
Pleasure (?) Travel in the Olden Time	22
Lake and Mountain Scenery	26
White Mountains	29
Lowell and Nashua Depot, Boston	34
Hooksett Falls, N.H.	36
Tilton, N.H.	38
Laconia, N.H.	40
View near Lake Village, N.H	41
Weir's Landing, Lake Winnepesaukee, N.H.	42
Steamer "Lady of the Lake"	44
Ragged Mountain, and Long Pond	45
Plymouth, N.H.	46
Pemigewasset House	47
Livermore Falls, N.H.	48
Owl's Head and Moosilauke Mount, N.H.	50
Littleton, N.H.	52
Lancaster, N.H.	54
Bethlehem Station, N.H.	56
Maplewood Hotel, N.H.	57
Twin Mountain House, N.H.	58
Fabyan House, N.H.	60
Ammonoosuc Falls, N.H.	61
Mount Washington Railway, N.H.	62
Boar's Head, N.H.	68
Portland, Me.	74
Falmouth House, Me.	75
United States Hotel, Me.	76
White Head Cliff, Portland Harbor	77
Portland Light	79
Sebago Lake	81

LIST OF ILLUSTRATIONS.

Conway Elms	85
Mt. Kiarsarge, N.H.	86
Diana's Baths, N.H.	87
Silver Cascade, N.H.	89
White Mountain Notch	90
Old Colony Railroad	97
Boat-House Landing, Newport, R.I.	98
Coast Scene	100
Trout Pond	101
Nantucket Wharf	102
Newport, R.I.	104
Saloon, Old Colony Steamer	106
Steamer "Bristol," passing East-River Bridge, N.Y.	107
Steamboat "Martha's Vineyard," passing Oak Bluffs	108
Riding Out the Storm	110
Sea-View House, Mass.	111
Seaside Cottage	113
Mattakeset Lodge, Mass.	114
Yachting	115
Sea-View Boulevard, Mass.	117
Providence Depot, Boston	126
Grand Central Depot, New York	129
Harbor Scene, New York	131
Fisherman's Cottage	141
Lehigh Valley, Mauch Chunk, Penn.	144
Coal Vein	146
Lehigh Gap	147
Mauch Chunk, Penn.	148
Mansion House, Mauch Chunk, Penn.	149
View from the Mansion House, Mauch Chunk, Penn.	150
The Flagstaff	151
Mount Pisgah Plane, Penn.	152
Onoko Station, Penn.	153
Coal Breaker	156
Lehigh Valley, Penn.	157
Prospect Rock, Penn.	159
Cloud Point, Penn.	160
Glen Thomas, Penn.	161
Solomon's Gap, Penn.	162
Lackawanna Valley House, Penn.	164
Nayang Falls, Scranton Gorge, Penn.	165
Switchback Railroad, Moosic Highlands, Penn.	166
Jones Lake	167
Wyoming House, Penn.	168
Cave House, N.Y.	172
Summer Life at North Mountain House, Penn.	183
Wild Woods	185
Ganoga Falls, Penn.	187
Mountain Stream	188
North Mountain View	190

LIST OF ILLUSTRATIONS.

Forest Life	192
Thousand Island House, N.Y.	199
Moonlight	203
Falls Village Bridge, Penn.	212
Columbia Bridge, Penn.	214
Schuylkill River, Penn.	215
Valley Forge, Penn.	216
Schuylkill River, above Pottstown, Penn.	217
Mount Carbon, Penn.	220
Little Schuylkill River, Penn.	221
Mahanoy Plane, Penn.	222
Brookside, Penn.	223
Herndon, Susquehanna River, Penn.	224
Mainville Water Gap, Penn.	227
Catawissa, Penn.	228
Marine View, — Coal Transport	230
Strawberry Culture, Del.	233
Peach Gathering, Del.	234
Cristfield, Md. : Oyster Shipment, Del.	235
Ridley Station, Md.	236
Residence of F. O. C. Darley, Esq., Md.	237
Mount Ararat, Md.	238
Hygeia Hotel, Va.	240
Marine View	241
Coatesville Bridge, Penn.	242
Connecting Railroad Bridge, Penn.	243
Bryn Mawr, Penn.	244
International Exhibition Building, Penn.	245
Fairmount Park, Penn.	246
Hestonville, Penn.	247
Ardmore Station, Penn.	249
Juniata River, Penn.	251
Logan House, Penn.	252
Mountain House, Penn.	253
Washington, D.C.	255
Baltimore Tunnel, Md.	256
Night Train	257
Susquehanna River	258
Harrisburg, Penn.	259
Renova House, Penn.	260
Dutchman's Run, Penn.	261
Empire Fall, N.Y.	263
Rainbow Falls, N.Y.	264
Hector Falls, N.Y.	265
Watkins Glen, N.Y.	266
Seneca Lake, N.Y.	267
University of Virginia	269
Earthworks on the Chickahominy, Va.	270
Commissary Department	272
Mountain Tunnel, Va.	275

Rockbridge Alum Springs, Va.	276
Griffith's Knob, Cowpasture River, Va.	277
Falling Spring Falls, Va.	278
Greenbrier White Sulphur Springs, W. Va.	280
Start down the Greenbrier, W. Va.	282
Richmond Falls, New River, W. Va.	283
Whitcomb's Bowlder, Va.	284
Running New River Rapids, W. Va.	285
Miller's Ferry, Va.	286
Charleston, W. Va.	287
Huntington, Ohio	288
Storm at Sea	300
Green Lake, Wis.	327
Pleasant Valley	339
Night Express	338
Giant Geyser, Mon.	342
Crystal Cascade, Mon.	343

SUMMER RECREATION.

The custom of setting apart a few weeks or months of the year as a respite from labor is fast gaining popularity; and each season adds to the number of those who leave their daily cares behind, and seek rest and recuperation for mind and body among the hills and deep green woods of the country, or at the sea-shore, bathing in surf or sunlight, and cooled by the invigorating breezes of the sea.

Those whom fortune has favored can devote the season to travel, visiting in succession the rare natural wonders with which the country is stored. Such have only to select the points of interest, and the most pleasing routes by which they may be visited. A far larger number, however, choose some desirable and healthful locality where they may secure the desired change and rest, at cheaper rates even than they could remain at home. Both classes will find in this book abundant directions for their guidance. But the following chapter is devoted to another class, who, from economy or adventure, choose more freedom in their movements. I refer to the pedestrian, equestrian, and camping-out party.

There is a certain age when young men glory in pedestrianism, and see in it a source of great pleasure. A few years later the same parties will prefer a horse to facilitate their movements; and, later yet, a carriage will be required to complete their happiness. As our army of young men is constantly recruited from the ranks of the *home-guard* of boys, and as the "wheels of time" as surely graduate these youth into manhood, it is proposed to treat briefly each of these sources of recreation.

PEDESTRIANISM.

When the place for "camping-out" has been determined on, the mode of travel will come up for discussion: this, of course, will be largely determined by the place chosen for camp, and the length of time it is to be occupied. If it is to be permanent, it matters little how the place is reached. But there is another manner of camping-out, combining with it pedestrianism or equestrianism, by which the advantages of tour and camp are combined; and this, when the party are physically able to endure it, will be found a source of great enjoyment, particularly if an interesting country be selected. Eight or ten miles per day, at early morning and late evening, can easily be made, which will sum up quite a trip during the season. There is a romantic novelty connected with an excursion of this kind, which commends it to the adventurous. Much of the enjoyment, however, will depend upon the similarity of tastes, and physical endurance of the party. The "best fellow in the world," socially, would soon become unendurable if he "broke down" every day on the march.

TROUT FISHING.

How delightful the sport in early morn, when the clear air resounds with the songster's happy note, and a roseate hue tips the mountain top; or at quiet eve, when the last rich golden rays of the sun struggle to pierce the overhanging boughs, — to cast the alluring fly, and land the gamey trout! And what a tempting dish it forms, caught and cooked by your own hands, seasoned by a good appetite and the excitement of adventure!

Pedestrianism determined on, it is of the greatest importance to reduce the stock of clothing and equipment to actual necessities, for "every ounce becomes a pound" at the end of a long jaunt. A frequent error of the novice in tramping tours, is to choose new and elaborately equipped knapsacks, heavy rifles or fowling-pieces, with patent ammunition and fishing-tackle, the accumulation of which soon becomes burdensome, and the pleasure of the excursion is spoiled. Fortunately nearly every one can get valuable hints on the subject of out-of-door life from men of army experience. Strong shoes and clothing are important. A change of socks and underclothes, and a rubber and woollen blanket, will be required. The three sides of a shelter tent, divided among a party of three, can be taken. A wire bread-toaster for broiling game or meat, a coffee-pot, with tin plates and drinking-cups, knives and forks, a hatchet, and pocket-compass, complete the outfit; and all weigh but a few pounds. Select a light game-bag or haversack, in place of the heavier knapsack. It is not only lighter, but can be carried with more freedom. A cape made of oiled silk, or glazed muslin, reaching to the hands, will be found very serviceable. It is light, and, when not in use, takes but little more space than a handkerchief, and is a complete protection against showers by day, or dampness at night. Milk, bread, &c., can always be purchased from the farmers; coffee and crackers at the stores. Unless hunting be intended as a specialty, rifles or fowling-pieces should be left at home. A Smith and Wesson sixteen-inch pistol with a detached stock weighs but a few ounces, and for ordinary tours will answer all requirements. Fish hooks and lines should be taken, but a rod can be improvised for the occasion. A valise containing additional changes of clothing can be forwarded from point to point. The expense of an excursion of this kind will be found to be surprisingly small.

EQUESTRIANISM.

In many respects the comforts and pleasures of an excursion will be increased if the party are mounted. Uninteresting sections can be quickly passed, and additional clothing and equipments can be taken. With a coil of rope to picket the horse, he will secure his own living, with the addition of oats purchased by the way. The camp should be pitched where cedar or hemlock boughs for beds can be had. Another very popular plan with pedestrians is to hire a horse and wagon to draw the baggage, cooking utensils, tent, blankets, clothing, &c.; and a larger supply can then be taken. Each of the party may become driver in turn. It also saves delay in case of the illness of one of their number, who may still be able to play the part of driver. This plan possesses many advantages, and is worthy of consideration.

OUT-OF-DOOR LIFE.
On the line of the Erie Railway, near Rock City, N.Y.

WAGON RIDING.

There is still another mode of travel for summer tourists which combines the pleasures of those already described, and possesses many additional advantages; namely, a good roomy covered wagon with curtains to be rolled up at the sides, in which all can ride, take their cooking utensils, tent, valises, guns, fishing-tackle, sketching-materials, &c. A good pair of horses, and an experienced cook for driver and "man of all work," will complete the outfit. Such a party can be always at home, can camp for a few days on the banks of some beautiful stream or lake, or where the scenery, fishing, hunting, or berries invite them to tarry. Last summer the writer met such a party from Wilkes Barre, encamped on the shores of Highland Lake, on the summit of North Mountain, Pennsylvania (see illustrated description of North Mountain), which had improved even upon this plan. In addition to tent and equipments, they had a light boat in their wagon, which could be readily launched. It might be slung beneath, and become the receptacle of small packages. This party were dressed in a unique uniform of blue flannel. They remained encamped several days at North Mountain, where the fishing is good, with fine shooting in the neighborhood; the ladies, meanwhile, dividing their time between their tent and the North-Mountain House. The most convenient and ornamental tent which we have seen for camping-out parties, or for the lawn, is manufactured by Gale & Co., 15 and 16 Faneuil Hall Square, Boston. It is pentagonal, and opens like an umbrella; is covered with striped canvas; the walls can be wholly or partially removed at pleasure. It has a folding centre-table, if desired, and is made of three sizes, accommodating from six to a dozen persons; yet it can be pitched or struck in ten minutes, and weighs but fifty pounds. A valuable adjunct to this tent is the lately invented *camp-store* sold by H. L. Duncklee, 87 Blackstone Street, Boston, which for its compactness, and the many conveniences it combines, should have been named *multum in parvo*.

CAMP-STOVE (OPEN).

CAMP-STOVE (CLOSED).

The accompanying illustrations represent the stove and furniture, both open, and packed ready for transportation.

In addition to the stove and detached oven, with a capacity for baking a turkey or fifteen pounds of beef, is an eight-quart kettle, six-quart tea-kettle, two-quart coffee-pot, fry-pan, two square and one round pans, a dipper, gridiron, tent-collar, and eight feet of telescope funnel. The ware is so constructed that it *nests*, and packs in the oven. The oven and funnel pack inside the stove, still leaving room for a half-dozen plates, cups, knives, and forks, &c. The stove and furniture complete weighs but twenty-seven pounds, and may be purchased for fifteen dollars. With such a tent and cooking apparatus how cosily a camping-out party may live, either at the mountains or sea-shore! Such a scene is represented on the shores of "Jones Lake." (See Index.)

There is still another species of camping-out which should be mentioned. I refer to the real camp of the explorer, the hunter, or adventurer. There is, extending from Canada to Mexico, a border of wild partially explored country, which affords a field rich in adventure for all who choose to visit it. There the camp is a real necessity, the requirements of which it behooves all who propose such an excursion, to study thoroughly before embarking. There are so many interesting localities on our frontier to visit that one can hardly go amiss; perhaps, however, there are none more attractive than the Yellowstone region of the North-west, the great "National Park," as it has been aptly termed; or the new State of Colorado with its "Garden of the Gods," its mighty cañons, &c. Here will be found in perfection those elements of grandeur in outline, and sublimity in effect, which please the eye, and gratify the senses.

The comparatively small number that choose these distant fields of adventure will usually select their grounds from information outside this volume, while the thousands who go into camp nearer home may consult its pages. To such, a few additional suggestions may be in place.

The seacoast and lake shores present unusual facilities for this mode of summer recreation, inasmuch as the lines of railway which usually skirt the water's edge afford convenient and cheap transportation and means of access. It is no new feature that the writer seeks to introduce into summer amusements, for thousands practise it every season; but it is the desire to show those who have never tried it how simple and enjoyable this recreation is.

During the "season" of 1874, while visiting summer resorts in different sections of the country, we saw camping-out parties at and in the vicinity of a large number of them. On the shores and islands of Lake Winnepesaukee several parties encamped for the season; and both there and at Lake George they remained until the autumn frost had tinged the foliage.

And so at numerous other pleasant localities, and along the routes leading to them, parties were seen who had chosen this mode of recreation.

Beyond all question, the most delightful and healthful way to spend one's summer vacation is in "camping out," provided the weather is reasonably pleasant. A time of storm is gloomy enough, whatever the mode chosen for enjoyment.

CAMPING OUT

In "camping out," all the stiff formalities of conventional life are put aside. The body is left free for any sort of dress except fashionable styles; and the mind is in constant and cheery repose, and therefore able to enjoy life with the keenest zest. Health comes to the invalid, with its building-up force of a sharp and eager appetite; and the strong feel an electric energy, daily renewed, unknown in great cities and marts of trade. In fact, while the visitor to thronged summer-resorts often returns home worn and wearied, the sojourner of the camp comes back increased in his avoirdupois, his strength, and his sense of having had a "glorious" vacation.

"Camping out" means a sort of woodman's or frontier life. It means living in a tent; sleeping on boughs or leaves; cooking your own meals; washing your own dishes, and clothes perhaps; getting up your own fuel; making your own fire; and foraging for your own provender. It means activity, variety, novelty, and fun alive; and the more you have of it, the more you like it; and the longer you stay, the less willing you are to give it up. In fact, there is no glory for the summer tourist, to compare with the "camping-out" glory.

For preparation, you will first know where your camp is to be, and what it affords for your pleasure. And you will scarcely make your party less than three, nor more than five. If the number exceeds five, it will be better to pitch two distinct camps at some distance apart, and thus have pleasant "neighbors" to visit, and hospitable parties to give, each to the other. Guns and fishing tackle carefully prepared for use will, of course, be required for localities where game and fish abound; and few places would be selected where one or the other, at least, would not be accessible. Two grand essentials should be thoughtfully remembered, — plenty of dish-cloths, and a good hatchet. A good blanket, rough clothes, strong shoes, and a convenient knapsack, are absolute essentials; but don't burden yourself with needless things. In fact, while nothing is needed in the way of choice cravats and white kids, there should be careful regard to the little things you will need but cannot buy in the woods, even to a stout-bladed jack-knife.

It is impossible, and useless to attempt, to describe particular spots, which would tempt a "camping-out" party to prefer. They are numbered by thousands. If you would have large game on land, and salmon in the waters, a location must be chosen in the more wild and rugged regions of our remote borders; and in the right season, — say, in early May. Should a more quiet and subdued locality be preferred, you may push for the mountain sides and slopes of Pennsylvania or of Virginia. No more attractive beauties of nature invite the tourists of our land, than await those who may seek the elevated portions of "Old Virginny."

Prince Edward's Island is also unsurpassed in natural charms, in healthfulness, in its sources for camping-out pleasures, and the broad hospitality of its rural population. In due time, a great summer pilgrimage will set towards that garden of the sea.

Nova Scotia abounds in novelties to our own people, and in its fine lakes, filled with the most eager and gamey of trout.

The solitudes of New Brunswick, so strangely overlooked by travellers and writers, possess some of the loveliest as well as the grandest and most romantic attractions to be found on any portion of our continent. This picturesque region also will soon, no doubt, be opened by pleasure hunters and the writers of many books.

Northern Maine about Moosehead and the Rangeley Lakes, the hills and streams of Vermont, a great and grand region lying between the White Mountains and Canada, a lovely land around the head waters of the Connecticut River, not forgetting Mount Desert nor the Adirondacks, — these are some of the leading areas of our Northern climate, where camping out may be enjoyed, in all its delicious and inspiring fulness.

This is all that space will permit in this work, for remarks upon "camping out." As yet it is only here and there that the camping-out party is to be found, in the warm months. The attention of vacation takers has not been turned to this best of all modes of seeking one's comfort and ease, to the degree required to make it popular and general. But it cannot be commended too earnestly, nor pressed too persistently upon public notice. It is not absolutely required that the party camping out shall locate in a place remote from all civilization. On the contrary, a vast number of our more popular summer resorts and towns offer most inviting spots for a camp (see cut) to which the belles would delight to ramble, and where primitive hospitality can be liberally dispensed, even when young bucks of fashion may be compelled to act the parts of Bridget the cook, and Mary the maid of all work. It is a matter of surprise that this charming way to diversify the individual and family trip is not more generally remembered and practised. Let the reader, as he decides whither his summer flight shall be, ponder well this idea of "Camping Out."

Harbor and Coastwise Excursions. — One of the most delightful yet economical sources of summer enjoyment is the harbor excursion. From each of our great maritime cities, boats conveniently arranged leave daily for some of the most popular resorts. From Washington they sail down the Potomac to Mount Vernon; from Baltimore to the beautiful water retreats in the vicinity; from New York up the North and East Rivers, to Staten Island, and Harlem; from Boston to Nantasket Beach, Long Island, Gloucester, and Nahant. Indeed, every large town with a harbor front has its pleasant resorts; and the stranger has only to look in the daily papers for particulars.

Again: if the tourist would consult comfort and economy, if he would take sleep and rest while passing familiar or uninteresting sections of the country, he can frequently give diversity to his travels by an occasional trip on a coastwise steamer. In going east from Boston, the daily steamboat line to Portland, the "Star of the East" up the Kennebeck, "Sanford's Independent Line" up the Penobscot, the "Inside Line" from Portland to Mount Desert and Bangor, and the Halifax boats, are all first-class, and deserving the notice of tourists. The New York boats are described elsewhere.

PLEASURE (?) TRAVEL IN THE OLDEN TIME.

HINTS TO TOURISTS.

It is a matter of no little anxiety to the devotee of summer pleasures, or the seeker after new wonders, to choose the field for his examination which shall yield the richest harvest of pleasure. In days gone by, it was largely the custom, as a matter of course, to visit those popular localities of most convenient access; and so it happened that each succeeding year found the same familiar faces returned to the haunts of past enjoyments. But, with the remarkable improvement in railroad and steamboat travel, new resorts have been opened, and fresh wonders present their claims for examination. This has induced a new feature in summer travel. Experienced tourists no longer choose the shortest line to an objective point, regardless of the scenery through which it lies; but, by judicious selection, with slight detour, they can embody such routes as lie through new or pleasant places, which can generally be done with trifling addition to the expense. The stranger will find much assistance in the selection of desirable routes and localities, by an examination of these pages, which, in the present instance, have covered largely the older and better-known routes of travel; but much remains to be done. The task set for this work is not completed by dwelling only upon the well-known and much-visited "resorts" which have secured popular favor, and are of ready and easy access. The perspective of other and less-regarded charms of Nature comes into our future picture. As population advances to its hundred millions, pushing the wave of frontier invasion on, and still on, until it shall break at last upon shores of the remotest sea, new objects of delight, new charms and beauties and wonders of creation, will be embraced in the "popular resorts" of this vast empire. It is not only the places of gayety and fashion and luxury, which appeal to summer migration. There are the zones of soft and bland climates, equable in temperature, and pure of air and water. There are the pharmacopœias of Nature's healing springs, formed from her own great recipes. And there are other regions, on mountains or in valleys, to all of which the sufferers from bodily ills will finally resort in great multitudes in the years now coming.

As yet, the stream of travel during the heated term may be said to move slowly toward the States lying upon, or contiguous to, the great lakes and rivers of our **North-west**. And yet Michigan, Wisconsin, Minnesota, Indiana, Iowa, and Illinois constitute a region replete with every degree of picturesque variety, excepting the wild grandeur of lofty mountain ranges. For invalids with debilitated systems and low vitality, especially consumptives, Michigan, Wisconsin, Minnesota, and Iowa are now esteemed by many as decidedly preferable, and more promising of cure than the soft and palliating atmosphere of Florida. In Michigan, the island of Mackinaw, in the straits of that name, rising in a line two hundred feet above the water; and the wonderful "Pictured Rocks," extending twelve miles, and having an elevation of three hundred feet from the water's edge, — are famous already. Wisconsin abounds in remarkable evidences of having been once inhabited by a now extinct race. One of these is said to be in the form of a recumbent man, a hundred and twenty feet long, and thirty feet across the body. (?) Another is in the shape of a huge turtle. There are falls, gorges, and wildly torn rocks and hills, the rivals of the most famous in our land.

Minnesota even surpasses some of our most favorite regions of charming scenery. Save great cataracts and lofty mountains, the State is of rarest wealth in natural glories; and no less in stupendous and yet not fully explored caves. The widely famed *St. Anthony's Falls* are in Wisconsin. And what is generally said just here of these three States applies no less to the sister States of that region. The traveller who would seek some new region for recreation, adventure, or health, may well make his summer campaign in the pleasant lands of the North-west.

Colorado is rapidly coming into popular consideration. Twenty years ago it was scarcely mentioned in gazetteers. Now it is a State of this Union. The advance of Colorado into public notice, as a place for emigrants to seek, had its origin in the California fever, and greed for gold. But the rapid and magical changes of a few years have revealed to our people a region of indescribable grandeur. Here every thing is laid out upon Nature's most gigantic scale. Rocks, piercing the clouds, rear their summits from cañons and vast gulfs and gashes that fill the soul with speechless awe and delight. All that imagination could conceive, of the stupendous, awful, and sublime, can be profusely studied in our new sister State of Colorado. Such a glorious panorama cannot remain unappreciated among such a travelling and voyaging people as ours. The Territory of **Montana** is another field of Nature's wonders, though hardly opened to the tourist. But **California** is a world by itself, rich in every variety of scene. Here Nature displays her grandest moods. To speak of **Florida** is to re-state what is now well known among all classes. The remarkably equable climate, at all seasons of the year to be found in certain portions of Florida, so novel and delightful to strangers, has invited a constantly increasing tide of visitors to that State, especially since the war. It is pronounced the most healthful section of our Continent. The modes of living, the hospitality of its citizens, its flowers, birds, and varied fruits, — but not its alligators, — are sources of sweet and placid enjoyment, *sui generis* in Florida. Other wonders are constantly presenting themselves, and claim our attention as "Popular Resorts." We now turn to New England.

The City of Boston. — If we depart from the general plan of this work, to take special note of a great city, it is because Boston stands out in marked distinctiveness from every other city on the continent, — perhaps it should be said, from all others of the whole world. For a certain class of vacation tourists, Boston contains as much to study and enjoy as is found by other classes in the peaceful woods, the mountain sublimities, or the ocean's grand moods. The claim to pre-eminence among all our cities, for lavish profusion and unstinted generosity in all matters pertaining to moral, intellectual, and philanthropic progress, is conceded to Boston, without dispute. Nor are these characteristics spasmodic or ephemeral. From the earliest history of the Puritanical settlements, this distinction has marked the history of Massachusetts, with Boston as the chief and centre of its manifestations. Institutions of learning; of moral and Christian teaching; of broad and comprehensive philanthropy; of art; of æsthetic culture; of hygiene; of all which tends to refine, purify, and elevate the race, — are not merely found here, but are full of progressive vigor. It is the innumerable systems of these classes, which induce many summer tourists to dwell for a season in Boston. We shall not delay to particularize these; for they would require a book to detail them.

For tortuous and narrow streets, lanes, courts, and alleys, no city of equal size can or would compete with Boston. Its plan, if it can be called such, may have been original with wandering cows and sheep; but no other design could ever have devised it as originally built since the late fire, however, many of the streets have been widened and extended, thus bringing some regularity out of seeming chaos; and the general architecture has been greatly improved, and in many cases is rich and elegant. The contrast of costly edifices, side by side with tumble-down ricketiness, is not to be found. What remain, even, of the older buildings are rapidly giving way to new. Widening of streets is progressing at enormous cost; and the demand for business facilities finds ample wealth to meet it. The city proper may be pleasantly studied.

Rare, beautiful, and refreshing to the eye as is Boston Common, the pride of Bostonians, the suburbs are even more attractive and grateful. We doubt if there is a city in the world with such a clustering zone of half city, half-country, — half nature, half art, — as adorns the environs of Boston. The peninsula being so much absorbed by trade, the population is forced to "roost" outside. Here, then, wealth and refined taste are free to combine and adorn. The stranger needs no special directions. Any course will suffice for the start; and the net-work of interlaced steam, horse-car, and carriage roads will permit one to study the whole of the delicious panorama, before finishing the day. Or, taking one

MOUNTAINS.

Mountains, lakes, rivers, and sea-shore form the principal resorts of New England. The three former are so interspersed and connected, that descriptions of them naturally blend one into the other. The sea-coast, fringed with some of the best beaches in the country, is thickly dotted with summer watering-places. High up in the unseen glens of the mountains the principal rivers take their rise, sometimes gliding a thread of silver to the valleys below; again leaping boldly from crag to crag, in a series of foaming cascades and waterfalls.

The mountains of New England form a marked characteristic of the scenery of that picturesque region. Bold and rugged in outline, grand in effect, clothed in the blue mystery of distance, and swept by an invigorating atmosphere, they embody the characteristics of popularity, and form a highly attractive point of interest for tourists.

While New England is proud of the fame of her mountains, she is equally pleased with her hundreds of beautiful lakes, sparkling in the sunlight of nature; but in the boldness of her mountains, and the beauty of her lakes, her waterfalls are eclipsed. In magnitude they would be buried in the spray of the Yellowstone, or lost in the grandeur of Niagara; yet they give life and interest to her scenery, and add variety to the pleasures of guests. They are subject to little change. The constant roll of waters through countless ages has, with few exceptions, produced little effect upon the texture of the primitive rock upon which they fall. The same jagged angles which broke their waters centuries ago, meet them now, and dash them into fragments of spray. It is far better as it is. The pleasure of travel would lose half its interest if one locality combined the excellencies of Nature. Florida without her unrivalled climate, Colorado shorn of her sublime cañons, or New Jersey bereft of her magnificent sea-shores, would fail to attract that tide of travel which now invades their borders.

There are really no waterfalls in New England of sufficient magnitude to draw visitors by their own attractions; but, as auxiliaries, they add to the variety of scenic characteristics otherwise interesting.

NEW HAMPSHIRE MOUNTAINS.

The early histories of nations show that mountains have always been objects of awe and veneration. Of this we have abundant proof in writing, both sacred and profane. Pre-eminent among the mountains of New England stand the **White Hills** of New Hampshire. Indeed, save in altitude, the remark might hold good for the country. It is well known that the natives held the White Mountains in religious reverence. They called them "Agiochook" (Mountains of the Snowy Forehead, and House of the Great Spirit), always approaching them with the greatest deference; seldom venturing far up their sides. From the settlement of the country, they were a source of great interest. They were visited by Derby Field only twenty-two years after the landing of the Pilgrims, notwithstanding an unbroken forest intervened; and it is even claimed that a party of Englishmen visited them ten years previous to that date. The "Notch" by which travellers can pass through to the country beyond, and through which the route of the Portland and Ogdensburg Railroad lies, was discovered in 1772; since which, improvements have been pushed to meet the wants of visitors, until now the hotels of this region are held among the best in the country. Unless mountains are of great repute, like the White Mountains, they are seldom visited as a specialty; but their attractions more frequently form an auxiliary to some neighboring resort, whose enterprising proprietor improves the paths of approach, and furnishes conveyance. Hence it is safe to infer, that good hotel and transportation accommodations may be had in the vicinity of most of our prominent mountains whose attractions are advertised to the public.

Hilly and mountainous regions, like New England, New York, and the North-west, are generally interpersed with ponds and lakes, which give a pleasing variety to the landscape. Should the mountains take the form of ranges, however, as in Pennsylvania, the surface is drained through the valleys; and lakes, which are so common in some sections of the country, are in such regions almost entirely unknown. New Hampshire has been aptly termed the "Switzerland of America." Her granite hills of rough primeval rock rear their bald and stately peaks high above the surrounding plains.

The scanty though productive soil at their base, formed by washings from the disintegration of ages, supports dense forests of hardy trees, which, as you ascend, become dwarfed and twisted by the winds; yet, when mere Liliputian in size, their proportions remain. When the line of vegetation is passed, mosses and lichens alone clothe the nakedness of the rocks.

The Black Mountain of North Carolina is higher than the White Mountains, but it is difficult of access, and its surroundings lack the grand scenic effects with which the former are clothed. The Sierra Nevadas, though loftier in altitude and grander in effect, are as yet comparatively unknown; while the ease with which the Highlands of New Hampshire are reached will always assure their popularity in the estimation of the travelling public.

THE WHITE MOUNTAINS.

The White Mountains, geographically known, comprise a large portion of that part of New Hampshire lying north of Lake Winnepesaukee, embracing an area of more than five hundred square miles. Through this region are located many of the summer houses for which New Hampshire is famed. These are frequently a long distance apart: five, ten, fifteen, twenty, or even thirty miles may intervene, — which results in frequent annoyance to the stranger, who, having been preceded by friends, expects to meet them at their hotel as readily as he would at Newport, Long Branch, or Saratoga; whereas they may have approached from an opposite direction, and their hotel may be twenty miles away. At a distance, all of this section of the State is termed the White Mountains: with the inhabitants, different localities have local names by which they are known. If, for instance, a citizen of Concord, the capital of New Hampshire, should announce his intention to visit Mount Belknap, it would be understood that he would go to Guilford. If he would visit Red Hill, he would go to Centre Harbor; if

WHITE MOUNTAINS, N. H.
From Milan, N. H.

Mount Kearsarge, he would stop at Potter's Station, on the Northern Railroad. If Mount Kiarsarge were to be visited, he would go to North Conway; if Mount Chocorua, or Ossipee Mountains, he would stop at West Ossipee; if Moosilauke, he would go to Warren; if the Franconia Range, he would continue to Littleton, and thence by stage. And to many other mountains popular as resorts, known at a distance as a part of the White Mountains, his routes would be equally divergent: yet a visit to neither of these would take him to the White Mountains as understood by the citizens of the State. It is therefore advised, that the visitor to that region procure a good map, or, better yet, a copy of "Eastman's White-Mountain Guide," a most complete and reliable book. The White Mountains proper, of which Mount Washington forms the crowning centre, are approached by four great natural thoroughfares, or valleys, up which run superior carriage or rail ways, traversed by excellent coaches or elegant cars. To these valleys, from every direction converge the various lines of New England. By either road the tourist will be taken to some portion of the White Mountain region along a route replete with interest. No two are alike; the scenery differs decidedly on each; and, as much of the enjoyment of a tour depends upon the pleasure *en route*, it behooves the traveller to use care in his selection. By going one road, and returning on another, the pleasure may be increased. If the design of the tourist were an objective point, like Niagara Falls, the Mammoth Cave, Watkins Glen, Cape May, or Mauch Chunk, he might well select the most direct route, and save himself for the anticipated pleasures in store; but the White Mountain trip *will pay from the start*. The scene changes incessantly, and the whole excursion is a panorama of interesting views. The person who has no love for the beautiful in nature, or who fails to appreciate its charms, and expects to find the great source of pleasure in store at the end of his journey, will be quite likely to return *disappointed*.

The eleven great peaks which form the White Mountain group proper are, Mount Washington, with an altitude of six thousand two hundred and eighty-five feet; Adams, fifty-eight hundred; Jefferson, fifty-seven hundred; Madison, fifty-four hundred; Clay, fifty-four hundred; Monroe, fifty-four hundred; Franklin, forty-nine hundred; Pleasant, forty-eight hundred; Clinton, forty-two hundred; Jackson, forty-one hundred, and Webster, four thousand. Connecting and adjoining these are many others of nearly equal altitudes. These mountains are generally accessible. A bridle-path from the south-west extends from the Crawford House, near the White Mountain Notch, over Mounts Clinton, Pleasant, Franklin, and Monroe, to the summit of Mount Washington; and, following as it does the crest of the mountain range, it unfolds a panorama of the grandest views east of the Rocky Mountains. A carriage-way, commencing at the Glen House, has been constructed up the northeastern slope of Mount Washington; and on its western face a railroad connecting by a short turnpike with the Boston, Concord, and Montreal Railroad, has been built to the top, by which the fatigues of the ascent have been overcome. Ample hotel accommodations will be found, of a superior character, on the summit.

ROUTES OF APPROACH.

The main routes of approach to the White Mountains are up the four great valleys, through which flow the waters from this region, — the Merrimac and Pemigewasset, the Saco, Androscoggin, and Ammonoosuc, a tributary of the Connecticut. Through each of these valleys railroads have been constructed; some extending to the mountains, others continued by stages, while each connects with tributary roads leading from distant parts of the country. Visitors from Boston and vicinity have the choice of four routes. The Boston, Concord, and Montreal Railroad extends up the Merrimac, Pemigewasset, and Ammonoosuc Rivers to the base of Mount Washington. This is also the direct line to Bethlehem, Franconia Mountains, Plymouth, and Lake Winnepesaukee *viâ* Weir's. The Conway Branch of the Eastern Railroad conveys passengers to North Conway, where intersection is made with the "Portland and Ogdensburg," by which they continue to the mountains. Tourists can also visit Lake Winnepesaukee by the Wolfboro' Branch of this road.

Passengers from Boston and Portland, by the Boston and Maine Railroad, reach the White Mountains *viâ* the valley of the Cocheco to Lake Winnepesaukee; thence by a delightful steamboat sail to Wolfboro' and Centre Harbor, continuing by stage and rail through the valley of the Saco, as above.

Another route is to continue to Portland by the "Eastern," "Boston and Maine," or by steamer, and thence by the Portland and Ogdensburg Railroad, up the Presumpscot and Saco valleys to North Conway, where the train receives tourists by the Eastern and Boston and Maine Railroads, and continues up the valley of the Saco, through the "Notch" to the Crawford and Fabyan Houses,* where connection is made with the Boston, Concord, and Montreal Railroad.

The approach through the Androscoggin valley, from Portland and the East, is by the Grand Trunk Railway to Gorham, and thence eight miles by stage to the Glen House. Travellers from New York to the mountains *viâ* Boston will take one of these routes. There are, however, inside lines connecting with the several Sound boats, which intersect the

---

* At the date of writing, the road is completed to Bemis Station, within eight miles of the "Notch," with flattering prospects of being finished before this meets the eye of the reader.

"Boston, Concord, and Montreal," without passing through Boston. Tourists by the Norwich or Stonington Lines, or by all rail from New York, can also go by the Worcester, Nashua, and Nashua and Rochester Railroads, intersecting the Boston, Concord, and Montreal, Boston and Maine, or Eastern Railroad routes to the mountains. If the trip be made by all rail from New York, travellers can take the Connecticut Valley Railroad at Springfield, continuing to Wells River, where intersection is made with the Boston, Concord, and Montreal Railroad, and passengers will reach the mountains by that line. In approaching from Montreal or Quebec *viâ* Grand Trunk Railroad, tourists can also strike the mountains on either the east or west side. If the former is desirable, continue on the Grand Trunk to Gorham, and thence by stage eight miles to the Glen House. If the west side is preferred, change from the Grand Trunk to the "Boston, Concord, and Montreal," at Northumberland, by which you are taken direct to the Twin Mountain or Fabyan Houses. Visitors to Bethlehem or the Franconia Range must make this change. Montreal and Quebec passengers may also go to the White Mountains *viâ* St. Johns, Canada, and continue thence by the South-eastern Railroad to Newport and Wells River. Those from Lake George go by Burlington and Montpelier to Wells River. From Saratoga, they can cross the lake to Burlington, or go by Rutland to Bellows Falls; and in either case intersect the Boston, Concord, and Montreal Road at Wells River, and reach the mountains by that line. Each of the above routes has its individual attractions. The *termini* of the several railroads are connected with each other, and with the Summer Houses, by lines of coaches. Parties can also secure private carriages for transportation throughout the mountains, which will be found one of the most enjoyable features of the excursion.

FRANCONIA MOUNTAINS.

The group of which Mount Lafayette, having an altitude of fifty-two hundred feet, is the central figure, is locally known as the Franconia Range. These mountains are situated about thirty miles south-west from the White Mountains proper, and, by their many points of scenic interest, successfully rival their more pretentious neighbors. **Mount Lafayette** commands a magnificent prospect. It is reached by a bridle-path, but the ascent is arduous. Here, also, is located that remarkable phenomenon, **Profile Mountain** (elsewhere alluded to), which is unquestionably the most wonderful natural curiosity in the country; while many other attractive features combine to render this a resort of great popularity. As before observed, the scenery on each of the above railway lines is entirely different, as shown by the accompanying illustrated pleasure routes.

ILLUSTRATED PLEASURE ROUTE No. 1.

Boston, Lowell, Worcester, Nashua, Salem, Lawrence, Manchester, and Concord to Lake Winnepesaukee, Franconia and White Mountains, Bethlehem, Lake Memphremagog, Montreal, and Quebec.

BOSTON, CONCORD, AND MONTREAL RAILROAD.

THE increase of travel to the White Mountains during the past few years has been something remarkable. The ease with which the trip can now be made, even by the aged or by invalids, has wrought this change. Cars of the most approved styles, equipped with all modern improvements, are run through without change from Boston and from the New-York boats. Hotels furnished with the comforts and luxuries of home spring up from the depths of the forest, and even crown the rocky summit of Mount Washington. A commendable emulation has actuated the several railroad companies, each striving to excel the other by adding to the comforts and conveniences of tourists.

To-day the **Boston, Concord, and Montreal Railroad** leads the van by placing its patrons at the end of their journey with the least effort to themselves. Its rails stretch to the base of Mount Washington; nay, by the patronage of this road, the cars now climb to the crest of that grand old peak, where they deposit travellers on the platform of an excellent hotel which has been built to shelter them. With the exception of a short ride from the Fabyan House to the Mount Washington R.R. Depot (six miles), there is a continuous line from Boston to the top of Mount Washington. This route receives more patronage, and distributes its patrons through more connecting lines, than any other.

Among the most prominent roads which contribute to swell the travel on the Boston, Concord, and Montreal, are the Portsmouth and Concord Railroad, Boston and Maine, Manchester and Lawrence, and Concord, with passengers from *Boston, Lynn, Salem, Lawrence, Manchester,* and the East; the Boston, Lowell, and Nashua, with guests from those cities; Framingham and Lowell, and its connections, with passengers from *New Bedford, Newport, Taunton, Fall River,* and *Providence,* and the *New-York* and *Stonington* lines of steamers; the Worcester and Nashua, with its local and *New-York* travel; and the Connecticut-River and Passumpsic Railways, with their numerous branches and connecting lines. Each of these roads must send its White-Mountain travel over the rails of the Boston, Concord, and Montreal.

Passengers from Boston take the cars at the Boston, Lowell, and Nashua Depot, or go by the "Boston and Maine" from Haymarket Square to Lawrence, and thence via the Concord Railroad to Manchester, where the train connects with that from Boston by the Boston, Lowell, and Nashua Railroad, and the two united continue to the mountains.

The Boston, Lowell, and Nashua Road is, perhaps, the legitimate route. This is made popular by the excellence of its equipments and running stock, and the promptness and regularity of its express trains, arranged at hours calculated to meet the wants of pleasure travel. Its depot, on Causeway Street, being one of the finest structures of the kind in the United States, is not only very popular with the travelling public, but it has come to be an object of interest to strangers.

LOWELL AND NASHUA DEPOT,
Causeway Street, Boston.

A large share of summer patronage is that in transit from New York and the South, through Boston, to the interior resorts of New Hampshire. The New York trains reach Boston early in the morning, at six or half past. The express train for the mountains leaves, by this road, at eight, A.M.; and the object was to construct a depot so ample in its appointments, so thorough in its equipments and conveniences, as to preclude the necessity of going to a hotel; and travellers will find, at the Lowell Depot, drawing rooms as elaborately furnished and conducted, and restaurants where meals are as well served, as at first-class hotels generally; thus affording to the wearied, travel-worn tourist, abundant facilities for toilet and rest before the resumption of his journey.

The ease with which the Lowell Depot is reached by public conveyance is also a feature of interest to the traveller. It is within fifty feet of the Eastern Railroad Depot, and within two minutes walk of the "Fitchburg" or "Boston and Maine;" while the depots in the southern portion of the city are connected with this by horse-cars, which pass the door every few minutes.

In the construction of this building, the architect has not only sought to combine all the advantages possible for the accommodations of the railroad, and the convenience of its patrons; but in its design and execution the city has secured an ornament in architecture of which its citizens may well be proud. The accompanying cut entirely fails to convey an idea of either its magnitude or finish. The "train-house," 565 feet in length, has been entirely ignored by the artist; but its ample proportions, completely covering the trains upon the several tracks, will, in the protection it gives, be appreciated by the public.

The principal tower is 148 feet high, the central dome 113 feet, and the east tower 104 feet. The central front of the building is occupied by the ladies' grand reception-room, 54 by 25 feet, which is elaborately finished and richly furnished. Upon its right and left respectively are also waiting-rooms for gentlemen and ladies. At the left of the main tower is the dining-room; and upon the same floor are baggage-rooms, barber-shop, toilet rooms, drinking fountains, and every convenience required by travellers.

The entrance is through archways at either end, of sufficient capacity for foot-passengers and carriages, a great convenience in stormy weather.

From these the entrance is to the main court, where the first impression of the magnitude of this building is felt. This court is 92 feet long by 52 feet wide, and extends from floor to ceiling, past three stories, a distance of 76 feet. The floor is covered with marble tiles in unique design. Ten pilasters at the sides, and six at the ends, continue to the roof, and, standing one above the other, support the balconies on the sides which lead to the various offices of the company. Between each pilaster is an arched-top doorway or window, opening to rooms beyond, an arrangement which gives great richness to the finish. The doors on the side open from the main entrance, and to the dining and baggage rooms; those at the south enter the reception-rooms, and those at the north lead to the trains; and between these is the ticket-office. The entire finish is in ash, elaborately carved. The ceiling is glass and stucco. From this depot the trains are made up for Lake Winnepesaukee, the Franconia Mountains, Northern New Hampshire and Vermont, Northern New York, Canada, and the West.

Passengers from *Boston* can also take the cars at the Boston and Maine Depot. These trains unite at *Manchester, New Hampshire*, and continue through *Concord* to the **Fabyan House**, at the base of Mount Washington. This route is made pleasant and interesting by the many streams and bodies of water along which it passes, among which may be mentioned the *Charles* and *Mystic Rivers*; the *Merrimac*, along whose banks it follows for many miles; the *Suncook* and *Winnepesaukee Rivers*; *Lake Winnesquam, Little Bay, Lake Winnepesaukee, Waukawan Lake, Long Pond, Pemigewasset*, and

Engraved expressly for Batchelder's "Parental Homes and Homes to Remember"

HOOKSETT, N.H.

1. Suncook
2. Suncook Valley Railroad
3. Hooksett Falls
4. Concord Railroad

Baker's Rivers; the Connecticut, Wells, Ammonoosuc, and Israel's Rivers; and many other smaller streams and ponds.

We strike the Merrimac at Lawrence or Lowell, following it past Manchester and Concord, crossing and re-crossing it at times. The beauty of its course is frequently varied by picturesque falls, affording more improved water-power than any river in the country. The falls at Lawrence, Lowell, Amoskeag, and Hooksett are particularly noticeable.

The accompanying cut, representing Hooksett Falls, also shows the Suncook Valley Railroad, which leads to Pittsfield, N.H., a thriving and beautiful village nestled among high hills, which are dotted with farm-houses, that are fast becoming popular with boarders from cities.

This route also leads through the heart of the cotton-manufacturing interest of New England; passing Lawrence, Lowell, and Manchester, besides many smaller manufacturing-towns. Concord, the beautiful capital city of New Hampshire, possesses many features which make it a favorite resort during the summer and autumn months. It contains about 12,500 inhabitants; yet all are so comfortably domiciled, that it is frequently remarked by strangers "Where do your poor live?" The shaded concrete walks of the city add much to the comfort of visitors.

The State Capitol stands in the centre of a small but beautiful square, handsomely laid out, and ornamented with broad-spreading trees. The structure is of pleasing architecture, built of native granite, for which the vicinity is noted, the whole surmounted by a lofty dome. Immediately fronting the State House, on the main street, is the **Eagle Hotel**, a fine brick structure, which is widely known as a first-class house. The "Eagle" receives much of its foreign patronage from parties, who, after starting for the mountains, prefer to spend a few days at Concord before leaving for the season; and particularly from those returning in the autumn, driven in by the early frosts, who always find here and in the vicinity a few weeks of charming weather.

Among the other public buildings may be named the Court House, Churches, Schools, City Hall, State Prison, and State Asylum for Insane. The two latter institutions are in fine condition. The Prison, unlike those of many States, is made a paying institution. The Asylum has been built 32 years, and is very successfully conducted.

Concord is somewhat celebrated for its manufactures, particularly of carriages and coaches, which are shipped extensively to all parts of the world. As a railroad centre, Concord presents admirable facilities for intercourse with various sections of the country.

The line proper of the Boston, Concord, and Montreal Railroad starts from Concord, though its cars and those of the Boston, Lowell and Nashua, the Framingham and Lowell, and the Worcester and Nashua Roads, run through from those cities, and continue to the mountains.

Engraved expressly for "Bachelder's Popular Resorts, and How to Reach Them."

TILTON, N.H.

1. Methodist Seminary
2. Winnepesaukee River.
3. Boston, Montreal, and Concord R.R.
4. Belknap Mountain.

A few miles above Concord, the road again crosses the Merrimac, and leads away towards Lake *Winnepesaukee*.

Tilton is the first town of interest. This was formerly known as *Sanbornton Bridge*. It is a thriving manufacturing village, and forms the centre of a large agricultural region. The Methodist Seminary located here, a good view of which appears in the engraving, has been long and favorably known. *Tilton* possesses an unusually fine water-power, not computed by its volume alone, but by its great regularity. The river which runs through the place is the outlet of *Lake Winnepesaukee*, in which large reservoir the water is held in reserve by the water-power company which owns it, to supply the cotton-manufactories at Lowell and Lawrence during the droughts of summer.

There is a charm in this whole region for summer life. Not only the town of Sanbornton, from which Tilton is an offshoot, but Canterbury and Meredith, Belmont and Gilmanton, all furnish desirable summer homes. The topography is particularly adapted to promote the health of its inhabitants. The land is generally high and rolling, and has been so long cleared that the climate is fully established.

A stage leaves Tilton, on the arrival of the morning train, for **Gilmanton Academy**, passing through *Belmont*, formerly known as Upper Gilmanton. Gilmanton is becoming popular as a summer residence for persons from Boston and New York, who build here houses for the warm season. It possesses the advantage of good and long-established institutions of learning, and is sought by persons having a family. Although supplied with daily mail and stage connections, there is no railroad within its borders; and, with a society cultivated by its fine schools, it possesses much of that pristine character which characterized New England towns of earlier days. *Lower Gilmanton* is reached by the "Concord" and "Suncook Valley Railroad" to Pittsfield, and Gilmanton Iron Works by the "Boston and Maine" to Alton, and thence by stages. From Tilton the road follows the *Winnepesaukee River*, and the shores of *Little Bay* and *Winnesquam Lake*, past *Union Bridge*, to *Laconia*.

Although undeveloped at present, this region possesses many features calculated to make it popular with the seeker after health and pleasure. Winnesquam Lake is some twelve or fifteen miles in length, is beautiful in form and surroundings, and, but for its more pretentious rival Winnepesaukee, would have, ere this, received the attention which its merits deserve. It has long been known as the home for the lake trout, and somewhat famed for its piscatorial advantages. Two small summer houses furnish accommodation for visitors,— the *Winnesquam* at the lower end of the lake, and the *Bay View*, which is admirably located in the suburbs of Laconia. The cars pass between it and the lake, and leave passengers when requested.

Engraved expressly for "Batchelder's Popular Resorts, and How to Reach Them."

LACONIA, N.H.

1. Lake Winnesquam.
2. Sandwich Mountains.
3. Boston, Concord, and Montreal R.R.
4. Bay View House.
5. Mount Belknap.

The route, which from Concord lies through an uninteresting country, now fairly enters the lake and mountain region. The scenery does not possess the grandeur of the White-Mountain section; yet it is marked by many elements of picturesque beauty. Its water-views are fine; a distant line of mountain-peaks cuts the horizon. It is only five miles, over a good country road, to *Mount Belknap*, which is easily accessible, and from whose barren summit may be had one of the finest landscape-views on the Atlantic slope. It varies from the *Red-Hill* prospect by having *Lake Winnepesaukee* and the entire group of the *White* and *Franconia Mountains* in the same view. The steamer "Mount Washington" can be distinctly seen soon after it leaves *Alton Bay*, and traced on its way for more than twenty-five miles to *Wolfboro'* and *Centre Harbor*. The steamer "Lady of the Lake" can also be followed in its tortuous course from *Weir's* to Centre Harbor and Wolfboro'. From this elevated position a much better idea of the great number of islands is obtained than while sailing on the lake. Beyond its placid waters the mountain ranges rise in successive peaks; and towering above all is the well-known "Presidential Group," of which *Mount Washington* is the commanding centre.

Mount Belknap is at present but little known to the travelling public; but its pleasant approach, easy access, and magnificent "View" must eventually bring it into great popularity. Visitors to Laconia will find the **Bay View House** (seen on the right of the engraving) delightfully located, and surrounded by beautiful scenery. It is noted alike for its good table, home-like atmosphere, and reasonable terms.

LAKE VILLAGE, N.H.
Boston, Concord, and Montreal R.R.

Lake Village is but a mile and a half from Laconia: their suburbs meet; and the towns are seemingly one. The views in the vicinity are very like those near Laconia. Indeed, the same mountain ranges may be seen in the distance, though the water foreground is different. There is nothing grand in the scenery as the train steams away towards the lake, but it is very picturesque. At *Weir's Station* passengers change for *Wolfboro'*, *Centre Harbor*, and *Conway*. Within the past year a Methodist camp-meet-

Engraved expressly for "Bachelder's Popular Resorts, and How to Reach Them."

WEIR'S LANDING, N.H.

1. Camp-Meeting Ground.
2. Sandwich Mountains.
3. Boston, Concord, and Montreal R.R.
4. Ossipee Mountains.
5. Lake Winnepesaukee.
6. Steamer "Lady of the Lake."

ing ground has been dedicated in a delightful grove adjoining the station, which bids fair to increase the popularity of this charming spot. Several commodious buildings have already been erected: lots for cottages have been secured on a site commanding a magnificent view of the lake, with fine boat and railroad accommodations. Those who desire to visit *Wolfboro'*, *Centre Harbor*, or *Conway* will find the commodious little steamer **Lady of the Lake** awaiting them at the landing. Arrangements have been made to run the boat from Wolfboro' to Weir's, and *vice versa*, to accommodate tourists to and from the Franconia Mountains. The distance to Wolfboro' is twenty miles, and to Centre Harbor but half that, although the latter route seems to combine all the beauties of the lake. When the steamer leaves the wharf, the jutting points of the adjacent islands would seem to bar our progress; but, as it speeds its way, the view unfolds, the channel opens; and we wind our pleasant course among the islands, at times so near that the overhanging branches almost sweep the boat. The lake is from twenty-five to thirty miles long, and varies from one to eight miles wide. It contains about sixty-nine square miles, and nearly three hundred islands, on many of which are fine farms, and several are used for grazing. Its surface is 472 feet above the level of the sea. The numerous islands which dot its bosom, the beautiful hills which hem it in, and its many points and inlets, combine to make Winnepesaukee one of the most pleasing inland resorts in the country. The sedative influence and peculiar quiet of the scene, during the charming days of an Indian summer, with the bright tints of an autumnal foliage, graduating to the soft haze of the mountain blue, reflected in its waters, is most wonderful. At Centre House or Wolfboro' for days and weeks the tourist lingers, forgetting, among the quiet beauties of nature, the cares of a business-life. The excursion to Centre Harbor also forms one of the most delightful *day-trips* from Boston. Leaving the city at 8 o'clock in the morning, via the Boston, Concord, and Montreal R.R. and steamer "Lady of the Lake," the visitor will have an hour for dinner at Centre Harbor, returning by the steamer "Mount Washington," and Boston and Maine Railroad, to Boston the same evening, thus passing through the cities of Lowell, Nashua, Manchester, Concord, Dover, Haverhill, and Lawrence, with the intervening towns, and traversing the entire length and breadth of Lake Winnepesaukee, by both routes, in a single day.

It would be easy to introduce pages of description from the pens of visitors; but all are embodied in the following quotation from that eminent writer, EDWARD EVERETT.

"I have been something of a traveller in our own country, — though far less than I could wish, — and in Europe have seen all that is most attractive,

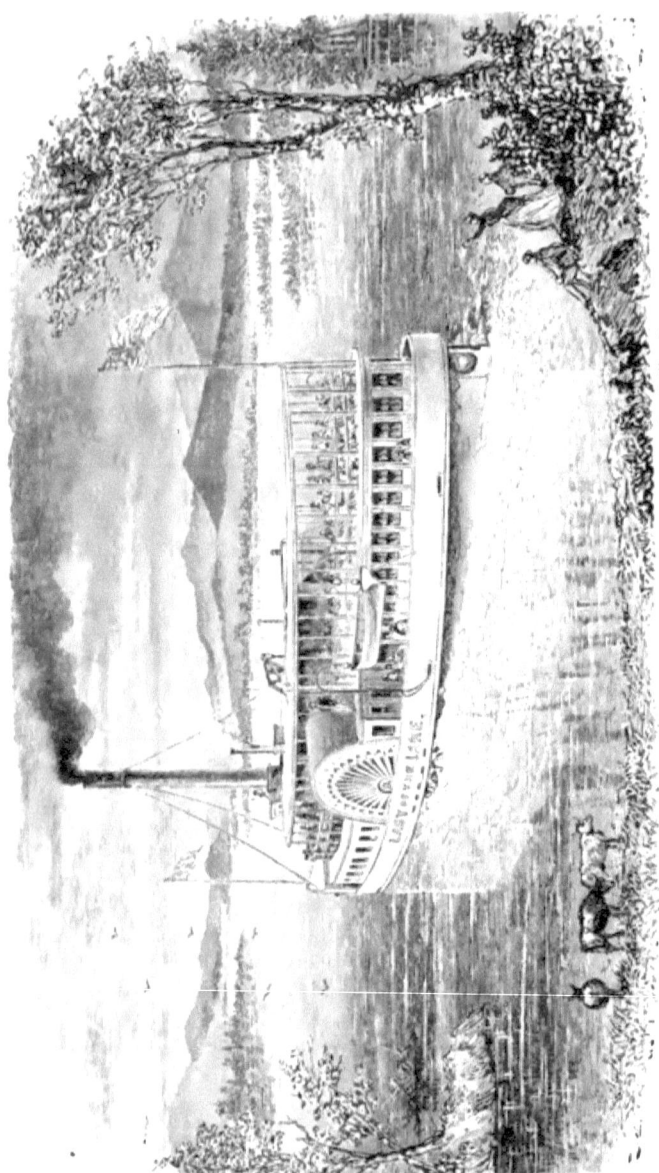

Engraved expressly for "Bachelder's Popular Resorts, and How to Reach Them."

STEAMER "LADY OF THE LAKE."

Connecting Weir's Landing, Boston Concord, & Montreal R.R., with Wolfboro' and Centre Harbor, N.H.

from the Highlands of Scotland to the Golden Horn of Constantinople, from the summit of Hartz Mountains to the Fountain of Vaucluse; but my eye has yet to rest on a lovelier scene than that which smiles around you as you sail from Weir's Landing to Centre Harbor."

From *Weir's Landing* the train continues northward past *Meredith*, a pleasant village located on the shores of the lake, from which steamers run to other villages during portions of the year. Above Meredith the route leads for four miles along the south shore of *Waukawan Lake*.

RAGGED MOUNTAIN & LONG POND, MEREDITH, N.H.
Boston, Concord, and Montreal R. R.

Long Pond on the right is the next body of water passed, the train gliding safely under the shadow of *Ragged Mountain*, whose rocky sides have been blasted away to give passage to the cars. This scenery and that around *Ashland* is very fine; and many a tourist artistically inclined will be lured from the cars to visit it. The *Pemigewasset* and *Squam* Rivers, which unite here, furnish many landscape "bits" of artistic beauty. The course of *Squam River* is not along our route; but the lover of the beautiful who would follow it three miles, to its source in *Squam Lake*, will be amply repaid.

Squam Lake has already been alluded to in a visit from Centre Harbor; but no single description can exhaust its picturesque beauties. Indeed, this whole region possesses peculiar charms for the liberated citizen of our larger towns, where weeks or months may be quietly spent; and, if he makes up his mind in advance to take the accommodations as he finds them, he cannot fail to be pleased. Unfortunately, no large hotel has yet been built here;

On entering the Pemigewasset Valley, at Plymouth, the scenery assumes beautiful combinations of lines, and scenic effects. The whole region, both on the river and inland, is made up of grand panoramic views or choice "bits," from which the artist really fills his sketch-book.

Engraved expressly for "Bachelder's Popular Resorts, and How to Reach Them."

PLYMOUTH, N.H.

1. Francoma Mountains.
2. Pemigewasset House.
3. Boston, Concord, & Montreal R.R.
4. Pemigewasset River.

but it is but a short drive from Centre Harbor, where all the quiet comforts of home will be found.

Ashland was formerly known as *Holderness*, and is remembered by members of the Episcopal denomination as one of the first places where that society flourished in this part of the State. Above Ashland we enter the valley of the *Pemigewasset*, which we follow to *Plymouth*.

The stranger will be particularly struck by the purity of the water in the wayside streams flowing from springs on the mountain sides. They furnish admirable nurseries for the speckled trout with which they generally abound.

The approach to Plymouth is very picturesque. The line of the road is along the banks of the river, which meanders its course through rich meadows, shaded here and there by broad-sweeping elms. On either side are high wooded hills, which, by gentle grade, sweep down to the

PEMIGEWASSET HOUSE.

valley below; while beyond in the blue distance are the *Franconia Mountains*. As you are whirled rapidly into the town, the **Pemigewasset House**, which in the distance seemed a mere speck among the trees, rises invitingly before you.

On reaching town, the train stops immediately in rear of the hotel; and, with an evident knowledge of the good things within, the passengers

soon fill the long dining-rooms of the house, or the restaurants attached
to it. Ample time is given for dinner, full thirty minutes, before the
conductor cries, "All aboard!" But here we find that many of our companions have left us, though the number is made good by others, who
have been spending a few days at this enjoyable place. *Plymouth* is
deservedly one of the most popular resorts in New Hampshire. It is a
compact village, with several fine churches, schools, county buildings,
railroad offices, &c. But tourists visit Plymouth for its delightful surroundings, pleasant drives, and magnificent scenery, and no less for
the popularity of its noble hotel, the **Pemigewasset House.**

This elegant and spacious hotel is delightfully situated on the banks
of the Pemigewasset, near its confluence with Baker's River. The halls,
parlors, and dining-rooms are large, light, and handsomely furnished.
The chambers are high and well-ventilated. There are bath-rooms with
hot and cold water, and all the modern conveniences of a first-class
house. It is under the patronage of the Boston, Concord, and Montreal
Railroad, and is frequented by persons of culture and taste, some
having secured rooms for nine consecutive years. An air of refinement
pervades its atmosphere, which is immediately *felt* by visitors.

LIVERMORE FALLS, PLYMOUTH, N.H.
Boston, Concord, and Montreal R.R.

LIVERMORE FALLS are on
a wild turbid stream, which
forces its way along a rugged
bed of shattered rocks. The
road-way crosses by a light,
airy bridge immediately below
the falls, affording an opportunity to view them without
leaving the carriage. Where
there are so many pleasant
drives as in the vicinity of
Plymouth, it is difficult to
particularize; indeed, with the
fine turn-outs furnished at
the hotel, one can scarcely go
amiss. The drive around *Plymouth Mt.* is very highly spoken
of; and a longer excursion by private conveyance up the valley of the
Pemigewasset to *Franconia Notch* is delightful.

Mount Prospect is much visited. A carriage-road leads to its summit,
which is 2,963 feet above the sea. It commands a landscape view of
rare beauty, embracing the Franconia and White Mountains, and this
entire lake-region, of which *Winnepesaukee* is the most noted. There
are also several elevations in the immediate neighborhood of the village,
which pedestrians will delight to visit.

STAGE-ROUTE FROM PLYMOUTH TO THE FRANCONIA MOUNTAINS.

There are those who would find their visit to the mountains unsatisfactory without a stage-ride; to such the writer can recommend the route from *Plymouth* to the **Profile House**. It is over a good road, and through one of the most picturesque regions of New Hampshire. Artists do not generally spend their summers at the mountain-houses, but select some desirable field for their labors. The route from Plymouth to the Profile House passes through *Compton* and *Woodstock*, which is emphatically a *field for artists*, where, through the months of summer and autumn they gather the choice *bits* which occupy their winter months, and delight their friends at home. A more interesting drive can scarcely be conceived. The road passes near the *Flume*. This is a wonderful freak of nature,—an upright fissure in the rocks, which have been forced asunder by some mighty convulsion; while high up their sides is held in unyielding grasp a huge bowlder, beneath which a wild mountain torrent dashes its feathery spray. The *Pool* is a curiosity scarcely less interesting, and should be visited by the tourist. An impetuous stream, shaded by forest trees, walled in by precipitous ledges, escaping from the thicket above, leaps from the rocks into the deepening gloom below. The *Basin* is passed at the road-side, and is an exceedingly attractive feature. Here a mountain torrent rushes obliquely into a rocky caldron, around which for ages past the waters with dizzy whirl have polished its granite sides. The *Old Man* of the *Mountain* is seen on the left a half mile before reaching the **Profile House**; and it is better to visit it late in the afternoon, with the bright sky behind it. It requires no stretch of the imagination to detect the cold, sharp outline of the human profile chiselled in colossal proportions by the hand of nature. This is unquestionably the most remarkable natural curiosity in this country, if not in the world. The likeness is formed of three blocks of granite, high up the mountain-side, located rods apart; yet when viewed from *one* particular spot the profile is perfect. It is 70 feet from chin to forehead; yet the lines are softened by distance. The beautiful lake at the foot of the mountain is known as the *Old Man's Washbowl*. *Echo Lake*, near the **Profile House**, is also one of the points of interest.

The ascent of *Mt. Lafayette* is made from here, and is scarcely less interesting than that of Mt. Washington, although much more difficult and fatiguing, as it must be done on horseback, unless the tourist is a good pedestrian. This locality can also be visited with a quarter-part the stage-coach ride by keeping the cars to Littleton. Resuming our seats in the cars at Plymouth, the train for twenty miles continues up the valley of *Baker's River*. There is nothing striking in the scenery;

but the mountains and river present varied combinations of forms in which the tourist will not fail to be interested. In the vicinity of *Warren* the mountains become bolder and more rugged; and the time is not far distant when this locality will be largely frequented by lovers of fine scenery. Even now the small hotel in the village, and many private boarding-houses, are well patronized. A wild mountain

OWL'S HEAD AND MOOSILAUKE, WARREN, N.H.
Boston, Concord, and Montreal R.R.

stream in the suburbs has several waterfalls and pretty cascades, which are well worth visiting. A good carriage-road leads to the summit of *Moosilauke*, five miles away. This mountain is 4,600 feet high, and commands a magnificent prospect. Visitors will find accommodations at the **Summit House**. A fine view of *Moosilauke* may be had from the right of the cars, while going northward. A few miles above

Warren, is a high barren cliff, called *Owl's Head*, which rises precipitously above the surrounding forests. This locality presents many points of interest, particularly for a pedestrian, who, with fishing-tackle or gun, may while away a few weeks in autumn.

The rugged form of *Owl's Head*, combined with *Moosilauke*, and the green meadows which surround them, make a beautiful landscape. Indeed, the scenery is all fine along this section of the route. *Haverhill*, a few miles farther on, is a pleasant village: the public buildings of Grafton County are located here. The line of the road has led us gradually towards the *Connecticut*. On our left are the rich bottoms which skirt its borders; and the thriving village of *Newbury, Vt.*, can be seen across the river.

The train crosses the *Connecticut* at *Woodsville* to *Wells River*, where connection is made with the Passumpsic, Montpelier, and Wells River Railroads. After receiving their White Mountain passengers, the cars re-cross to the east bank, and continue up the *Ammonoosuc*. This is indeed a pleasing stream. Its course is broken by falls and rapids; and its waters are swept by the branches of overhanging trees. The next village passed is *Bath*, which is charmingly situated on the bank of the river, and presents a very picturesque appearance. *Lisbon* is but a few miles farther on. This is a very interesting village, and pleasantly located. The discovery of a gold-mine has given it interest. "Passumpsic" passengers for Newport and Lake Memphremagog diverge here.

Littleton is the largest and most populous village in this section of the State. It contains several hotels and boarding-houses, among which **Thayer's** is the best known. The scenery at *Littleton* presents many artistic combinations. The village is built mostly on the right bank of the river, extending up the hillside. From the upper portion of the town is had an excellent view of the *White Mountains*, flanked by the *Franconia Mountains*, and other ranges equally interesting.

Littleton contains about 2500 inhabitants, and is well supplied with churches, schools, banks, and printing-offices. Indeed, it seems a miniature city, yet so small that ten minutes' walk in any direction will take you into the delightful suburbs, where all the pleasures and amusements of the country may be enjoyed. During the summer months the number of inhabitants is largely increased. The atmosphere is exhilarating, and the water pure, for which so many come heree, wher more home comforts can be enjoyed, in preference to going to the mountain-houses. From Littleton, tourists can easily visit the more important points of interest. *Mount Washington*, the *White Mountain-Notch*, *Pool*, *Flume*, *Profile*, and many other interesting places, can be visited in a day, and return the same night. Stages to the *Profile House* and *Franconia Mountains* leave here twice daily. No tourist to the mountains can afford to pass the *Franconia Notch*, without a call. Indeed, it is one of the few

Engraved expressly for "Bachelder's Popular Resorts, and How to Reach Them."

LITTLETON, N.H.

1. Oak Hill House.
2. High School.
3. Ammonoosuc River.
4. Mount Washington.
5. Boston, Concord, and Montreal R.R.

places where the traveller lingers. The **Profile House**, near the Notch, is one of the largest and best appointed in New England.

Seven miles above Littleton the *Wing Road* branches to the right, and continues past *Bethlehem* and **Twin Mountain House** to the **Fabyan Hotel** at the foot of Mount Washington.

The next station of importance on the main line is *Whitefield*, extensively known for its lumber operations, but more recently as a summer-resort. In the neighborhood are some fine views of mountain scenery; and the place is fast growing in popularity, which may also be said of *Dalton*, the next station on the line.

Lancaster is one of the most beautiful villages in Northern New Hampshire. It is well laid out, has concrete walks, and fine shade-

LANCASTER HOUSE,
Lancaster, N.H.

trees ornament its streets. The architecture is good; and tasteful gardens are everywhere to be seen. There are six churches, a public library, and other public buildings; and throughout the town pervades an air of taste and refinement. The village is built in an immense amphitheatre, surrounded by hills and mountains, which are reached by excellent roads, affording some of the most delightful drives in the State. The view from *Lunenburg Hills, Vt.*, is unsurpassed. Israel's River passes through, and the Connecticut near the town. There are several hotels, the most prominent being the **Lancaster House**.

The **Lancaster House** is superior to most houses situated so far from the centres of trade. It accommodates conveniently 150 guests.

LANCASTER, N.H.

Engraved expressly for "Bachelder's Popular Resorts, and How to Reach Them."

1. Lunenburg Heights.
2. Connecticut River.
3. Boston, Concord, & Montreal R R
4. Lancaster House
5. Mount Lyon.
6. Stratford Peaks.

The rooms are large and high-posted. From the ample cupola which surmounts it, the view of the *White Mountain Range*, *Stratford Peaks*, *Starr King*, *Pilot Range*, *Mt. Lyon*, and the green hills of Vermont, is unsurpassed. Parlor-cars run through the village from Canada, Boston, Fall River, Newport, New London, and Worcester. Lancaster is a town well calculated to please the visitor who would make it his home during the summer or autumn months.

The Waumbec House is but eight miles away, and can be reached by stage. This and other houses in *Jefferson* are in a romantic locality, which will well repay a visit.

From Lancaster the train continues to *Northumberland*, where connection is made with the Grand Trunk Railroad for *Canada* and the West.

Percy or **Stratford Peaks** are northern outstanding spurs of the White Mountains; bold and rugged in outline, grand in effect, yet less popular with the tourist than those forming the principal group of the White Mountains proper, or those which are farther south. These mountains are visited from the village of Northumberland, where the Boston, Concord, and Montreal, and Grand Trunk Railroads intersect. A charter has been obtained by the Boston, Concord, and Montreal Company to construct a railroad to the Canada line; this will open up a new and varied field for the tourist, happily divided into rich, arable land, and wild, interesting scenery.

Dixville Notch. — In the extreme northern section of New Hampshire, near the Canada line, there is a barren region, sparsely inhabited, yet rich in picturesque grandeur. It is a favorite resort of the few, though but little known to the general tourist. The altitude is high, the atmosphere clear and dry, and the water pure and sweet.

The streams abound in speckled trout. This is near the head-waters of the Connecticut and Androscoggin Rivers; the former flowing southward through New Hampshire, Vermont, Massachusetts, and Connecticut, into Long Island Sound; the latter bearing eastward, skirting the base of the White Mountains, and moving through New Hampshire and Maine into the broad Atlantic. Dixville Notch has many objects of interest. A rugged pinnacle, five hundred and sixty-one feet above the carriage-road, approached by a rough stone stairway, is called *Table Rock*. *Jacob's Ladder* is the name which has been given to the path that reaches it. Another path near at hand leads to the *Ice Cave*, a protected gorge where the snows lodge in winter, and remain throughout the summer. Dixville Notch, which is a mile and a half long, like the White Mountain Notch, has its towering rocks, grand *Flume*, turbulent stream, and its snowy cascades.

The **Connecticut Lake,** lying to the northward, is also a feature of this unfrequented region. It may be reached from Colebrook, twenty-five miles, by stage. A pleasure steamer has been placed on its waters. The principal lake is but five and a half by two and a half miles; a second, four miles away, reached by a forest path, is but half this size; a third and fourth, both small, are beyond. These lakes are well stocked with fish, whose unfamiliarity with the sports of civilization make this a desirable locality for a "camping-out" party, though comfortable hotel accommodations can be had. This is similar to the Lake Umbagog region described elsewhere.

Resuming our route on the Wing Road (see index), we follow up the

RAILROAD STATION, BETHLEHEM, N.H.
Boston, Concord, and Montreal R.R.

banks of the *Ammonoosuc;* though for several miles there is nothing in the character of the scenery to indicate to the tourist that he is rapidly approaching one of the most celebrated summer-resorts in America.

Bethlehem Station is the first stopping-place. The village of *Bethlehem*, two miles from the station, is one of the favorite summer residences in the mountain-region. With the increasing popularity of White Mountain travel come large numbers as sight-seers and pleasure-seekers; still, there are many who visit the highlands of New Hampshire for the water pure from its mountain springs, and fine invigorating atmosphere which sweeps the hills, and after a few months' sojourn feel that they have renewed their lease of life. The extent of country thus visited occupies an area of more than 500 square miles, embracing every variety of surface and surroundings, from the green meadow, the rolling upland, the high mountain-peak, to the dense primeval forest. The village of Bethlehem is built on a plateau or ridge of deep, rich soil, which connects the White and Franconia ranges of mountains, and commands striking views of both. *Its altitude is greater than that of any other village east of the Rocky Mountains.*

Some years ago a Boston merchant, overtaxed by business cares, and suffering from loss of health, was recommended to try a season at Bethlehem. He returned in the autumn well, — completely invigorated and restored; but each season finds him with his family at their mountain home. Thankful for this marvellous and unexpected restoration, with his ample means he determined to prepare accommo-

MAPLEWOOD HOTEL.
Bethlehem, N.H.

dations where others could have the comforts of home without the expense of fitting up an establishment of their own.

A valuable farm of five hundred acres was purchased, and thoroughly stocked with improved breeds of horses, cows, and sheep, and large numbers of poultry of the most approved kinds; while the house was enlarged and placed in perfect repair, bowling and billiard saloons erected, and other games and amusements improvised for the entertainment of guests. The farm was placed under the charge of a competent person, for whom a commodious farm-house, barns, dairies, stables, and extensive out-buildings, were erected. From this farm guests are daily supplied, during the summer, with the *very best of every thing* fresh from the fields. Green

TWIN-MOUNTAIN HOUSE.
Boston, Concord, and Montreal R.R.

1. Ammonoosuc River. 2. White Mountain Range.

corn, pease, beans, and garden-sauce, growing at one hour, are bountifully served upon the table the next. Fresh cream, butter, and eggs, of home production, are furnished, not at fabulous hotel-prices, but at fair and reasonable rates. The verandas at the **Maplewood** are shaded; and the grounds are ornamented by a fine growth of sugar-maple, forming delightful play-grounds for children, and a cool and cleanly out-of-doors resort for adults. One of the finest and most picturesque views of *Mount Washington*, and others of the White-Mountain group, is from the veranda of this house; while the vicinity abounds in delightful drives. The admirable drainage, secured at great expense, renders this a healthful and desirable summer residence. It is supplied with never-failing spring water, and is but one and one-fourth miles from the depot.

From the station at Bethlehem the train continues along the bank of the *Ammonoosuc* to the **Twin Mountain House** and to the **Fabyan House**, at the base of *Mount Washington*. The **Twin Mountain House** until the present season was the terminus of the railroad. It has been extensively patronized, and will be pleasantly remembered by its patrons, as a most free and social summer home.

The buildings of this extensive summer resort were erected and furnished new in 1869-70, on a spot long occupied as a hotel, and popular with the public. The vicinity not only commands fine and varied views of the White and Franconia Mountains, but has better facilities for water amusements than any hotel of the mountain region. The house stands high on a commanding bluff, which overlooks the *Ammonoosuc*. So near its head, this stream is not usually suitable for boating; but here it is held by a dam, thus affording an admirable opportunity for that healthful and fascinating amusement. The water is fringed with trees of most delicate foliage, among which guests have constructed rural seats and arbors. This is a romantic spot, where lovers and those socially inclined do love to congregate. The forests about the "Twin Mountain" are very charming, and the shrubs and ferns fresh and varied. But the chief and practical excellence of this locality is in the entire absence of hay-fever, that disagreeable disease indigenous to so large a portion of the country. The following extract from "The New York Ledger" is from the pen of the Rev. Henry Ward Beecher, who has long been afflicted with this distressing malady, and who now spends his summer and autumn months at this health-giving place: "Meanwhile another year warrants me in saying that a resort hither is almost certain relief; not one per cent of patients failing to obtain essential if not entire relief. We can go out into the sun, stand in mud morning and evening, and in spite of dust, rain, or chill, we are well."

From the Twin Mountain House, the route continues to the Fabyan House, the terminus of the Boston, Concord, and Montreal Railroad; from

Engraved expressly for "Batchelder's Popular Resorts, and How to Reach Them."
FABYAN HOUSE.

1. White Mountain Range.
2. Mount Washington Turnpike.
3. Ammonoosuc River.
4. "Notch."

which point stages convey tourists six miles, to the Mount Washington Railway, by which they are taken to the summit the same evening.

Carriages will also be found in waiting, to take passengers from every train to the Crawford House five miles distant, White Mountain Notch, and other points of attraction in that neighborhood, described under the head of Pleasure Route No. 5."

The stage-ride from the Crawford and Fabyan Houses to the Mount Washington Railway Station is one of the most exciting features of mountain travel.

AMMONOOSUC FALLS.

The falls of the Ammonoosuc are passed by the wayside, and are well worth a visit. Here the rocks have been worn by the action of the water into a thousand fantastic forms. The road leads through a primeval forest: luxuriant vines laden with fruit and berries spring from the virgin soil, often tempting the visitor from the carriage. We occasionally catch a glimpse of the grand old mountain, as it raises its granite head above the clouds. The ascent of Mount Washington was once a feat of rare occurrence, accomplished only by the daring hunter or adventurous traveller; but the industry and perseverance of man have smoothed the way; and the route has been made easy, safe, and pleasant.

To accompany an aeronaut, to look out upon the surrounding world, has been the desire of many, though enjoyed by few. Here the "iron horse," guided by the hand of genius, climbs triumphantly to the dizzy height of 6,285 feet, more than a mile in the air, where the "storm-king," riding on the wings of the whirlwind, have hitherto reigned supreme; and yet all this is done in absolute safety, and with as much ease as the same distance could be accomplished over any road in the country.

The ascent should be made the subject of some preparation. To attempt it improperly clothed would risk the pleasure of the excursion. You *may* not meet a snow storm, or find icicles hanging from the roof in the morning; but you are *liable* to any month in the year. Ladies, particularly, should not rely upon a shawl alone for protection, but add a full suit of winter extra under-clothing. You will find the house on the summit heated by steam, and a cheerful fire in the grate; but you should not, for want of proper clothing, lose the opportunity for out-of-door pleasures.

The views while ascending and descending are supremely grand. To stand upon the summit of *Mount Washington* is the one desire of every visitor to the mountain-region. Here, from the highest point on the Atlantic slope, he can look down upon this vast panorama of hills and valleys, cities and plains, dotted with a thousand silvery lakes blended into one harmonious whole. Without putting foot upon the ground, he is lifted step by step up this rugged steep, to the very doors of the Hotel, which, bound with chains to the barren cliff, has been built and furnished to receive him.

The **Mount Washington Summit House** accommodates conveniently one hundred and seventy-five guests, though more than two hundred have been entertained. Several thousand persons visited it during the past season. Its appointments are very complete. Lighted by gas, and heated by steam, with all modern improvements, a liberal table, and good attendance, the visitor can be made comfortable for any length of time. Stages run from the railroad depot to all the prominent houses, enabling tourists to return by any route they choose.

PLEASURE ROUTE No. 2.

Boston and Vicinity to the Summer Resorts of Northern Vermont, Lake Memphremagog, Montreal, and Quebec.

PASSUMPSIC AND SOUTH-EASTERN RAILROADS.

TOURISTS go by Pleasure Route No. 1, by Lowell or Lawrence, Manchester, Concord, Plymouth (twenty minutes for dinner), to Wells River Junction. A short distance south from Wells River, at Newbury Station, the visitor will find **Newbury Sulphur Springs.** They are located near the village, and are a favorite resort of invalids and travellers, being well recommended by the medical faculty.

The village, which is rurally pleasant, commands a fine view of the Connecticut River, and the rugged hills of New Hampshire beyond. *Mount Pulaski* is near at hand, and the whole surface is charmingly diversified. If from this point we continue up the river by the "Passumpsic" due north, along a route rich in landscape beauties, passing *en route* the thriving village of St. Johnsbury, after an hour's ride we reach the quiet station of *West Burke,* from whence by stage a half dozen miles, we are taken to **Willoughby Lake,** which, although inferior in size to many lakes in New England, has no superior in picturesque variety and beauty. It is favorably known to the travelling public as a summer resort, and is situated between two high mountains, *Annanance* (Willoughby) and *Hor,* which rise abruptly from its shores. The water is of remarkable depth: over six hundred feet sounding has failed to find bottom. Willoughby Lake is six miles long, and about two in width. A pleasure-drive has been constructed along the east side. There are many points of rare interest in the neighborhood, among which are Mount Annanance, Mount Hor, Silver Cascade, Point of Rocks, &c.

Mount Annanance, or *Willoughby* as it is generally called, thirty-eight hundred feet high, derives its name from that of an Indian chief of the St. Francis tribe, who here made his home. This elevation is generally visited on foot, the path leading up through a beautiful forest with occasional glimpses of the lake, whose quiet waters lave the rocks below. It is not, however, until the summit is reached that the visitor realizes the richness of the scene about to be spread out before him. A fine panoramic view, embracing the Connecticut Valley, the Franconia and White Mountains of New Hampshire, the nearer peaks of Mansfield, Camel's Hump, Killington, and Jay; Owl's Head in Canada, and the Adirondacks of New York, in the distance; while near at hand the beautiful waters of Willoughby Lake, and the bold outline of Mount Hor which rises beyond,—serve to form one of the most effective scenes in New England.

CANADA AND THE PROVINCES.

Lake Memphremagog is the connecting link between the summer resorts of New England and Canada, more than half its surface being in the latter country. It is the next point of interest after leaving West Burk Station. It is an unusually pleasing sheet of water, about thirty miles long, hemmed in by bold, rugged mountains, traversed by pleasure-boats, from whose decks, as they steam along its quiet surface, rich views of the surrounding scenery may be had. The lake varies from two to four miles in width, its bold shores and numerous islands contributing wonderfully to the interest of the scene. The village of *Newport*, Vt., located at the southern extremity, is already an inland watering-place of considerable repute, and its fine summer houses are rapidly increasing in fame and popularity.

Bolton Springs in Canada, fourteen miles distant; *Clyde River Falls*, two miles; *Mount Morrill*, two miles; *Bear Mountain*, seven miles, and *Prospect Hill*, close at hand, — these are among the points of interest near Newport. From Prospect Hill a charming view is presented, not only of the lake and mountains near by; but Mount Annanance, Jay Peak, Orford, Elephantis, and Owl's Head, are all visible, and, with the lake and its picturesque islands, combine to form an exceedingly interesting landscape.

Owl's Head, Canada, a conical peak of singular formation, having an altitude of nearly three thousand feet, rises from the west shore of Lake Memphremagog. Steamers *en route* from Newport to Magog make a landing at its base. The ascent is made from the *Mountain House*, a half-mile distant, by a footpath. The view from its summit is unusually fine, and well repays the hard climb to secure it. *Round Island*, *Minnow Island*, and Skinner's Island, with its cave of legendary fame, are among the local attractions.

Mount Elephantis guards the western shores of Lake Memphremagog. It is not visited as much as Owl's Head; but its attractions are enhanced by a sparkling lake, two miles in length, far up its side, which is noted alike for the crystal purity of its waters, and the abundance of trout it contains.

Mount Orford is one of the attractions of the village of Magog. It is reached by a pleasant drive, and a carriage-road extends to its summit. It is the highest peak in the vicinity, and commands a fine prospect. The village of Magog is also visited daily by stage, sixteen miles from Sherbrooke on the Grand Trunk Railway.

Pinnacle Mountain and **Lake** are about ten miles from Stanstead, but have not yet attained great popularity. The mountain is the most singular feature, rising with great precipitousness from the lake below.

Lake Massawippi, on the line of the Passumpsic Railroad, affords rare sport for the fisherman. It is only about a mile and a half in width, yet extends for nine miles. It is overlooked by *Blackberry Mountain*, which rises from its eastern shore.

Lachine Rapids are among the most thrillingly fascinating attractions in the vicinity of Montreal; a visit to which is usually made by the cars, and the return by steamer.

"The Lachine Rapids are visited by taking the seven, A.M., train (at the Bonaventure Station) to Lachine, where a steamer is in waiting, by which the tourist returns through the rapids to Montreal, arriving about nine, A.M. After taking a pilot from the Indian village of Caughnawaga, the steamer passes out. Suddenly a scene of wild grandeur breaks upon the eye: waves are lashed into spray and into breakers of a thousand forms, by the submerged rocks which they are dashed against in the headlong impetuosity of the river. Whirlpools, a storm-lashed sea, the chasm below Niagara, all mingle their sublimity in a single rapid; now passing with lightning speed within a few yards of rocks, which, did your vessel but touch them, would reduce her to an utter wreck before the crash could sound upon the ear. Did she even diverge in the least from her course, if her head were not kept straight with the course of the rapid, she would be instantly submerged and rolled over and over. Before us is an absolute precipice of waters: on every side of it are breakers, like dense avalanches thrown high into the air.

"Ere we can take a glance at the scene, the boat descends the wall of waves and foam like a bird, and in a second afterwards you are floating on the calm, unruffled bosom of 'below the rapids.'"

The **Falls** of **The Shawanegan** are visited from the city of *Three Rivers*, usually in canoes. Notwithstanding the magnitude and grandeur of these attractions, the difficulty of reaching them has detracted materially from their popularity.

They are on the *St. Maurice River*, thirty miles above the city of Three Rivers.

The towering rocks which set their bounds rise on either side, between which the stream makes a fearful plunge of one hundred and fifty feet.

PLEASURE ROUTE No. 3.

Boston, Lowell, Lawrence, Lynn, Salem, Newburyport, Dover, and Portsmouth, to Portland and the East, Lake Winnepesaukee, North Conway, and the White Mountains.

EASTERN RAILROAD.

THE **Eastern Railroad**, aside from leading directly to the popular watering places in Massachusetts, New Hampshire, and Maine, and being the through line to Bangor, St. John, and the Provinces, is also one of the principal routes of approach to Lake Winnepesaukee and North Conway; and, in its intersection with the "Portland and Ogdensburg," to the White Mountains proper, through the picturesque valley of the Saco, and the famous *White Mountain Notch*.

Of the four depots in the northern section of the city of Boston, the "Eastern" holds a central position, and is in the immediate proximity of each. It is also connected with those of the southern portion of the city by horse-railways, rendering it easy of access for strangers. This road extends from Boston to Portland, and Boston to North Conway, with a branch to Wolfboro'. There are also several other branches; viz., Saugus, Swampscott, Marblehead, Salem and Lawrence, South Reading, Gloucester, Essex and Amesbury, and Portsmouth and Dover. No route in New England possesses more varied charms for the tourist and pleasure-seeker.

Twenty-eight seashore-resorts are reached by the Eastern Railroad and its branches; the more prominent of which are Chelsea or Revere, Nahant, Swampscott, Marblehead, Lowell Island, Gloucester, Rockport, Pigeon Cove, Plumb Island, Salisbury, Hampton or Boar's Head, Little Boar's Head, Rye, Isles of Shoals, Kittery, York, Wells, Cape Arundel, Old Orchard, Mount Desert, and all others in Maine. Most of these are directly on the main line of this road, and all in close proximity to it.

Chelsea, Mass., was one of Starr King's favorite spots for sight-seeing. "Powder-horn Hill," with its remarkable pictures extending over a circle of miles, was a special object of frequent visits with the scholarly and enthusiastic young preacher, the rising or the setting of the sun being his chosen times for studying its wonderful beauties. The name has been arbitrarily altered to "The Highlands," which is more pretentious, certainly; but old names, after all, cling closest to historic associations. The "Eastern Railroad" and horse-cars pass through the city.

Chelsea Beach (**Revere**) is no part of Chelsea proper: it possesses many points of interest, and its proximity to Boston makes it a place of great resort in the hot months. A line of horse-cars connects it with Boston.

Nahant, Mass., is among the mature celebrities of the New-England coast. It is one of those rare combinations of natural and remarkable beauties which assert their superiority without the need of art or special praise. Yet Nahant is a lesson. It teaches the fickleness of human fancy, and the uncertainty of popular favor. If this really charming spot were only located a hundred or more miles from the leading marts of New-England trade, it would scarce find a rival in fashionable and public approval. It is too near Boston and other cities, too easy of access, and too comfortable generally, to attract the great multitude, who prove that "distance lends enchantment to the view" of a summer trip and life, by seeking remote and out-of-the-way places in preference. Yet Nahant is so delightfully located, so varied in its scenery and surroundings, so dotted with wonderful curiosities of nature, so graced with romantic and ever-varying specialties, and so readily reached, that the number of its summer residents and brief visitors will ever be very large. It was the chief resort of the wealthy and the gay only a few years since; but the worshippers of fashion now travel to other shrines.

Lynn, Mass., is a busy and thriving city, famed as the leading shoe-manufacturing place on the continent. A lofty and commanding eminence called High Rock, from which a singularly picturesque view is obtained, is the chief point of attraction to the traveller. Swampscott and Nahant, popular resorts, are contiguous. Trains on the Eastern Railroad, and horse-cars from Boston, pass through the city.

Swampscott, adjoining Lynn, is a favorite with the wealthy classes of Boston and neighboring cities, and has numerous costly and elaborate summer residences. Its comfortable boarding-houses have attracted many strangers for a summer's sojourn. The chief industry of the place is fishing; and a very clever addition to the season's profits is made by letting rooms and dwellings during the summer months.

Gloucester, Mass., is the great centre of the New England fishing interests. Thousands of her hardy population pursue their perilous avocation at all seasons of the year, and upon all the great fishing-grounds, especially upon the Banks of Newfoundland. No season passes without its sad tragedies among the vast fleet which leaves the harbor of Gloucester. The sources of pleasure and of cultivated intercourse located around Gloucester are worthy of an elaborate detail, and are full of agreeable surprises and rare delights. Great numbers take the cars of the Eastern Railroad, or boats from Boston direct, in the travelling season.

Rockport, Mass., was once a part of Gloucester. This place will not attract a great deal of attention from sight-hunters, although its extensive granite-quarries will richly repay a visit.

The famous and justly popular resort called **Pigeon Cove** is close by Rockport. This and other spots of novel and rare curiosities form a group of too much interest to be overlooked; and it has long been a fixed centre for a very large summer attendance. Few places on the New-England coast afford greater gratifications to visitors.

Newburyport, Plumb Island, and **Salisbury Beach** possess a local fame, and receive considerable patronage from the towns adjacent, but cannot be recommended to the general public. The bathing at Plumb Island is treacherous; Salisbury Beach is better, but the loose sands prevent driving.

The town of **Hampton, N.H.**, has little to distinguish it from towns of modest pretensions generally; but its beach — Hampton Beach — is renowned in every quarter. *Boar's Head*, a bold and commanding promontory, projecting a quarter of a mile from the mainland directly into the sea, is the hospitable castle which "lords it" over the adjacent beaches. Here the admirer of the murmuring sea can find full scope for his admiration. The views from this lofty eminence are numberless and varied. The origin of the name is somewhat shrouded in mystery. Tradition says it was given by fishermen, from the similarity of its foam-laved rocks, when lashed by the fury of the waves, to the enraged boar.

HOTEL

This summer resort has been long and favorably known. The house stands on the crest of a rocky promontory, which rises gradually to the height of eighty feet, against whose jagged base for ages past the waves in ceaseless roll have dashed their whitened spray. On either side, stretching for miles away, extend beautiful beaches, whose waters furnish rare facilities for bathing, and whose hardened sands present a surface for driving not excelled along this coast.

Little Boar's Head, North Hampton, N.H., is a connecting link between Hampton and Rye Beaches. It would be famous but for the superiority of its great rival, Boar's Head. It is a projection also into the sea, but of a lesser altitude. These marked spots, adjacent to such grand beaches as Hampton and Rye, are assured of a constant popularity. This is a favorite summer resort for families of taste and refinement. At present there is no hotel, but many excellent boarding-houses.

Rye Beach, N.H., half a century ago had an occasional straggling admirer, or possibly a company from the back country, in the summer season, to appreciate its beauties, and enjoy its lonely solitude. But it has since acquired a distinctive fame. At present its popularity is widely established, and thousands make it their resort for recreation and rest. It is animated and exhilarating in "the season," and is able to maintain its partial preference against all rivals of the coast. It is abundantly supplied with every source of enjoyment, — city, country, sea, and fashionable elegances and refinements, and all modes and moods of life, to suit all tastes.

Portsmouth, N.H., has proved an admirable place from which to emigrate. It has one of the best harbors, rears the smartest of men and most charming of women, but the city persists in not growing in population. It is a grand centre or starting-point, however, from which to visit a vast number of famed and delightful spots; and it wears a thronged and busy air during the hot months. It has, in the *Rockingham House*, a first-class hotel, which in all its appointments exhibits an air of elegance and comfort, and is convenient as a "roost" for travelling birds.

Frost's Point, near Portsmouth, N.H., is a very pleasant place, and has a local popularity.

The Isles of Shoals, off Portsmouth Harbor, have risen to wonderful fame within twenty years. Fifty years ago it was one of the places to visit, and have a chowder, and was noted for its wild and rugged features, even in those prosaic days. It is now a fixture in popular favor, and is visited by multitudes, who make a marked stay there in summer time. Its chief interest lies in its remoteness from the land, and its home in the sea. The entire scene is wild, grim, and barren, excepting the homelike comforts which enterprise and money have supplied.

We have written of "The Isles of Shoals" as "it," although there are half a dozen islands in the group; but we have always associated the places with the idea of but one. And old people still call them "Isle of Shoals;" and this is not far from correct. The eccentric Leighton, who really laid the foundation for the present great fame of this resort of pleasure, faithfully believed, that no person coming there, however sick, could die of disease if the invalid remained. Mr. Leighton, although living to a good old age, now rests with his fathers.

New Castle is one of the marine suburbs of Portsmouth, three miles distant. It has a new summer hotel, pleasantly located.

Kittery Point. — This quaint old Maine town, recently rejuvenated for a summer resort, possesses, in its historic associations, its admirable location, and its cool, exhilarating atmosphere, many attractions for the visitor who desires quiet and repose. Its principal hotel — the "Pepperell House" — occupies an elevated and commanding site, and the harbor offers unusual facilities for boating and fishing.

The government navy yard is near at hand, Portsmouth is across the harbor, and the Isles of Shoals in full view but a few miles away. Take cars on the Eastern Railroad to Portsmouth or Kittery Station.

York Beach, Me., and, beyond that, **Bald Head Cliff** (a wild, stern, defiant-fist rock, in almost constant battle with the waves of the sea), are places of interest, and when better known will command their share of patronage. The run to these points can be made by stage a half dozen miles from Portsmouth.

Wells and **Old Orchard Beaches** can also be reached by this road; the former by stage, six miles from *Wells Station*, and the latter, five miles by stage from *Saco*. There is little of interest between Saco and the city of Portland. For a description of the city of Portland and vicinity, see separate article.

TO THE INTERIOR.

Leaving such of our friends as we have directed to the numerous watering-places on the coast to their own amusement, we return to the "Conway Branch," to accompany those who desire to see that charming inland resort, Lake Winnepesaukee, to visit North Conway and the intermediate points, or go with us to the White Mountains proper.

The "Conway Branch" diverges from the main line a few miles east of the city of Portsmouth, and continues northward past Berwick, Great Falls, to Conway and North Conway; another road branches from this to Lake Winnepesaukee, where connection is made by boats to all points on the lake, and with the Boston, Concord, and Montreal Railroad beyond. At West Ossipee the stage from Centre Harbor, with tourists that have crossed Lake Winnepesaukee, from the Boston, Concord, and Montreal, and Boston and Maine Roads, intersects with this railroad with which its passengers continue.

Centre Harbor, Wolfboro', and **Alton Bay** are the three prominent places on Lake Winnepesaukee. The former is mentioned in Pleasure Route No. 1. Wolfboro' is pleasantly located on the east shore of the lake, at the terminus of the Wolfboro' Branch Railroad, and has several hotels. Alton Bay is at the lower end of the lake, reached by the Boston and Maine Railroad and the steamer "Mount Washington."

Mount Chocorua, N.H., 3,358 feet high, is most easily reached from West Ossipee Station. The trip to Chocorua, eight miles distant, is tedious, but amply repays those physically able to make it. This mountain is more Alpine in its character and outline than any in New Hampshire. The beautiful **Chocorua Lake** is passed *en route*, from whose borders one of the finest views of the mountain may be had. The surface, from far below its summit, is completely bare of vegetation. High overhanging rocks seem ready to topple from its craggy peak.

The view from Mount Chocorua is a singular combination of the beautiful and grand. Hundreds of lakelets dot the landscape, increasing in size to the charming Winnepesaukee, from whose placid bosom spring myriads of leafy islets; while northward the mountains rise tier above tier to Washington and the "Presidential Group." The **Ossipee** and **Sandwich** Mountains are also visited from this place, which is surrounded by picturesque scenery, and must eventually become a popular resort. Indeed, this entire region, extending from Centre Harbor to Wolfboro', embracing the towns on the eastern shore of Lake Winnepesaukee, is filled with interesting localities, and is beginning to be annually frequented by persons looking for the quiet, substantial requirements of summer life.

Ossipee Lake is usually visited from Bank's Hotel at this station. The road to it passes many attractive points; among which may be named an Indian mound, from which various articles of interest have been exhumed, including arrow-heads, implements of various kinds, and human bones. The locality also has its historic associations. The remains of Lovewell's Fort (built by Capt. Lovewell's band in 1725) are still pointed out. The lake once reached, a feeling of sequestered quietude reigns. It is less than half a dozen miles long, and the close proximity of Chocorua and the Ossipee Mountains increase the interest, and add to the beauty of the scene. **Ossipee Falls**, in Moultonboro', within a few years have received many visitors, and as an auxiliary attraction to other resorts, are very fine.

Passaconaway Mountain, N.H., 4,200 feet high, is one of those bold, unfrequented peaks lying west from Conway, and north-east of *Whiteface* of the Sandwich Mountains. At present it is but little known to tourists, and will only be visited by the adventurous spirit desirous of studying Nature in her primeval state. This mountain bears the name of a famous Indian sachem of the Pennacooks, a warlike tribe whose territory embraced this region at the time of its discovery by the English. Go by the West Ossipee and Centre Harbor stage to Sandwich, from West Ossipee to Centre Harbor.

Whiteface Mountain, N.H., 4,100 feet high, is the most noted of the group known as the Sandwich Mountains. The number of visitors has increased within a few years. The fine prospect from its summit is said to amply repay the arduous climb to secure it. The view of Lake Winnepesaukee is particularly fine. Stages between Centre Harbor and West Ossipee which connect with the lake boats pass through Sandwich, intersecting the railroad at West Ossipee, for North Conway and the White Mountains. This whole country in this neighborhood is exceedingly fine.

North Conway is a pleasant little village located upon the banks of the Saco River, and is the central feature of the charming Saco valley.

There are several fine hotels, ranging from the first class to the comfortable country inn; indeed, nearly every house is a boarding-house. The inhabitants are largely migratory, going and coming with the birds, and through the hot months of summer resting in the shades of Conway's famous elms, sauntering along her sparkling streams, or climbing her rugged mountains. North Conway has one noted mountain. —

Pequawket, or *Kiarsarge*, as frequently called, though by this name it is confounded with Mount *Kearsarge*, in the western part of the State. Mount Pequawket is three miles from the village, has a good bridle-path; and a public house on the summit furnishes refreshments or lodgings for those who desire.

Diana's Baths, Hart's Ledges, Echo Lake, and Artist's Falls are among the attractions of the place.

At North Conway the Eastern Railroad intersects the Portland and Ogdensburg Railroad, by which tourists continue to the Crawford House, White Mountain Notch, or by stage to the Glen House on the eastern side of the mountains.

The stage line to the *Glen House* is still in operation, and many prefer this route of approach. *Goodrich Falls* are passed *en route;* and it also leads through "Pinkham Notch," one of the natural thoroughfares to the mountains.

Glen Ellis Falls are among the most attractive in New Hampshire. They are in the woods at the right of the road, six miles before reaching the Glen House. *Crystal Cascade*, of equal height, is at the left, a mile farther on. *Thompson's Falls* are still nearer the house.

(For continued description to the mountains, see Portland and Ogdensburg Railroad.)

PLEASURE ROUTE No. 4.

Boston, Lowell, Lawrence, Haverhill, and Dover, to Lake Winnepesaukee, Wells, Cape Arundel, and Old Orchard Beaches, Portland, Mount Desert, North Conway, and the White Mountains.

BOSTON AND MAINE RAILROAD.

WHILE the Eastern Railroad has run along the coast, distributing tourists at the various seaside resorts east of Portsmouth, the Boston and Maine, whose depot in Boston is in Haymarket Square, at the head of Washington Street, has kept inland, passing through several cities and villages to Dover, whence a " Branch " leads up the Cocheco Valley to Lake Winnepesaukee, which is crossed thirty miles by steamer to Wolfboro' and Centre Harbor, and thence by stage and rail to North Conway, continuing to the White Mountains over the line of the Portland and Ogdensburg Railroad, to which description the reader is referred for a continuation of the route.

Returning to Dover, the road soon leads gradually towards the coast, which it touches at **Wells Beach,** an old and well-known summer watering-place. Coaches at the depot take visitors to the hotels, two miles distant.

Kennebunk is the next regular station. Three miles away by stage is **Cape Arundel,** a new resort just springing into existence. The beaches here are short, although very good for bathing. Cape Arundel will be eschewed by the fashionable tourist whose pleasure comes from " drives " and " hops," and whirls of excitement; but the lover of rocks and foaming spray, the student of nature in her wildest moods, should not pass it by.

Old Orchard Beach is the next point of interest to the tourist.

At no place along the New England coast has nature done so much, or planned such a magnificent beach, as here. For nine miles the surface is level and hard, and it has few equals in the country. As much cannot be said of the hotels. Good fair accommodations are furnished, nothing more, nor are the prices high. But with such hotels as those at Newport, Long Branch, or Cape May, Old Orchard might become the queen watering-place in the land.

From Old Orchard to Portland there is little to interest the tourist. From here he may go by steamer to Mount Desert, continue by the Maine Central Railroad to Rangeley or Moosehead Lakes, to St. John and the Provinces ; or he may go to the White Mountains by the Portland and Ogdensburg, or Grand Trunk Railways. See description of Portland.

PORTLAND AND VICINITY.

Portland, the chief city of the State of Maine in point of population and commerce, is situated at the southerly extremity of Casco Bay, and contains about thirty-five thousand inhabitants. It is of considerable importance as a railroad centre, being the terminus of six different roads converging at that point, and of numerous lines of steamers constantly plying between Portland and New York, Boston, Eastern Maine, and the British Provinces, and is the winter port of three lines of European steamers. Its facilities for communication are excellent; there being despatched daily eight trains and one steamer to Boston, connecting there with points farther south and west, besides several trains daily for Montreal and Quebec, Bangor, St. John, N.B., and Halifax, N.S., weekly; also tri-weekly steamers for Bangor, Mount Desert, St. John, Halifax, &c., and steamers four times a week for New York direct.

FALMOUTH HOUSE.

As a pleasure resort, Portland is becoming well known throughout the country; and the number of visitors to the city and its vicinity is increasing with each summer's return. Not only in lines of communication, but in hotels and other accommodations, is she particularly fortunate. The "Falmouth," "United States," and "Preble" are the leading houses; each of these possesses its peculiar merits.

The "Falmouth" is a first-class house in all its appointments: it is the largest, and is the only hotel in the State that has an elevator, a luxury that the weary traveller can appreciate.

The tourist who concludes to "do" Portland will find rich materials for his sketch-book and his notations. The vicinity is not only rich in landscape scenery, but the climate is delightfully cool, the heat of summer being tempered by the pleasant sea-breezes from the ocean, three miles distant. which combine to enhance its charms.

The United States Hotel is centrally located, and has been recently rejuvenated and put in condition to accommodate guests. It stands on Market Square, at the junction of Middle and Congress Streets, the principal thoroughfares of the city.

Portland is built upon a small peninsula jutting into Casco Bay; and a ridge of land through its centre, sloping on both sides to the water, affords excellent drainage. At the east and west extremities are high elevations, known respectively as *Munjoy* and *Bramhall*, with fine driveways and promenades, which command grand and extensive views of the surrounding country. The observatory on Munjoy, which no one should fail to visit, is provided with a powerful telescope, sweeping the horizon in every direction, by aid of which nearly every summit of the White Mountain Range, eighty miles distant, can be easily distinguished. The view from this point seaward is magnificent, embracing as it does the numerous islands of Casco Bay, the surrounding coasts, and an uninterrupted view of old Ocean, extending more than thirty miles from land, dotted with sails, and flecked with foam. The shady streets and attractive suburbs invite to charming walks and drives through their quiet avenues. It is visited yearly by families and parties who spend their entire summer vacations hereabout, interspersed with short-trip excursions in the harbor, to Mount Desert, Fryeburg, Sebago Lake, Winnepesaukee, North Conway. or the White Mountains, all of which are within a few hours of Portland, and of easy access.

UNITED STATES HOTEL.

Casco Bay, with its fifteen-score of islands, their rocky promontories and pleasant coves, their green fields and forests rivalling in romantic beauty the archipelagoes of Greece, is much visited during the pleasure season. The islands nearer the city contain a considerable population; and their many good hotels and boarding-houses are well supported in the summer. Four steamers of good capacity and accommodations run daily between them and the city, making several trips each way, and are largely patronized. The round trip is made in from one to two hours, and forms a delightful afternoon's recreation.

Casco Bay, the musical waters whereof sing sweet songs, even up to the margin of the discordant city, — Casco Bay seems expressly formed for the lovers of the romantic, the beautiful, and the wonderful. Here Nature has nestled the charms of the sea and of the land in almost every variety, — in miniature continents, rivers, hills, valleys, bluffs, beaches, wild rocks, soft verdure, fragrant flowers, and birds of richest

WHITE HEAD CLIFF, PORTLAND HARBOR.

plumage and sweetest song. Indeed, the dullest nature is moved with unwonted stirrings, approaching the poetic; and the man of sensibility feels a pleasure rarely found, all the more keen that the noisy and disorderly crowd have not yet invaded this undefiled paradise of the sea and shore.

More extended trips are also daily made which are assuming a national popularity. Mount Desert, Old Orchard, Cape Arundel and Wells Beaches, Fryeburg, North Conway, and the White Mountains are all within a few hour's ride.

PORTLAND, BANGOR, AND MACHIAS STEAMBOAT COMPANY.

The steamer "Lewiston," after receiving passengers from the Eastern, Boston and Maine, Portland and Rochester, Portland and Ogdensburg, and Grand Trunk Railroads, and Boston Boats, leaves Portland from Railroad Wharf every Tuesday and Friday evening, at ten, P.M., touching at Rockland at five o'clock the next morning, Castine at seven, and arriving at Mount Desert at noon. One has but to glance at the map of Maine to become impressed with its wonderfully serrated coast, its numerous bays and headlands, sand and gravelly beaches, rocks, coves, and outstanding islands, embracing some of the most delightfully rural resorts in the country. Many of these are passed *en route*, and the morning view from the steamer's deck is enchanting.

Mount Desert is the rising star of Maine's attractions for the summer-home seekers. Apart from any little side-shows which may have been put up in the papers from time to time by shrewd and calculating capitalists, Mount Desert has gifts that are all its own, and such as will continue to swell its fame as years progress, and its charms become revealed. Its area is reckoned at about a hundred square miles; and it is, therefore, quite a little world in itself. The island might aptly be likened to a lot of marbles dropped from a pocket of a giant, provided a giant's marbles were small mountains. At least, the more notable and striking portions of the island are made up of a group of mountains huddled together, of a singularly wild grandeur. Upon one portion there is a sheer and almost vertical descent of rock, nearly a thousand feet from the brink to the deep water below; and the progress of the explorer is constantly met by changes and surprises of panoramic and kaleidoscopic beauty. Much fine soil is found, which is considerably cultivated; but the inhabitants are chiefly absorbed in fish-catching. Portions of Mount Desert are still primeval in their solitudes; and Nature yet prevails in her simplicity and peculiar sovereignty. Hence wild game may still be hunted; and sylvan streams are enriched by great numbers of the gamey trout. The indications within a few years are unmistakable that Mount Desert will take a rank among the families of the wealthy and fashionable second to none on the coast, or even on the continent. As soon as the needed capital decides to invest, and the newspapers open their trumpet throats to proclaim Mount Desert and "all about it," the armies of summer pilgrims will commence the mighty march to grand and glorious Mount Desert. Go by steamers of the Portland, Bangor, and Machias Steamboat Company, or by boat from Rockland, connecting with the Knox and Lincoln Railroad. Good accommodations at *South West Harbor* may be had at the *Ocean House*, and visitors to *Bar Harbor* will be well entertained at the *Rodick House* and others.

UP THE PENOBSCOT.

The scenery along the Penobscot is unusually fine. Summer hotels are springing up; and when this interesting region is fully understood, through descriptions and illustrations, it must become popular with tourists. In full view are the **Megunticook Peaks** (altitude 1,355–1,457 feet). They are near the pleasant river village, *Camden*, and form one of its attractions. Although of less elevation than many mountain resorts, they command views of rare landscape interest, combining mountains in the north; the serpentine windings of the Penobscot, bounded by cultivated farms and thriving villages; and, south-east and east, Penobscot Bay with its thousand "sea-girt isles." Camden is passed by two lines of steamers, Portland and Bangor, and Sanford's Independent Line from Boston, both of which touch here.

PORTLAND LIGHT.

Near the head of Penobscot Bay, and within twenty-five miles of Bangor the boats pass **Fort Point**, a place of much scenic interest, and famed for its many historic recollections. The ruins of *Fort Pownal*, built by the English in 1759, still remain. Near by on the "Point," a summer hotel, the *Massaumkeag*, has been recently erected, which forms a commanding feature in the landscape, and which, from its cool location, fine scenery, and good fishing and boating in the vicinity, is receiving its full share of patronage.

PLEASURE ROUTE No. 5.

Portland to Sebago Lake, North Conway, and the White Mountains.

PORTLAND AND OGDENSBURG RAILROAD.

No more attractive route to and from the White Mountains is afforded to tourists than this from Portland, Me., up the beautiful valleys of the Presumpscot and Saco Rivers, by the lovely Sebago Lake, and through the famous intervales of Fryeburg and Conway, and the wonderful Crawford or White Mountain Notch to Crawford or Fabyan Houses. In former days it was considered the most interesting of the many stage-routes to the mountains; the gradual and distinctly marked transition from the lowlands of the coast, to the higher and grander elevations of the mountain region, invests it with untiring interest for the traveller from the beginning to the end of the journey. The recently constructed Portland and Ogdensburg Railroad follows almost without deviation the old and popular stage-route, and without losing the attractions of the old highway, has opened new scenes of beauty; and, for the short time that it has been open to travel, has already acquired an enviable reputation as a pleasure route, and will, as it becomes better known, have an honored place in the itinerary of every visitor to the mountains, as the mountain tour is not complete if the Portland and Ogdensburg Railroad is overlooked, especially that portion of it which passes through the "Notch."

The many lines of railroad and steamboat conveyance, centring at Portland from Boston and the South and West, make this city a most available point from which to commence a mountain trip. Portland, of itself, possesses many attractions, and is becoming celebrated as a summer resort. Its fine hotels, beautiful walks and drives in and around the city, the magnificent views of ocean and inland scenery to be obtained at various points within the city limits, and its proximity to many popular resorts of the seashore and country, make it a desirable stopping place for the pleasure-seeker. But those wishing to avail themselves of a ride over the Portland and Ogdensburg Railroad are by no means obliged to stop over at Portland, unless arriving in the city late at night; in which case a good night's rest will prepare them to enjoy all the more heartily the trip from Portland to the mountains. The Portland and Ogdensburg trains connect closely with the Eastern and Boston and Maine Railroads from Boston, and the new through line from New York direct, viâ Worcester and Nashua, and the Portland and Rochester Railroad, and also with the fine steamers of the Portland Steam-Packet Company, which leave Boston every evening during the summer season; and arrive in Portland in season to take cars for North Conway and the mountains, which will be reached the same evening; passengers meanwhile enjoying the near approach to the mountains by sunset, the most beautiful hour of the day. The time from Portland to Crawford or Fabyan's is but little over three hours; and parties going by way of Portland from Boston will arrive as early as by any other route. It will thus be seen, that, while rivalling all others in attractiveness, this route is as expeditious as any to the mountains.

SEBAGO LAKE,
Portland and Ogdensburg Railroad.

A brief description of some of the more prominent points of interest will serve to show in a slight degree the pleasure to be enjoyed in a trip through its charming scenery; and the personal experience of all who travel this way will relieve us from any charge of exaggeration or partiality. **Sebago Lake** is the bright spot in our picture; but we will not anticipate.

On leaving the station in Portland, the road passes around the western boundary of the city, along the shore of Fore River, that forms the upper part of the harbor for which Portland is celebrated; and under the brow of *Bramhall*, a promenade much resorted to by citizens and visitors for its superb and extensive views of the country and the White Mountain range, ninety miles distant. It then crosses the old canal basin, and continues through an open country to Westbrook, five miles from the city. At this point we strike and cross the *Presumpscot River*, along the banks of which beautiful stream the road lies for some seven miles, through the farming town of Windham; affording many picturesque views of the river with its numerous water-powers, mostly unimproved, but awaiting the advent of enterprise, when their energy shall be harnessed to the wheels of industry, making this one of the busiest as it now is one of the most powerful of the rivers of New England. Again crossing the Presumpscot, and passing through a portion of the town of Gorham, we come to an uncultivated tract with deep cuttings, whose bare walls of sand and gravel only serve as a reminder of the labor which was expended in preparing for our convenience and pleasure; here also, in a natural depression of the land, nestle the lovely *Otter Ponds* in calm and retiring beauty; another moment, and, without any previous intimation of their nearness, the broad waters of *Sebago Lake* roll at our feet as we dash out from behind a projecting bank upon the shores of the lake, taking in at a glance its wide expanse and distant shores with their outlying hills.

Lake Sebago is fourteen miles long, and eleven miles wide in the widest part; and its deep waters, noted for their purity, supply the city of Portland, seventeen miles distant. Through its natural outlet, the Presumpscot River, it also furnishes a series of most valuable water-powers which never fail even in the dryest seasons, nor are subject to disastrous freshets. The lake with its appendages forms a navigable water extending from the *Lake Station* on the line of the railroad, and at the southern extremity of Sebago, thirty-four miles, to Harrison at the northerly end of **Long Pond**. This latter sheet of water is connected with the lake by *Songo River*, whose narrow and tortuous channel twists and turns through a course of six miles to overcome a straight distance of only a mile and a half; a lock near the outlet of Long Pond raises the steamers and other craft plying upon these waters from the level of the lower to the upper lake. Two steamers, in summer season, perform the round trip between *Sebago Lake* Station and *Harrison*, starting from either terminus daily, and connecting each trip with the Portland and Ogdensburg trains; and the excursion thus offered is replete with charming attractions, and is rapidly increasing in popularity. The trip to Harrison and return, including landings at Naples, Bridgton, and North Bridgton, is made in about eight hours; tourists may then go to Portland, North Conway, or the Mountains the same evening.

Sebago Lake, with its far-reaching landscapes and distant mountain views, the passage through the serpentine windings of the silent and beautiful *Songo*, and the fair shores of *Long Pond* dotted with pretty hamlets and thriving villages, offer an excursion of varied and exciting interest, never to be forgotten by the favored ones who participate in its enjoyment.

After leaving the Lake Station, the railroad skirts the shores for some three miles, then in almost unbroken tangent strikes across the country to **Steep Falls**, twenty-five miles from Portland. At this point we reach the *Saco River*, through whose lovely valley the road is laid until we reach its source at the summit of the mountain pass.

Here, too, the tourist will observe that the land assumes a more hilly character: the surface which until now had the general undulating features common to New England begins to rise in abrupt elevations increasing in height as his journey extends, constantly changing to new forms of beauty and grandeur, and gradually closing in around him on either side as if to oppose his farther progress.

From Steep Falls onward an endless succession of enchanting views gives an untiring interest to the journey. Ever-varying pictures of many-sided nature, harmonious and complete in all the elements of beauty, unfold themselves to the delighted gaze of the traveller. Mountains rising grandly from the plain in gentle slope or more precipitous inclines give grace and dignity to the landscape; and the panorama, gradually changing from the broad acres of the lowlands to the narrow valleys of Conway and Bartlett shut in with mountain walls, finally culminates in the grand and inspiring passage through the wonderful Crawford Notch of the White Mountains.

Passing through the town of Baldwin, after leaving Steep Falls we cross the Saco, and enter the town of Hiram; but just before reaching Hiram Bridge two pictures of the river are obtained in such quick succession, and at the same time in such wide contrast with each other, as to vividly impress them upon the memory. After leaving the West Baldwin station, and crossing the short high trestle known as *Breakneck*, you will see on the left through an opening in the woods the Great Falls of the Saco, white with foam, plunging over a solid ledge a total descent of seventy-two feet. As if guarding the seclusion of the scene, there rise on either side high hills thickly wooded from base to summit, whose deep-tinged foliage and overhanging shade give a sombre character to the view. It is a picture of wild and solitary beauty, and is viewed to best advantage from the railroad. When we next see the river, it is in softer mood. The waters lie seemingly motionless beneath the shade of gracefully impending branches, its surface flecked with dancing sunbeams which have pierced the leafy canopy ; and as we obtain a glimpse of the stream in perspective, retiring far within the deep and pleasant shade, it seems the very home of the water-sprites as memory recalls the fairy tales of childhood. But the picture is fleeting ; and with regret that we cannot linger, and enjoy its surpassing loveliness, we hurry on to other scenes; past Hiram and its hilly surroundings, and through the wide meadows of Brownfield, overlooking which stands Mount Pleasant, a mountain situated in the town of Denmark. We soon reach the beautiful village of **Fryeburg**, whose broad, shady avenues, enchanting scenery, and delightful climate, make it a growing rival to the more celebrated North Conway.

Good hotels and boarding-houses are found here: and the peculiar charms of Fryeburg, with its proximity and ease of access to various resorts among the mountains, make it a most desirable spot in which to spend the summer vacation. Before proceeding farther, a short notice of **Mount Pleasant** will not be amiss. This mountain standing alone, and rising in graceful outline to an elevation of nearly three thousand feet, is reached by a short drive of two or three hours from Fryeburg or Bridgton. A carriage-road has been laid out to the summit, and a commodious hotel erected upon the highest point of the mountain, where visitors are entertained in excellent style. The hotel has been opened only within two years past, but has been largely patronized; and Mount Pleasant has already attained a high place in the estimation of pleasure-seekers. Its situation, nearly midway between the loftier peaks of the White Mountains and the seacoast, renders the outlook from its summit one of great variety and interest. On the east side is seen the ocean with its white sails and crested waves, and the roofs and spires of Portland are also in view. On the west rise the stately domes of the White Hills from the symmetrical *Pequawket*, or Kiarsarge as it is sometimes called, to the towering peak of Mount Washington. In all directions the valley lies spread out in wide extent, adorned with lovely lakes and winding streams, and rich in all the charms of nature.

Leaving the village of Fryeburg upon our right, we shortly cross the western boundary of Maine, and enter the town of Conway in New Hampshire. A fine view of the easterly slope of the Rattlesnake Range is obtained on rounding the curve just out of Fryeburg; and the broad interval, with its graceful elms, standing singly or in groups of pleasant shade, forms a beautiful foreground to the picture. On the left, as we near the station of Centre Conway, is seen the distant *Chocorua*, its bold, precipitous ledges sharply outlined against the sky. After leaving Centre Conway we recross the Saco, and, rounding the southerly spur of Rattlesnake Range, are quickly arrived at renowned and beautiful **North Conway**.

NORTH CONWAY, N.H.

When the days begin to lengthen, and the sun runs high in the heavens; when the short nights fail to cool the heated streets of a dusty city, and man feels that he must have respite from the care and excitements of business; when the mother rises in the morning unrefreshed,

CONWAY ELMS.

and the children grow languid for a change; when, finally, the family council decide that a few weeks' vacation must be spent in the country, — no place can be found where the cool mountain air blows fresher, where the crystal streams flow purer, or where Nature wears a lovelier garb, than at *North Conway.* Since the early settlement of the country, the praises of Conway's rich meadows have been sung. To-day her broad-sweeping elms and luxuriant gardens indicate the strength of their rich alluvial soil.

MT. KIARSARGE FROM NORTH CONWAY.

North Conway has an additional advantage, wholly and peculiarly her own. She has a mountain, — Mount Kiarsarge, — which forms an admirable objective point for her visitors. It is only two miles distant, and three more to its summit; yet the prospect from it combines some of the finest panoramic scenery in New England. A small hotel on its crest affords refreshments and shelter for those who desire.

The climate of North Conway is free from mists and fogs; and with its pure air, and dry and invigorating atmosphere, it is one of the most desirable points in the whole White Mountains for those who may be seeking health or pleasure. "And then the sunsets of North Conway! Coleridge asked Mont Blanc if he had 'a charm to stay the morning star in his steep course.' It is time for some poet to put the question to those bewitching elm-sprinkled acres that border the Saco, by what sorcery they evoke, evening after evening, upon the heavens that watch them, such lavish and Italian bloom. Nay, it is not Italian: for the basis of its beauty is pure blue; and the skies of Italy are not nearly so clear as those of New England. One sees more clear sky in eight summer weeks in Conway, probably, than in the compass of an Italian year."

North Conway is not only noted for the beauty of its scenery in the spring and summer months, but later in the season the bright tints of its autumn foliage make it more lovely than ever; and nowhere can the magnificence of the autumnal forest scenery of New England be seen to better advantage than on her hills and mountains.

DIANA'S BATHS.

Few localities are better or more favorably known to the "artist world" than North Conway. The variety of the scenery is particularly noticeable; while it possesses some of the broadest landscape and mountain views, it is celebrated for the beauty and artistic value of its choice "bits." Although midsummer is the most fashionable season at North Conway, we have our own opinion that it yields more pleasure when nature is fresh and redolent with the sweet breath of spring, or in the later season when valley and hillside are resplendent with the crimson and gold of autumn; and the same is true of the entire mountain tour. To be sure, the chilly nights must be prepared for; and, if properly provided for in that respect, one need fear no discomfort.

At this point passengers who left Boston by the Eastern or Boston and Maine Railroads for the White Mountains join our train.

Six miles from North Conway is *Glen Station*, where stages are in readiness to convey passengers to Jackson (two miles), or to the *Glen House* fourteen miles distant by way of **Pinkham Notch**, one of the three great highways which nature itself has hewn through the mountain wilderness. Six miles beyond we reach Upper Bartlett, shut in on all sides by high mountains, the situation possessing many elements of beauty, in some respects resembling North Conway and its surroundings.

The railroad is already built from this point to *Frankenstein Cliff*, nine miles distant; with every indication that early in July the entire trip from Portland to the Crawford and Fabyan Houses will be accomplished in the cars.

Leaving Upper Bartlett, the road keeps to the valley for some six miles, delaying till the last moment the steeper ascent it must soon commence; and the mountains, in seeming displeasure at this bold attempt to penetrate their hiding-places, draw closer together as if to dispute the passage. But onward we ride into the dark shadows of the hills, crossing and recrossing the Saco, or leaping torrents rushing from the mountain side to join the larger stream. Just before reaching Bemis Station, we cross *Nancy's Brook*, of memorable fame. At Bemis we shall see the old Crawford Homestead with its pretty clearing, and the residence of Dr. Bemis, built of granite quarried from these hills. From this place stages will connect temporarily for the Crawford House. Here begins our ascent of the mountain side; and we shall reach the **Gate of the Notch**, nine miles distant, by a continuous climb of a hundred feet and more in each mile of the journey. The character of the scenery does not change materially after leaving Bemis, until nearing *Frankenstein Cliff*, when the most magnificent view in the entire mountain region rewards our waiting eyes.

Towering high above the beholder, stands Mount Washington in grander proportions than can be observed at any other point. Spurs of high mountains ranging from its sides, like strong arms held out towards us, protect in their rough embrace the valley of the *Mount Washington River*, of which we have a splendid view in perspective from its confluence with the Saco to its source far up the side of Mount Monroe. This glorious picture which words fail to describe continues in full view from the road, and is not lost sight of until we have passed under the precipitous cliff of Frankenstein, continuing our journey up the side of *Willey Mountain*. The railroad is most advantageously located for viewing the conformation of nature in this remarkable mountain-pass. Built upon the side-hill hundreds of feet above the old carriage-road, it overlooks the entire valley, and commands an unobstructed view from mountain-summit to river-bed throughout the six miles of the "Willey Notch." The advantage of the open observation cars is now realized. The lofty battlements of *Mount Webster*, scarred and torn by the storms of ages, guard the valley upon the east, a silent witness to the terrors of the flood which ravaged its fellow guardian of the west, and, beating down the rugged escarpment of the mountain, prepared the pathway for commerce which we in later days are utilizing. Across numerous water-courses, and over deep and rocky gorges, or clinging to the sides of precipitous ledges, we ride, each moment revealing to us some new phase of mountain scenery. Just before emerging from the Notch, the line passes under the brow of Mount Willard, which stands a stern sentinel at the head of the valley, and overlooks the whole.

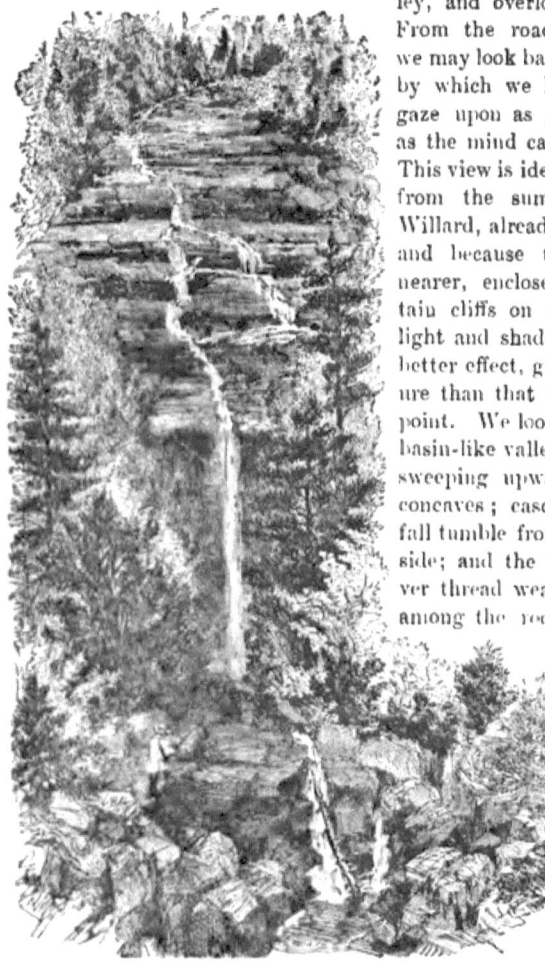
SILVER CASCADE.

From the road at this point we may look back over the path by which we have come, and gaze upon as grand a picture as the mind can well conceive. This view is identical with that from the summit of Mount Willard, already so celebrated; and because the prospect is nearer, enclosed with mountain cliffs on either side, and light and shade are seen with better effect, gives more pleasure than that from the higher point. We look down into the basin-like valley with its sides sweeping upward in graceful concaves; cascade and waterfall tumble from the mountain side; and the Saco like a silver thread weaves in and out among the rocks and through the green texture of overhanging foliage; while at our feet lie the waters of *Dismal Pool*, secluded and still as if sullenly reluctant to move into the sunlight beyond.

Across the valley the sparkling waters of **Silver Cascade** are seen, a mere thread of silver. Down from the mountain top it springs near a thousand feet in successive leaps, a rushing, boiling, foaming mass, till it joins the waters of the Saco below.

Turning to the left, while yet under the shade of Mount Willard we reach the summit of our ascent, and, bursting through the narrow **Gate of the Notch**, are soon at the hospitable doors of the Crawford House.

WHITE MOUNTAIN NOTCH,
From the Crawford House.

It is four miles from Crawford's to the Fabyan House; and the railroad, having passed the source of the Saco in the pretty lakelet between the Gate of the Notch and the hotel, follows the valley of the Ammonoosuc River for the remaining distance. After leaving the Crawford, and before we reach the Fabyan, another grand view of the Mount Washington Range is had, including the entire western slope from Clinton to Clay, and the summits of Washington and Jefferson, the highest elevations of New England. The completed portion of this division of the Portland and Ogdensburg Railroad terminates at the doors of the Fabyan House, where connection is made with the "Boston, Concord, and Montreal;" and with the Mount Washington Turnpike and Railway, for the ascent of Mount Washington; but the road is located from this point through the pleasant towns of Carroll and Whitefield to the Connecticut River, passing within easy distance of Jefferson Hill, the favorite resort of Starr King.

From the Connecticut the line, already mostly built and in operation, will cross the splendid farming country of Northern Vermont to the head of Lake Champlain, with branches to Montreal and Burlington. There are many attractions near the Crawford House, besides those already named, that will command the attention of the tourist, — *Gibbs Falls, Beecher's Cascades, Mount Willard, Willey House, Sylvan Glade*, and many others.

For the lover of the wild and picturesque, the tourist will find ample opportunities to gratify his taste by a visit to Gibbs Falls, particularly if the stream is traced to its source at the base of the mountains. This locality was examined and the falls sketched by the writer in 1857, and named for the (then) proprietor of the Crawford House. Beecher's Falls have been popularized by the interest taken in them by their noted namesake.

The "Notch" is the great natural gateway to the White Mountains proper. Mounts Webster and Willard form its outstanding pillars. The scenery is grander than by any other approach. Nowhere can this be so well realized as from the summit of Mount Willard, which, with its admirable carriage-way, must always be one of the most popular resorts in this region. The "Gate of the Notch," flanked by perpendicular ledges, is but twenty-four feet wide; through which passes the carriage-road, and flow the waters of the *Saco*, which rises a short distance above. The *Crawford House* is also the starting-point of the only bridle-path to Mount Washington, which no person physically able should fail to visit. The path enters the forest at the house, through which it winds its way by a rough course to the summit of Mount Clinton; thence continuing by a rugged pathway over (or around) Mount Pleasant, Mount Franklin, Mount Monroe, to Mount Washington. The route follows the crest of the mountains, and affords a combination of the finest views in the region, — one of the grandest of which embraces that stupendous gulf, **Tuckerman's Ravine,** which falls sheer down a thousand feet. This not only forms one of the wildest retreats about the mountains, but it generally contains an individual feature of interest, the snow-arch. During the winter months, the north-west winds completely fill this chasm with snow, which, packed by the driving storms of wind and sleet, by the warm rains of spring and the hot sun of summer, settles to a firm, compact mass. As the swollen streams pass beneath, the snow is melted. The massive bowlders which fill the valley become the base of so many ice-pillars, which remain and uphold the enormous snow-arch above. On the 12th of August, 1857, the writer entered this cavern to the distance of three hundred feet, and, by estimate, found the snow still twenty-five feet thick. It all passes away, however, by the last of August or the first of September. Tuckerman's Ravine can be visited from the Summit or from the Glen House.

The bridle-path excursion, about nine miles, is frequently made by pedestrians. But no one should attempt it without being well shod; and the sudden accumulation of clouds and mists on the mountains renders an experienced guide indispensable. The "Crawford" has always been noted for its admirable *cuisine*, and will be found withal one of the most desirable houses at the mountains. It is also reached by the Boston, Concord, and Montreal Railroad, to the Fabyan House, and thence by Crawford House coach.

PLEASURE ROUTE No. 6.

Portland to the White Mountains, Lake Umbagog, and Rangeley Lakes, Montreal, Quebec, Thousand Islands, Niagara Falls, and the West.

GRAND TRUNK RAILWAY.

THE Grand Trunk Railway connects in Portland with all roads running east and west. The approach to the White Mountains from Portland and the East by this line presents many points of individual interest. This is also a favorite route to Lake Umbagog, and the Rangeley Lakes; and when properly understood, through description and illustrations, must prove an excursion of great popularity. In this connection a few of the more important objects of interest will be described; although general illustrations must be deferred to a future edition. Leaving Portland, the line of the road soon enters the valley of the Androscoggin, up which it continues surrounded by the beautiful scenery for which this region is noted.

Bethel, a fine summer resort of local fame, possesses the elements of popularity, and, when better known, will be sought for its many attractions. It has much the character of North Conway, and each year increases in popularity.

Lake Umbagog is reached by stage twenty-six miles through scenes of various interests, and, like most of these inland sheets of water, the fishing is good.

Rumford Falls in the hilly town of Rumford, Me., on the Androscoggin River, for height, rugged grandeur, and picturesque beauty, have no superiors in New England. In three rapid and successive leaps the river makes a quick descent of over one hundred and fifty feet. The interest in the falls is greatly enhanced by the wildness of the surroundings. Reached by the Grand Trunk Railroad to Bryant's Pond Station, and thence by stage.

Near the station at Gorham, N.H., are located several mountains of considerable fame. One of the first points of interest is **Mount Hayes**, one of the north-eastern spurs of the White Mountains, an elevation of twenty-five hundred feet. It rises from the eastern bank of the Androscoggin River; and, before the Alpine House was destroyed by fire, a visit to this mountain was a popular excursion for its guests. A foot-path leads to the summit where is obtained a fine view of Mount Washington and neighboring peaks.

Mount Moriah, N.H., having an altitude of forty-seven hundred feet, rises from the valley of the Androscoggin, near Mount Hayes, north-east from the White Mountains proper. In a region where mountains were less common, this would be worthy of more attention. A bridle-path once led to the summit; but it is now neglected, and out of repair.

Mount Surprise is another name for one portion of Mount Moriah. This mountain is generally visited for the fine view of *Pinkham Notch* which it affords. It has had less visitors since the destruction of the Alpine House, but it has always been one of the favorites with tourists to the mountains. It has a good bridle-path; but, although saddle-horses can be always obtained, it is much visited on foot. The summit commands a very fine view of the White Mountain group. It is reached from the Gorham Station of the Grand Trunk Railroad. Visitors to the Glen House go by stage eight miles from Gorham Station; from thence a carriage-road extends up Mount Washington.

The Glen is one of the largest summer resorts in the White Mountain region. This fine establishment occupies a most picturesque location in the beautiful valley of the Peabody River, within a few rods of Mount Washington Summit Carriage-Road (which is one of the best constructed roads in the country), commanding a fine view, from base to summit, of Mounts Washington, Jefferson, Adams, and Madison, head of Tuckerman Ravine, and the Carter Range, — forming one of the finest panoramas to be obtained in the whole mountain region of New Hampshire.

Other points of especial interest in the vicinity are Glen Ellis Falls, Crystal Cascade, &c.

Berlin Falls, N. H. — Whoever shall have made the tour from Gorham to the Rangeley Lakes, and sailed across the charming Umbagog, where the wild Androscoggin gathers its head-waters preparatory for a race to the sea, and neglected to visit Berlin Falls, will have missed one of the most attractive features of that enjoyable excursion. Here, indeed, for the space of a mile, is one continuous rush and roar of waters, one wild, foaming cascade. Walls of adamantine rock crowd the Androscoggin into a narrow space through which it rushes, and in its mad career falling, tumbling, boiling among the rocks, a mass of glittering spray. They are but a half-dozen miles distant from Gorham Station.

Dixville Notch, Connecticut Lake, and Stratford Peaks, described in Pleasure Route No. 1, are also reached by this, which is also one of the principal thoroughfares to Montreal, Quebec, and the West.

PLEASURE ROUTE No. 7.

Portland to Lewiston, Farmington, Rangeley Lakes, Bath, Augusta, Moosehead Lake, Bangor, St. John, and the Provinces.

MAINE CENTRAL RAILROAD.

THE Maine Central Railroad is a continuation of the Eastern, and connects in Portland with the Boston and Maine, Portland and Ogdensburg, Portland and Rochester, and Grand Trunk Railroads, and Boston, Mount Desert, Machias, and Bangor boats.

The **Rangeley Lakes** are perhaps, at this time, receiving more attention from tourists than any other section of Maine; not only from their individual attractions, but from their wild surroundings. The complete tour, with its boating and portage, possesses just enough adventure to give romance to the occasion. This remarkable chain of waters consists of several distinct lakes, connected by narrows and streams, yet forming one continuous water communication for a distance of over forty miles. Each has its individual name, but they are collectively known as "The Rangeley Lakes." There are, probably, few places in the country where trout-fishing can be more successfully enjoyed. They are generally reached by the Maine Central Railroad to *Farmington*, a place of growing popularity as a summer home, and thence thirty-six miles by stage. You will spend the night at *Phillips*, seventeen miles from Farmington, where excellent quarters will be found at the Barden House; and, rested and refreshed, the traveller will better enjoy the remainder of the journey. Indeed, so comfortable are the accommodations, so fine the surrounding scenery, and so excellent the speckled-trout fishing in the neighboring streams, that, in spite of itself, Phillips is growing rapidly to assume the character of a summer resort, in every thing save exorbitant prices. Many tourists arrange to spend several days or weeks here, for rest and recuperation. The proximity of Mounts *Blue*, *Abraham*, and *Saddleback* not only gives a pleasing variety to the landscape, but form objective points for short excursions.

Mount Blue, Me., with an altitude of four thousand feet, is but a few miles away, and is famed for the wonderfully diversified views its summit affords. The region is mountainous and picturesque, and is noted for the pure water of its streams, and for its invigorating atmosphere. Mount Blue is reached by the Maine Central Railroad to Farmington, thence by stage to Weld, fifteen miles, or Phillips, eighteen miles; usually the latter.

Saddleback Mountain in altitude is but little lower than Mount Blue; but, although affording fine views, is not yet so favorably known. Go by stage from Farmington to Madrid, or from Phillips direct.

Damariscotta Lake, located in Lincoln County, is a resort just springing into existence. It is twelve miles long by three wide, at its greatest extremities; although its remarkable irregular shores give great changes in its apparent size. It is beautifully interspersed with islands; its coast is divided into sand and gravelly beaches, and rough rockbound shores, frequently overhung and shaded by groves of leafy trees, which form delightful resorts for "picnic grounds." A summer hotel has just been erected, and a pleasure steamboat placed on the lake for the entertainment of guests. Reached by the Knox and Lincoln Railroad, from the "Maine Central."

Moosehead Lake is the largest and most important of the inland waters of Maine. It is thirty-five miles long, and varies from four to twelve miles in width. It is somewhat larger than Lake Winnepesaukee, but falls far behind it in notable characteristics and attractions. It is nearly surrounded by dense forests, and is much frequented by sportsmen. Deer and moose were formerly plenty, and are now occasionally met. The special amusement, however, is trout-fishing. Still-fishing is also good. Moosehead Lake is a resort of increasing attractions.

Mount Kineo is a bold, rugged bluff, rising precipitously from the east shore of Moosehead Lake, beneath the shadow of which, on a protruding point of land, the *Mount Kineo House* is situated. This elevation is easily ascended, and affords fine views of the lake. The village of Greenville, at the foot of the lake, is reached by several distinct routes. Take the Maine Central Railroad to Skowhegan, thence by stage fifty miles; or by stage from Dexter Station, on the same road; or by rail or steamer to Bangor, thence by the Bangor and Piscataquis Railroad to Guilford, and from there by stage twenty-three miles to Greenville, where the boats connect with Mount Kineo.

Lakes Chesuncook, twenty miles long by two wide, **Caucomgomuc, Caucomgomosis,** and **Allagash,** are a chain of lakes on the head waters of the Penobscot River, beyond the lines of civilization, and only interesting to tourists as camping-out fields for sportsmen. They are passed by canoes successively, from Moosehead Lake to Mattawamkeag or Old Town. A week's supply of provisions, with experienced guides, will be required. Mount Katahdin can be visited *en route.*

Mount Katahdin, 5,385 feet high, is the most elevated mountain in Maine. Its altitude is but nine hundred feet below Mount Washington; yet it is seldom visited, and but little known to tourists. It rises from a dense primitive forest intersected by streams, ponds, and lakes. The region is without public highways and hotels, and lacks those homelike conveniences and accommodations necessary for a popular resort; yet it possesses the very elements desired by the angler and hunter, and is much frequented by sportsmen.

Mount Katahdin is reached from Bangor to Mattawamkeag, fifty-eight miles, *viâ* European and North-American Railway, thence by stage thirty-eight miles to Patten, the nearest public conveyance, from which place guides are necessary.

THE EASTERN PROVINCES.

From Bangor, the tourist bound to the Eastern Provinces — New Brunswick, Nova Scotia, and Prince Edward's Island — by rail makes his first stop in the city of St. John, where there are several fair hotels, the best being the Victoria. Fredericton, the capital of New Brunswick, lies some distance above on the River St. John, and is a much cleaner and pleasanter place. From here the tourist can proceed to Shediac, and take the steamer for **Summerside**, Prince Edward's Island, a charming spot, from whence he can stage it through a beautiful and romantic country to Charlottetown, forty miles distant. The only watering-place proper of the island is a little village eighteen miles distant, called Rustico. From Charlottetown steamers can be taken to Pictou, the centre of the coal region. From there the distance to Halifax is made by rail, passing through Windsor and Truro, two important towns. Halifax is a quaint old city, with a few fine buildings and many dingy ones. It is an interesting place, however. Before leaving Pictou for Halifax, a visit to Sydney and the site of the ancient city of Louisburg in Cape Breton would be time and money well spent.

On the return, the tourist would either go by the way of Truro, and thence to Monckton on the Intercolonial Line through the pleasant towns of Sackville, Amherst, and Dorchester; or, by taking the Annapolis Railway at Windsor Junction, pass through the most delightful portion of the Province, south of the Basin of Minas, and through the Annapolis valley, the scene of Longfellow's pathetic poem of "Evangeline." Here are the famous dike marshes extending uninterruptedly for ten or twelve miles, the remains of ancient orchards, and the stone chimneys of ancient habitations. From Annapolis the steamer can be taken to St. John; or stage can be taken for Yarmouth, a large and important town at the Southern extremity of the Province, from which port there is regular communication with Boston by steamer.

ILLUSTRATED PLEASURE ROUTE No. 8.

Boston to New Port, R.I., Duxbury, Plymouth, Provincetown, Hyannis, Woods Hole, Nantucket, Oak Bluffs, and the Coal Towns of South-Eastern Massachusetts.

THE OLD COLONY RAILROAD.

Who has ever examined the map of Eastern Massachusetts, with its long arm stretching out into the sea, — the land of the "Pilgrim Fathers," — without feeling a desire to visit it? From Boston to Narraganset Bay, the coast is filled with interesting localities, many of them identified with the early history of the country. The rockbound shores of Cohasset are noted for their grand marine views; the beautiful harbor of Plymouth is surrounded by scenes replete with historic memories; the barren coast of Cape Cod is made interesting by the beautiful summer resorts which line its borders; the sail through Vineyard Sound, "the great highway of commerce," to Newport, with *Falmouth Heights* and the *Elizabeth Islands* on the right, and *Martha's Vineyard* with *Oak Bluffs* and *Gay Head* on the left, is truly delightful, and cannot fail to please.

To all these localities the Old Colony Railroad leads. Its branches and connecting lines of ●steamers unite all these prominent points of interest with Boston and New York; and its admirable construction and superior equipment render it one of the most popular summer routes in the country. The original line extended from Boston to Plymouth thirty-seven and one-half miles, and from Fall River to Myricks, twelve miles. It now comprises a line of three hundred miles extending from Boston, the metropolis of New England, to Newport, R.I., and to all the principal cities, towns, and villages of South-eastern Massachusetts.

A glance at the map shows the two routes between Boston and Newport: viz., *via* Randolph, Stoughton, Easton, Taunton, Dighton, and Somerset; and *via* Brockton, Bridgewater, and Middleboro', which unite near Fall River; the line to Provincetown, the end of Cape Cod; that to Woods Hole, the mainland terminus of the Vineyard and Nantucket steamers; the line to "Plymouth Rock," passing through the Abingtons, Plympton, and Kingston; the Sea-Shore Line, through Quincy, Braintree, Weymouth, Hingham, Nantasket, Cohasset, the Scituates, Marshfield (the home of Webster), and South Duxbury, the American station of the "French Cable," to, and intersecting, the Plymouth Line at Kingston; also the suburban Branches, "Shawmut," "Milton," and "Granite."

BOAT-HOUSE LANDING, NEWPORT R.I.

No more beautiful summer routes can be selected for a day trip from the metropolis, than from these shore lines. Notwithstanding this was the earliest settled portion of New England, considerable sections of it still exhibit a primitive wildness. The following are among the noted summer resorts reached by the Old Colony Railroad: —

Wollaston Heights, in the northerly portion of the old town of Quincy, is situated upon a beautiful eminence overlooking Boston Harbor and the surrounding country. Good hotel accommodations.

Hingham. — This ancient town is a favorite resort for summer residents, is situated but seventeen miles from Boston, upon the southerly shores of Boston Bay, has many agreeable attractions, both for the tourist and the temporary resident.

Nantasket Beach is rapidly rising into popularity, although more of a local than a general character. The beach is long, and just like all the sandy frills of our Atlantic coast. The surface is hard, and admirably adapted to driving or bathing. It is but eighteen miles from Boston. Eleven light-houses can be seen from the shore; and it may well be pronounced one of the most delightful watering-places in the country. Sailing and fishing are without limit; and tens of thousands flock to enjoy the varied beauties of the scene, and the soothing temperature of the coast and sea air, daily, when the heated term is in full power.

Cohasset, which was sliced from Hingham, is a glorious spot for all lovers of the moody sea. Here are cliffy rocks enough, with a broken sea-margin, to insure a turbulent ocean even in a comparative calm; and, when old Neptune is in one of his fiery moods, the scenery around the shores of Cohasset rises to a degree of sublimity and grandeur that surpasses description. The wild, picturesque beauties of Cohasset rocks form an admirable subject for the artist's pencil; and here the lover of the beauties of nature delights to linger. **Marshfield**, the home of Webster, will also attract attention. **Duxbury** is thirty-nine miles from Boston. A walk to Captain's Hill, where a monument is being erected to the memory of Miles Standish, forms a pleasant objective point.

Plymouth. — Although the road to Plymouth passes through several thriving and interesting villages, the chief object of the tourist will be a visit to Plymouth itself, — a visit which cannot fail to interest him; and fortunately for his enjoyment, whether his sojourn is for a day or a week, he will find, at the Samoset and Clifford Houses, accommodations which will render his stay agreeable. Plymouth has a world-wide fame. If the orations delivered in honor of "The Pilgrim Fathers" were all printed in one book, it would make a volume fearful to encounter; for those famous "Pilgrims" landed there, as is generally known. Plymouth is a wholesome, steady, well-to-do town, with nothing remarkable about it except its historic notoriety. And yet the pleasure-seeker can find enough here to busy his hands. There is fair shooting at "the Point," some nine miles down the harbor; and sea and pond fishing are abundant. But the harbor is not good for commerce; and the place will depend chiefly upon the direction in which its capital is utilized. At one period Plymouth was of considerable maritime importance. It is but thirty-seven miles from Boston, and is reached in a little more than an hour's ride. The town is well laid out, and pleasantly located, on ground sloping to the water. Burial Hill, above, commands a fine view of the harbor. Pilgrim Hall, with its many curious relics; Plymouth Rock, Cole's Hill, and Clark's Island (where the Pilgrims "rested on Sunday, Dec. 10,

1620," before landing at Plymouth on the 20th), are all interesting places to visit. The vicinity of Plymouth, with its fresh-water lakes and fine drives, also presents objects of interest.

Scituate is really a very pleasant place, and has many agreeable attractions both for the tourist and the temporary resident.

The Cape Cod Division of the Old Colony Railroad extends to Provincetown, the extreme point of Cape Cod, one hundred and twenty miles distant from Boston. The fine harbor at Provincetown presents a refuge

for the storm-driven mariner. The place is inhabited largely by seafaring men; and its thrifty appearance is a fine illustration of what an enterprising community can drag from the sea. It is built on and surrounded by sand-hills; and the earth of its gardens, so green and beautiful, is mostly brought from the mainland. Branches lead from this road, from Cohasset Narrows to Woods Hole, and from Yarmouth to Hyannis, which was the terminus of the road before it was extended to Provincetown.

Hyannis is a pleasant village, and, next to Provincetown, the largest place on the Cape. It is prettily laid out, and ornamented by shade-trees. The inhabitants are engaged in a sea-faring life; and many retired sea-captains and merchants have made this their home. The railroad leads through the village a mile to the sea, where a fine wharf has been constructed. A growing seaside village, a short distance to the west, overlooks the harbor to Vineyard Sound and Martha's Vineyard beyond. A land company has erected a new hotel, for the accommodation of summer guests; and many fine cottages have already been built, and others are in course of construction. The serrated coast from Hyannis to Woods Hole is alternated with fine beaches, summer resorts, and wild lands, where the sportsman and fisherman delight to roam.

The name of "Cape Cod" is synonymous in most minds with sand, sea, and codfish. For the delicate and sensitive devotees of fashion these words have no charm; but, for the more hardy seekers for novelty and pleasure, they indicate shooting, fishing, and pure air. The characteristics of Cape Cod, although having a likeness to the whole coast family of attractions, are, after all, peculiar to itself. It is not an island; and yet it is as really in the arms of the Atlantic as though it were alone in its waters. To be on the shores of "Cape Cod" is to have the alternating humors of the ocean, as much as though it were Nantucket. Of course, Cape Cod is not a town nor a city nor an island; for it is "Cape Cod," and embraces towns, villages, islands, beaches, headlands, rocks, reefs, sand, salt, plover, ducks, coots, and codfish. It is of varied pleasures, found in numerous and peculiar places.

TROUT POND.

There is Cotuit Port, for example, nestled on high land, and in a charming location, almost romantic. It is also the rural home of many families of taste. Few summer resorts surpass it. It is reached by the Old Colony Railroad and connecting stages.

Falmouth Heights, a rising place, with rare attractions, now in process of development by a company of capitalists. It is a delightful location, and commands a fine view of Vineyard Sound. The prospect opens to the south; the ground is high, falling gently from a wooded crest to the bluff, which drops thirty feet to the beach below. Serpentine walks and drives permeate the groves of oak, in which cottages are pleasantly placed. A grand ocean avenue leads for miles along bluff and through the groves to the point of starting. A commodious hotel is open to guests in summer. The railroad which leads to Falmouth Heights and Woods Hole branches from the main road at Cohasset Narrows. This locality abounds in fine landscape and marine views; and summer residences will soon dot the scene.

Woods Hole is a picturesque hamlet of a hundred buildings, located on a promontory, on the extreme southern point of the peninsula which forms Barnstable County. The harbor is small, but affords good anchorage, and is well protected by outlying headlands and islands. This is the terminus of the road in this direction, and is the nearest point of railroad connection to Martha's Vineyard.

NANTUCKET WHARF. — THE ARRIVAL.

Thirty miles out at sea is a tufted sand-bank fifteen miles long: on it is the quaintest and most old-fashioned town in the Commonwealth. The zeal and energy of its hardy seamen, who pursued the whale in its arctic home, made Nantucket in earlier times familiar to the British Parliament. Nantucket is now coming to the front as a watering-place. The facilities for the rugged sports of the seaside which it offers, its bracing and genial air, the many attractions that surround it, the absence of fogs, and the home amusements presented, the easy sail of two hours, — all commend Nantucket to parties in search of out-door excitement and healthy recreation.

The town is specially commended to invalids who seek quiet and repose. Visitors can take their breakfast at the White Mountains, and sleep in Nantucket; men of business can reach New York or the White Mountains as easily as from Boston. The connections are swift and sure for travel east, west, north, or south. The town is peculiar for its quaint customs. Lectures, meetings, and arrivals are announced as of old by the bellman. Carts take the place of coaches, in which passengers stand. Young ladies invite their friends to a ride, back up a one-horse cart to the door, mount by means of a chair, and drive off with as much *sang-froid* as a fashionable city lady would enter the park in her phaeton.

Nantucket is full of surprises, and has many attractions. Its hotels and boarding-houses are ample : its athenæum, library, marine curiosities, its ancient houses of 1686, and the residence of the last of the Indians, are full of historic interest.

Oak Bluffs as a summer resort is an immense success, and its magical growth has fairly earned for it the appellation of *the Cottage City of America*. The sail from Woods Hole in the elegant steamers that ply between that place and the Vineyard and Nantucket, in connection with the Old Colony Railroad, is simply delightful. *Buzzard's Bay* is studded with beauty and with places of rare interest. The entire fleet plying between New York and the coast passes in view of the prominent hotels, often bringing into view a thousand sail.

This city of cottages has concrete drives, horse-railroads, a trotting-course, and all modern improvements. The great "Vineyard Camp-Meeting Association" holds its annual gatherings in the beautiful "Wesleyan Grove," adjacent to the place. Fully fifteen thousand visitors have been upon the island at one time. Hotels and boarding-places are numbered by scores.

A narrow-gauge steam-railroad connects Oak Bluffs with Edgartown, and with Katama, the beautiful new watering-resort, about nine miles southerly from the Bluffs. Katama is situated upon the magnificent bay of the same name, has a fine hotel, and delightful still-water and surf bathing, and attracts many visitors. The direct route to the Vineyard is *vià* the "Old Colony Line." Taking the cars at Boston, an hour is saved over any other route, and all sea-sickness is avoided.

Newport, R. I., is now the fashionable queen of all American watering resorts, for summer pleasure. With comparatively little of striking or romantic scenery, it has attractions peculiarly its own. Wealth and social distinction having approved of this really delightful location, the summer gatherings are of the gayest and most brilliant description. In elegance and splendor of outfit; in fame and beauty of its throngs; in all that invites the curious, the seekers after pleasure, the invalid's repose, and the glare and extravagance of fashion, — Newport is unrivalled. Indeed, this ancient and once renowned seat of commerce, after sinking into semi-oblivion, has been Rip-Van-Winkled into fame again, and is now in the bloom of a vigorous summer life, though still inclined to its winter drowse. The location of Fort Adams at Newport also adds to the attractiveness of the place.

NEWPORT, R.I.
Old Colony Steamboat Company's Docks.

From the south, Newport is reached by the Sound Steamers of the *Old Colony Steamboat Company*, "Fall River Line;" and from Boston by the "Old Colony" Road. Steamers also leave Providence for that city, stopping at all the leading places of interest along the shores of Narraganset Bay.

Perhaps in no particular has greater improvement been made in the last few years, than in the taste displayed in the construction of steamboats and railway cars, thus greatly lessening the fatigue and annoyance of travel. A journey may now be made without losing for a day the comforts of home. Cars and boats furnished with elegant parlors, inviting saloons, and luxuriant state-rooms, are now found on all the principal routes in America.

OLD COLONY STEAMBOAT COMPANY, "FALL RIVER LINE,"

Between New York and Boston, via Newport and Fall River.

The **Old Colony Steamboat Company** may well be said to occupy the front rank in this improvement; and its boats have no superiors in the world. Plying between the metropolis of the nation and the most fashionable watering-place on the continent, their saloons are constantly patronized by the *élite* of society.

Every afternoon long lines of carriages deposit their passengers at the company's wharf, PIER 28 (FOOT OF MURRAY STREET) NORTH RIVER, NEW YORK CITY; and at 5 P.M. in summer, and 4 P.M. in winter, the signal gun announces the hour of departure, and these magnificent floating palaces, crowded with human freight, glide into the stream. Martial music, by Hall's Boston Brass Band, enlivens the scene, as the gayly-dressed steamer majestically threads her way through the noble harbor, made rich in panoramic scenes by the marine of all nations. The twilight deepens as the stately vessel enters the East River, on her way to the placid waters of Long Island Sound. The scenery becomes beautified by the enchanting villas that line the shores, the homes of wealth and beauty; and nought is heard but the exclamations of delight from the assembled throng, the merry laugh of the promenaders, and the intoxicating strains of the reed and string music which have replaced the brass band. Thus into the night glides this living freight, — faith, comfort, and contentment resting in the minds of all.

The fleet of steamers formerly owned by the Narragansett Steamship Company (now by the "OLD COLONY") comprises the "Bristol," "Providence," "Old Colony," "Newport," all of which are well known to the travelling public.

If we had not been warned by the march of improvement in the past, we should be tempted to believe that steamboat building has reached its climax in the superb vessels "Bristol" and "Providence."

On crossing the gang-plank the visitor finds himself on a broad deck, surrounded by richly carved and gilded panelling. The deck itself is composed of alternate strips of yellow pine and black walnut. In extreme width, this main deck measures eighty-four feet. Surrounding that portion of it which we enter from the wharf are the various offices for tickets, luggage, &c. Large doors in the after bulkhead lead to the ladies' saloons and state-rooms, which are appropriately divided for the use of ladies travelling alone, and for families with children, the most complete accommodations being provided for all. The main deck is divided into two general divisions by sliding glass doors. The forward part is used for freight; and the after part, which has just been described, is devoted to the use of passengers. From this after part, stairways lead to the upper and lower saloons. These stairs, with their highly-polished brass steps and their carved and graceful mahogany balusters, are separated from the open deck by a semicircular partition of woodwork and glass, which prevents the too strong draughts which a head wind sometimes occasions. Entering this semicircular enclosure, we descend to the lower saloon and supper-room. Here, in long perspective, tables, glittering with cut-glass and silver, stretch away toward the stern of the boat. Just forward of the stairway are the china-closet and kitchen, where all the culinary operations required on the boat are performed, and whose neat array of shining cooking utensils would delight the heart of the most fastidious housekeeper.

MAIN SALOON
Of the world-renowned Steamers "Bristol" and "Providence."

Leaving the appetizing scenes of the kitchen and supper-room, we ascend two broad and easy flights of stairs to the main saloon, which runs fore and aft nearly the whole length of the boat, with rows of state-rooms on each side, and, in fact, overhead, — for there is yet another stairway, and another tier of state-rooms above us. The eight rooms which occupy the after part of the main saloon are for the accommodation of those who desire more luxurious surroundings than are sought by the public at large. These rooms far excel in elegance those of any first class hotel, and in size they are at least equal to the ordinary rooms of seaside houses. The other state-rooms, numbering in all three hundred, are large and well ventilated. They possess the peculiarity of having, in place of the ordinary fixed bunks, a kind of two-storied black walnut bedstead, which, being detached from the light woodwork, is comparatively free from the vibration commonly perceived when a steamboat is under way. Besides this provision against vibration, it will be observed that the partitions between the rooms are built diagonally; so that, instead of working with every revolution of the wheels, they form, in the aggregate, a powerful set of braces, adding much to the strength of the superstructure. The state-rooms of the upper tier are entered from broad galleries, which run around the saloon. These galleries unite at either end, and form spacious landings, on which are tables and chairs similar to those in the saloon; and the boats are lighted by gas.

STEAMER BRISTOL
Passing proposed Bridge between New York and Brooklyn.
Taken by permission from Warren's Geography, published by Cowperthwait & Co., Philadelphia.

At 5.30, P.M., daily (Sundays, during the summer at 6.30), passengers for New York leave Boston, from the depot of the Old Colony Railroad, connecting with steamer at Fall River, leaving there at 7.15, P.M.; and at 8.45 in the evening the boat leaves Newport, arriving in New York in season to connect with all through trains South and West.

Passengers from New York, the West, or South, for any of the above places, can purchase tickets and have their baggage checked to destination; and by branch roads will be taken direct to any of the delightful resorts to which it leads, without the necessity of going to Boston.

PLEASURE ROUTE No. 9.

EXCURSION TO OAK BLUFFS AND KATAMA BAY.

THE coast of New England abounds in beautiful harbors, charming bays, and quiet inlets, many of which are unknown to the public for want of means of communication. But the popular and increasing custom of spending the summer months at the sea-shore every year causes the development of new and delightful resorts, — localities whose quiet beauty frequently proves a surprise to the travelling public. Yet it has so often happened that a long time intervenes before such accommodations are prepared, that tourists have hesitation in visiting any but the beaten tracks of travel. Very fortunately, however, this difficulty is being largely obviated by the action of business men, who, with a quick eye for the beautiful, and sharp discernment for the wants of the public, do not hesitate, on discovering a desirable locality, to announce it with

STEAMBOAT "MARTHA'S VINEYARD" PASSING OAK BLUFFS.

hotel accommodations complete. Such is the case at KATAMA BAY, where the first building erected was a large and commodious hotel. The name of "Katama" is not a familiar one; and many will read it now for the first time. If such, however, will look at the map of Massachusetts, they will see off the eastern shore of Martha's Vineyard a bay or channel separating it from Chappaquiddick Island, which forms the point. This beautiful sheet of water is but eight miles south-east from Oak Bluffs, now so extensively known as a summer resort.

Tourists by rail to New Bedford connect with the fine side-wheel boats "Martha's Vineyard" and "Monohansett," of the Martha's Vineyard Line of steamers, for Oak Bluffs and Katama.

The excursion from New Bedford is one of the pleasantest on the coast. We have a fine view of Fairhaven as we sail down the harbor. This town is pleasantly located; and a number of beautiful private residences, half hidden by foliage, overlook the water. An old fort stands upon a rocky promontory at the east entrance of the harbor, opposite to which, upon an island, is the light-house. At this point we enter Buzzard's Bay, and cross directly to "Woods Hole," a dozen miles away.

Buzzard's Bay and Vineyard Sound are so protected by headlands and outlying islands, that the sail is delightful. While the coast turns back to the left, after passing the fort, on the right the main-land pushes out into the sea, forming a cape, on which Clark's Point Light-house stands, and Fort Tabor occupies a commanding position. From this the bay opens to the right; and the main-land stretches away in the distance to a marked promontory known as "Round Hill." "Dumpling" or Round Hill Light stands on an isolated rock off this point, beyond which is the main entrance to Buzzard's Bay. The Elizabeth Islands are on the opposite side of the channel. These are individually known by their Indian names, commencing with the westernmost, as Cuttyhunk, Pennikeese, Neshawana, Peskeneese, Naushon, Nonfameusett, and to complete the rhyme, mariners have added, Woods Hole, Quequonkesset. Turning to the left, the main-land, extending towards Cape Cod, can be seen in the blue distance, with Black Rock, a dangerous shoal, lying between. As we approach Woods Hole, the island of Naushon stretches six or seven miles away to our right. This is the property of Capt. John M. Forbes of Boston. His summer residence, and that of his son, form conspicuous features in the landscape. An extensive grove of beech and oak has been stocked with deer, where his friends are annually entertained with a genuine deer-hunt.

The entrance to Woods Hole is narrow and tortuous, with sunken rocks on either side, requiring great nautical skill in its passage. The harbor is small, but deep, and well protected. A hundred houses, perched upon the surrounding hills, many of them quite beautiful, comprise the town; the depot of the Old Colony Railroad being the most conspicuous feature. Visitors to Katama or Oak Bluffs who prefer to go to this place by rail can take the cars at the Old Colony Depot, Boston, and continue from here by boat. The light-house, with its beacon light, stands on a headland at the mouth of the harbor, and marks the entrance to Vineyard Sound. The sail across the sound is delightful. Martha's Vineyard lies directly before you. Gay Head Light is seen far away to the right, while Falmouth Heights are on the left.

VINEYARD HAVEN.

This town, formerly called Holmes' Hole, is approached between two headlands, known as the "East" and "West Chop." Its harbor is indeed a haven for the storm-driven mariner; and hundreds of sail frequently lay here for days awaiting a favorable wind. The village of Vineyard Haven rests upon a hillside, sloping gently to the water.

The town is old: several churches, a few newly erected residences, and an old wind-mill whose arms point to the past, form the conspicuous features in the landscape. By the formation of a natural dike across the southern portion of the harbor, a small lake, three miles long, has been separated from it, known as Lagoon Pond, which is noted for its fine oysters of artificial culture. A carriage-drive along this dike extends, via Oak Bluffs, to Katama.

RIDING OUT THE STORM.

As we leave Vineyard Haven, and the steamer rounds the "East Chop," we approach the locality of summer life for citizens from our large towns. Cottages and villas are scattered over the hillsides, which increase in number and beauty until we reach Oak Bluffs, where we have a *rural* city spread out before us, from which, standing in bold relief, rises a colossal structure surrounded by broad verandas, and surmounted by towering cupolas. — THE SEA VIEW HOUSE, — with the steamboat-landing directly in front.

THE "Sea View" is the prominent feature of the town, which lies beyond. On either side, overlooking the water, and extending for thousands of feet, is a broad plank promenade, with seats the entire length.

At the right is a building one thousand feet long by eighteen feet wide, built into the bluff, protected by a heavy bulkhead. This is used for amusements. The side is of glass, and opens to the sea. In front is a broad promenade, provided with seats. The roof is flat, covered with concrete, and is also used for a walk, over which pavilions are placed at intervals, the whole overlooked by beautiful cottages. At the left of the hotel are hundreds of bathing-houses, with pavilions and seats for spectators. Steamboats and yachts crowd the landing, while the wharves, the verandas, the balconies and bluffs, are filled with the life and gayety of the scene. Over sixty-nine thousand guests visited this renowned resort during the season of 1874.

SEA VIEW HOUSE.
Oak Bluffs, Martha's Vineyard.

If the boat remains at the landing long enough to allow it, a visit to the "Sea View" will amply repay the trouble. From the wharf, the entrance is made through an ornamental gate-house, which is devoted to offices. In the tower at the right is the baggage-room, with a general railway ticket-office over it. At the left is the wharfinger's office, over which is the office of the Oak Bluffs Company. The basement of the hotel is approached by a private entrance from the wharf, by which the baggage and stores are taken, and, by the steam elevator, raised to any part of the house.

Wide passage-ways extend through the basement, cutting each other at right angles. At the left of the entrance, opening to the sea, are the barber's shop, bath-rooms, and billiard-hall, beyond which is the engineer's and boiler room. &c. On the right are store-rooms, ice-house, chill-room, laundry, bakery, and servants' rooms. The house and promenades are lighted by Walworth's solar gas generator, with gas manufactured in an underground building, distant from the hotel.

The Sea View House is approached by a broad flight of steps, leading to a capacious veranda at the east end, twenty-six feet in width. This is an important architectural feature of the house. It is three stories high, giving beauty to the structure, and comfort and pleasure to the guests. It commands a full view of Vineyard Sound, the great "highway of commerce." Ninety-five thousand vessels are reported to have passed Gay Head Light in 1872. The ladies' reception-room is at the right of the entrance, with hat and coat and wash rooms, and stairway beyond. On the left is the gentlemen's reception-room, elevator, and office. Opposite the main entrance, the doors open to the dining-hall, which occupies the entire width of the building, with long windows opening to wide verandas on either side. The private dining-rooms are beyond. Broad stairways and the elevator lead to the stories above.

The public parlor, on the second floor, is pleasantly located across the southern end of the building, commanding, from its windows and balconies, an extended view in three directions. On either side of the passage are private parlors and sleeping-rooms; and the two stories above are similarly arranged, with walks the entire length, affording excellent ventilation. Balconies lead from every story. A tank of two thousand gallons capacity, placed under the roof, supplies the water for the house.

The peak of the south tower has three fine sleeping-rooms. The north tower has two, with a passage-way leading to the lookout-room above, from the four windows of which a most extensive marine and landscape view can be had.

The chapel is a beautiful structure, of ornamental design, and seats eight hundred and twenty-two persons. It stands on a rounded knoll, surrounded by a grove of oak. It is octagonal in form, with four doors, leading from opposite angles, and a gallery entrance on the side, in front of which stands the pulpit, with ante-rooms on either side. The seats on the floor and in the gallery occupy seven sections of an octagon, facing to a common centre. The inside is not plastered; but the painted tri-colored walls and ceiling, relieved by the ornamented framework, produce a singular and beautiful effect. The building is not confined to any particular denomination; but any stockholder can secure the pulpit for a friend on any unengaged day.

The free and harmonious use of a building so beautiful lends a charm to the exercises, and, furnishing as it does a regular place of worship, contributes largely to the popularity of the place.

Circuit Avenue commences at the wharf, and extends through and around the town. It has a paved drive-way, with concrete sidewalks. Other important streets are paved with concrete. The town is emphatically a *Cottage City*. Within an area of one mile stand six hundred and ninety-one cottages. Among the hundreds of beautiful houses, of exquisite architecture, scarcely two can be found alike.

SEASIDE COTTAGE OF HON. E. P CARPENTER.

Oak Bluffs is a delightful place, frequented by persons of culture and taste, which can only be partially described in this article, but will well repay a visit from the pleasure-seeker. Katama, on the eastern point of the island, is now attracting attention.

The scenery at Katama was so beautiful, the fishing and shooting so unusually fine, that in 1872 a few gentlemen of means determined to erect a hotel there, and secure a building site for others who, like themselves, should be desirous of leaving the heated streets of a town for a quiet cottage life by the seaside. The result was beyond the expectations of the most sanguine; and forty-three persons came forward the first season, and secured lots, with the agreement to build cottages on them. The place has been artistically surveyed by a landscape-gardener, streets and avenues graded, and parks laid out. Summer residences are being built; and a town is springing up as if by magic.

Several of the managers of this enterprise are the same who made Oak Bluffs a success; the same whose lots, placed in the market five years ago at $100 each, have since sold for $1,600; the same who, in six years, from a single house, have built "THE COTTAGE CITY OF AMERICA," with paved and gas-lit streets, — in a word, a complete town, clothed with taste, comfort, and picturesque beauty. And now the same energy and perseverance which succeeded at Oak Bluffs promise success at Katama.

It is a short but pleasant sail across the bay from Oak Bluffs to the village of Edgartown, the spires of which can be seen in the south-east. This place was once a town of some enterprise; but, with the decline of the whale fishery, the young men have generally sought employment elsewhere. From Edgartown we enter Katama Bay, a beautiful sheet of water, some five miles long. It is of itself an excellent harbor, affording not only good anchorage, but is well protected from the winds, making it a favorable resort for yachting and fishing clubs. On entering the bay, "**Mattakeset Lodge**," the model hotel of Katama, becomes the absorbing object of attention.

MATTAKESET LODGE,
Katama, Martha's Vineyard.

The house stands on a commanding bluff, at the opposite end of the harbor, its symmetrical towers cutting boldly against the southern sky. As you near the landing, immediately below the hotel, its peculiar structure and singular location become apparent. The surface at Katama is an extended table-land, broken by gentle undulations, but at the coast falling abruptly to the beach below. In a single instance, by some natural agency, a pathway has been grooved through the bluff to the water's edge, forming an admirable passage, of easy grade, from the beach to the plain above. Here a wharf has been built; and on the bluff, spanning the ravine, stands the hotel, under which the drive-way passes to the town beyond.

"Mattakeset Lodge" has been constructed to afford the *maximum* amount of comfort and pleasure. Numerous balconies and broad verandas, commanding a complete view of the landing and harbor beyond, surround the house; but the peculiar and favorite feature is an open gallery, which occupies the entire upper story. In the evening this is brilliantly lighted by gas, and is entirely devoted to promenading and dancing.

YACHTING.

It is, indeed, a novel feature. Here, in the hottest summer's day, a cooling breeze is always felt, and the roar of the breakers on the south side of the island, scarcely a mile away, fills the air. The whitened foam, as the waves dash upon the beaches, which stretch away like a ribbon in the distance, is always an interesting object of contemplation. From this elevation, high above the surrounding water, securely shielded from the sun's scorching rays, hundreds of whitened sail can be seen.

The admirable facilities for yachting, and the abundance of fish and birds in this vicinity, have induced the proprietors to make special arrangements for the convenience of sportsmen; while others, whose taste for these amusements commences after the game has passed the culinary department, will have reason to be equally well pleased. Fishing parties from Oak Bluffs have come to be daily affairs.

Few localities on the coast possess better facilities for fishing than Katama,—deep-sea fishing off Cape Pogue for cod, haddock, hake, whiting, pollock, and halibut; exciting sport in sailing or trolling for blue fish, striped bass, and Spanish mackerel; still fishing in the bay, within a gun-shot of the house, for scup, tautog, sea-bass, and sea-perch. But the sport in which Katama leads is in the serving of her unrivalled *Clam-bakes*; and, for the convenience of guests, a grand pavilion has been erected for their shelter.

THE MIGHTY CLAM-BAKE.

Tradition gives color to the claim, that the great genius of Mattakeset, the famous and powerful Indian chief of the primitive days, devised the art of preparing the delicious and now world-renowned "CLAM-BAKE." Charles Lamb relates the remarkable way in which "roast pig" was revealed to the "heathen Chinee." Doubtless the claim set up for Mattakeset is quite as authentic as Lamb's bit of tradition; but it is not as full in particulars. Clams of the very finest variety abound in the region around Mattakeset Lodge; and the formula for composing and compounding a clam-bake, in the style of a fine art, is naturally found here. For general satisfaction, we now describe the process:—

First, a huge saucer-like space is dug in the sand or ground, and is well paved over with stones. This may be called the bake-oven. To prepare the grand bake, the "oven" is filled with fuel, intermingled with goodly sized stones. This is fired (the combustible part): and after the stone portions are all thoroughly heated the coals are raked off. The "oven" is now ready. First, a layer of rock-weed is equally spread over the heated surface; next, from fifteen to twenty-five bushels of clams are thrown in, and then covered with another layer of rock-weed; and over that sea-weed is thickly placed. The heat of the oven is sufficient to raise a great cloud of steam from the water of the clams and the weeds; and in about half an hour, the capacious mound of savory bivalves is ready to be borne to the feast-board. Here, drawn butter, salt, pepper, and vinegar, or any of the more pungent relishes of the table, served in convenient dishes, are used to add zest to this notable and popular food. The instruments used to dislodge the clam from the shell, decapitate it, and submit it to the teeth, are simply fingers. Experts at this kind of feed are wonderfully dexterous in the work, and raise huge piles of shells around them in brief time. This is a clam-bake simple. But, to have a compound bake of appetizing temptations most excelling, as frequently served by mine host of "Mattakeset Lodge," lobsters, green corn, fresh fish, chickens, &c., are to be placed among the clams at the outset. With these additions, the rudely improvised feast becomes one not to be surpassed for lusciousness by the skill of a regiment of French cooks. In the opinion of many, the transcendent glory of Mattakeset Lodge lies in its unrivalled clambakes.

COMMUNICATIONS.

Katama possesses admirable facilities for communication by steamers, and the sail is delightful. A new and beautiful steam-yacht, of unrivalled speed, connects with steamers from Woods Hole and New Bedford, at Oak Bluffs; or you can go by rail, a new and elegantly equipped narrow-gauge road having been built from Oak Bluffs to Katama. And the citizens of Edgartown, alive to the requirements of the public, have laid out and constructed a splendid drive-way of twelve miles, extending from Katama to Vineyard Haven, viâ Edgartown and Oak Bluffs. Steamers will run daily. In a word, every thing which experience can dictate is being done to make this a popular summer resort, and a pleasant seaside retreat.

THE SEA-VIEW BOULEVARD.

This fine drive-way, commencing at Katama, extends twelve miles along the coast to Vineyard Haven. It has been constructed by the citizens of Edgartown, to meet the increasing wants of visitors. From Katama to Edgartown it continues along the table-land to and through the village. A couple of miles beyond, it leads down to the sea, approaching it between two smaller bodies of water, which lay contiguous to and parallel with the ocean. That upon the right, and nearer Edgartown, is known as Crystal Lake, — a beautiful pond, one mile in diameter. On the left we pass the foot of San-cha-can-tack-et Lake, which for miles is only separated from the sea on our right by a natural dike, evidently thrown up by some mighty convulsion, or by the action of the waves, beyond the present history of this region. The road-way has been built along this dike, which in places is so narrow that a stone could be tossed into the water on either side. This is a delightful drive; and, although of recent construction, it has attained a great popularity, affording as it does, to persons having objections to boating, the rare opportunity of securing an equally refreshing sea-breeze while riding in a carriage. Nor is this all. Midway between Edgartown and Oak Bluffs the dike has been cut, by the action of the waters, through which, with the tide, the current ebbs and flows.

This channel has been spanned by a bridge four hundred and fifty feet long, affording a rare opportunity for fishing; not merely small, worthless varieties, but blue-fish, bass, flounders, and others of large size, are taken in abundance. This was only needed to secure for ladies and children, or persons averse to boating, the full advantages of the exhilarating sport of fishing, shorn of the disagreeable annoyances of sea-sickness. The Sea-View House, at Oak Bluffs, is but three miles distant, and in full view from the bridge, which is one of its most favorite resorts. Indeed, the Sea-View Boulevard is one of the most enjoyable features of this popular watering-place.

San-cha-can-tack-et Lake is three miles long, and from one to two wide, and is a favorite boating and sailing locality. Cultivated farms rise to a wooded crest on the opposite shores.

The drive from the bridge to Oak Bluffs is along an undulating surface, which will soon undoubtedly be filled with cottages. It affords a fine variety of landscape views, with occasional glimpses of the town beyond, — the Sea-View House always forming the most imposing feature. Island Lake nestles quietly by the wayside, with a miniature island set like an emerald in its centre.

STRANGERS IN BOSTON.

Boston is well supplied with hotels, and, like every city, with cheap and expensive ones: but the Crawford House, which is under the same management as Mattakeset Lodge, will be found one of the most desirable for strangers. It is centrally located, in Scollay's Square, from which point all the city and Metropolitan horse-cars start. It is convenient to all the depots, and is a first-class house in every respect. It is kept on the European plan, with rooms from $1 to $4 per day; and with four dining-halls the proprietor is able to satisfy his guests in style or price. I have no doubt strangers will find this a desirable home while in Boston and vicinity.

PLEASURE ROUTE No. 10.

New York, Boston, and Vicinity to the Mountains and Springs of Vermont, Lake Champlain, Montreal, and Quebec, Thousand Islands, &c.

CENTRAL VERMONT RAILROAD.

THE southern termini of the great system of railroads known as the "Central Vermont" are Miller's Falls, Mass., where connection is made with the New London Northern Railroad; also at Chatham Four Corners, N.Y., connecting with the Harlem, Boston and Albany, and Hudson Railroads. It is approached from Boston and the east *via* Boston, Lowell, and Nashua, Concord, and Northern Railroads to White River Junction; and by the Fitchburg and Cheshire Railroads to Bellows Falls, and by the Boston and Albany Railroad to Chatham Four Corners. From these several termini the various routes and branches permeate the entire State of Vermont, reaching into New York and the Province of Quebec. At St. Johns, Canada, the northern terminus, twenty-two miles from Montreal, connection is made with the "Grand Trunk," and again at Ogdensburg, N.Y., the western terminus. At this point also the St. Lawrence and Lake steamers touch.

A description of the route from Boston to White River Junction will be found in *Illustrated Pleasure Route No.* 1, as far as Concord, N.H., where the Northern Railroad commences. The first noted point of interest beyond Concord is **Newfound Lake,** located about two miles north from the quiet and sequestered village of Bristol, N.H., on a branch of the Northern Railroad. This delightful sheet of water, seven miles long by three wide, is surrounded by mountains, which from every direction are mirrored on its quiet surface. *Sugar-Loaf Mountain* rises in the west, and *Mount Crosby* in the east. The Newfound and Pemigewasset Rivers unite near here, and add their attractions.

Mount Kearsarge, N.H. (2,461 feet high), is four miles from *Potter Place* Station, on the Northern Railroad, reached by stage. This mountain is not unfrequently confounded with another of similar name, *Kiarsarge* (see North Conway), which, in an air line, is sixty miles distant in a north-easterly direction. Mount Kearsarge is a place of considerable repute as a summer resort. Its isolated position affords a magnificent view of the surrounding country; the landscape is dotted by beautiful lakes; and far in the northern horizon are seen the Franconia and White Mountains with the Green Mountains of Vermont at the left. Visitors will find comfortable quarters at the *Winslow House,* which is located half way up the mountain side.

The steam frigate "Kearsarge" was named for this mountain.

VERMONT.

We strike the Central Vermont Railroad at **White River Junction**, a railway centre of importance. Considerable stop is made at this station, to allow passengers time for refreshment at the admirable restaurant in the depot.

After New Hampshire, the Green Mountains of Vermont rival in popularity those of any other of the New-England States; and of these **Mount Mansfield** (4,348 feet high), the crowning peak, is the most famed. This is an objective point from *Stowe*, one of the most fashionable resorts in the State, which, when better known by illustrated description, is certain of liberal patronage. Stowe is reached by the Central Vermont Railroad. From this place the excursion to Mount Mansfield (five miles) is made by carriage; the visitor's path is shaded by a hardy grove, but the trees gradually decrease in size. To meet the demands of travel, a hotel, the *Summit House*, has been erected near the top; which is largely patronized by those who would enjoy a sunrise from the summit. Mount Mansfield is the central peak of several mountains, each a point of interest to visitors.

Sterling Mountain (3,500 feet high), separated from Mount Mansfield by a gorge known as *Smuggler's Notch*, a wild, romantic pass which derived its name from incidents in the past history of the region. The drive from Stowe, nine miles, is picturesque and pleasant. The most impressive view of Smuggler's Notch is from Mount Mansfield. Here also is a profile rock called "The Old Man," curious in itself, yet lacking the stern grandeur of its namesake in the Franconia Mountains.

Bolton Falls, near Ridley's Station, form one of the interesting features of that favorite region for tourists, of which Mount Mansfield is the towering sentinel. More than four thousand feet below Mansfield's crest is a deep rugged ravine, overhung by frowning rocks, screened by foliage, and ornamented by mosses, lichens, and clinging vines. Here almost unseen this wild mountain stream boils and foams.

Camel's Hump (4,083 feet in altitude) is but little below Mansfield, and is in the same neighborhood. It is also reached by the Central Vermont Railroad to *Ridley's Station*, thence by carriage six miles, to the base of the mountain. From this point, a carriage-road has been built three miles. A good pedestrian can readily make the remainder; but, for those who desire, saddle-horses are to be had. A small summer-house near the summit furnishes refreshments, and serves for protection in case of storms. The view is fine.

Sheldon Springs have long been famous for the excellence of their waters, efficacious for rheumatism, erysipelas, &c. The scenery in the neighborhood also possesses many other attractive features.

The Portland and Ogdensburg Railroad, leading from Portland past Sebago Lake, through the White Mountain Notch, will cross the *Missisquoi Valley* Railroad at Sheldon.

Highgate Springs are located immediately at the station of that name, and form the central group of a number of medicinal springs, many of which have become noted for the virtue of their waters, and when better known will rank higher as fashionable resorts. Among those within a few-miles' radius can be named *Alburgh Springs, Missisquoi Springs, Champlain Springs, Sheldon Springs,* and *Vermont Springs. Highgate Springs* contain chloride of sodium, carbonate of soda, and sulphate of soda. The proximity of *Missisquoi Bay* which lies at the east, and **Highgate Falls** only two or three miles distant, add to the attractions of this region.

Champlain Springs and **Highgate Falls** are rival attractions at the village of the latter name. The waters of the springs are recommended for cancer, dyspepsia, and skin-diseases generally.

Alburgh Springs are also reached by the Central Vermont Railroad, and have won considerable attention. They are located north from Lake Champlain, and on the direct route from the "springs region" of Vermont to the Thousand Islands, Alexandria Bay, N.Y.

If we make an approach to Vermont by the Fitchburg Railroad, we can visit **Wachusett Mountain** of Princeton, Mass., having an altitude of 2,480 feet, which is very favorably known to tourists. It is a crowning eminence second only to *Greylock* in altitude, rising from an undulating yet highly cultivated country, and is located midway between Massachusetts Bay and the Connecticut River. It was occupied by Government officers on the coast survey, and forms a distinctive landmark for mariners approaching the coast. It has long been a popular resort; even the red men are said to have made it a rendezvous. Its beauties are enhanced by an attractive lake, *Wachusett,* which nestles under its shadow. The elevated situation of the village of *Princeton,* the pure water with which it is supplied, its invigorating atmosphere, and pleasing surroundings, would of themselves prove attractions sufficient to popularize the place with pleasure-seekers. To these are added the charms of *Lake Wachusett,* and Wachusett Mountain; the former noted for the purity of its waters, the latter for the boldness and rich variety of its landscape.

If the tourist has a desire to climb another of New Hampshire's famous mountains, this will be found a convenient time to visit **Mount Monadnock**, in the town of Jaffrey, which has an altitude of 3,450 feet. (Vermont also has a mountain of the same name located near the little village of Colebrook.) This mountain is known as the *Grand Monadnock*, and is in the extreme southern section of the State. Being the highest elevation in the vicinity, it is a distinctive feature in the landscape. The view from this mountain is magnificent, and differs from that of any other in the vicinity. This elevation seems the connecting link between the more undulating surface of Massachusetts, and the towering highlands of New Hampshire. In the east the historic shaft on Bunker Hill indicates the location of Boston. In the southwest, Wachusett, Holyoke, and Mount Tom form the conspicuous objects in the scenery; the green hills of Vermont become blue in the western horizon. In every direction, like gems set among the hills, sparkle the waters of picturesque lakes, the queen of which, *Contoocook*, with its charming little steamer, lies half enveloped in the shades of the overhanging mountain. From the north, however, the landscape draws its grandeur. Rising one above another are seen all the important peaks of this mountainous State.

A mineral spring in the vicinity adds to the attraction of Monadnock.

Contoocook Lake, set like a gem among the hills, as before mentioned, nestles under the shadow of the grand old Monadnock. It is a delightful retreat for the pleasure-loving guests of the summer hotel clinging to a shelf on the mountain-side. A small steamer plies on its waters. Go to Jaffrey, N.H., *viâ* Monadnock Railroad.

Bellows Falls, Vt. — Bellows Falls, or, as might more properly be said, cataract, forms an object worthy the attention of the sight-seer. It consists of a channel fifty feet in width, cut far down into the solid rock, through which the waters of the Connecticut River rush. A descent of fifty feet is made during the passage over which the river tumbles and foams. The scenery in the vicinity is fine, both in richness and variety. Bellows Falls may be reached by any of the railroad lines which strike the Connecticut River above or below. At this point connection is again made with the Central Vermont Railroad.

Black River Falls, Springfield, are little more than a wild cascade where the stream has worn a curious fantastic channel through the slate formation. In a descent of six hundred feet the river falls one hundred and ten feet, fifty of which is by a single leap. Reached by the Central Vermont and Cheshire Railroads.

The **Hoosac Tunnel** is so far completed that trains for merchandise passed through it for the first time on the date of writing this article. By this great engineering success the *Hoosac Mountain* has been pierced, forming a direct passage from the Atlantic seaboard to the west. Arrangements for tourists are not yet completed, but this must eventually become one of the attractive features of an already interesting region. Go by the Vermont and Massachusetts Railroad from the east, or Harlem Extension from New York.

If our approach to Vermont is by the Boston and Albany Railroad we shall pass near **Mount Holyoke**, a place long and favorably known as a popular resort. Although it has an altitude of only 1,120 feet, the prospect from it is remarkably fine.

Many other mountains are higher, yet few afford more interesting views. It has long been visited as a summer resort. As early as 1821 a hotel was erected on its summit, the same site now occupied by the famous *Prospect House*. It is ascended, partly by carriage, and partly by railway. Three hundred and sixty-five feet of the steepest portion is overcome by an incline six hundred feet, with cars drawn up by stationary engine. Over twenty thousand visitors are annually lifted to this commanding spot.

The mountain is but three miles from the thriving village of Northampton, with a good carriage-road. Visitors can also go by horse-cars to the mountain railway. Northampton is reached by the Connecticut River, and New Haven and Northampton Railroads.

Mount Tom (1,320 feet high), is also in this neighborhood, and is visited from Northampton (five miles). It commands a more extended view than Mount Holyoke, but from the difficulty of ascent it has never gained the same popularity.

Our next point of interest as we move westward is the **Berkshire Hills**, whose queen is the charming village of Pittsfield, Mass. One of the chief attractions is **Lake Ashley**, a quiet lakelet set on the summit of *Mount Washington* (1,800 feet high), near the village of Pittsfield. Its pure limpid waters supply the town.

Lake Onota, Pontoosuc Lake, Berry Pond (in Hancock), **Melville Lake**, or the Lily Bowl, **Lulu Cascade, Silver Lake**, and **Sylvan Lake** are also among the interesting attractions near Pittsfield, which is really one of the most delightful homes of the hill region of Massachusetts. It is reached by both the Boston and Albany, and Housatonic Railroads. Before continuing to Vermont, the tourist should not fail to visit *Greylock*, the crowning peak of the Berkshire Mountains.

Greylock is partially cleared on its summit, and commands a view pleasantly interspersed with every variety of landscape. Near by are the lower ranges of the Berkshire Hills, generally wooded to the crest; beyond are the hills and valleys of an undulating country, dotted with farm-houses, lakes, ponds, and villages, which are agreeably intermingled. Above and beyond stretching far into the blue distance may be seen the towering form of Grand Monadnock. Turning with the sun, Mount Wachusett, in the eastern section of the State, forms a notable feature. Holyoke and Tom are seen in the south-east, and the Catskills in the south-west. There are several paths by which the top of Greylock is reached. The ascent is somewhat tedious, but the cheering prospect repays the effort. Go *viâ* Boston and Albany Railroad.

New Lebanon Springs are reached by rail on the Harlem Extension Railroad. Here may be found a fine summer house, which is well patronized during the season. The medicinal qualities of the waters are highly recommended, having a temperature at 73°; the flow is large, and the fame of this spring is increasing. The surroundings are pleasant, walks good, and drives fine. The Shaker village is two miles distant.

Mount Equinox (3,706 feet high) is one of the popular resorts of Manchester, Vt., from which village a fine carriage-road has been constructed to the house on the summit. Of all the charming drives, for which the environs of Manchester are famed, that to Mount Equinox is the most desirable. The landscape view is extensive and exceedingly interesting; reaching from the far-off Monadnock in the east, the Catskills in the west, to all the prominent Green Mountain peaks in the north. Manchester, which is reached by the Harlem Extension Railroad, contains many other objects of attraction for tourists. Its fine hotels, the "Equinox" and "Taconic," are worthy of patronage. The village possesses many picturesque charms which make it popular with visitors. Its marble walks shaded by beautiful trees give the place a quiet air of elegance.

Clarendon Springs are located about six miles from Rutland, and not only form a delightful and much frequented resort for the citizens of that prosperous town, but are of themselves fashionable attractions.

Killington Peak, having an altitude of 3,924 feet, is best visited from Rutland, which is reached by the Central Vermont, Rensselaer and Saratoga, and Harlem Extension Railroads. The excursion is made by carriage seven miles. The ascent requires the usual amount of hard climbing; but the view from the summit is fine.

St. Catherine Lake, Poultney, is the central feature of many interesting points in that region. It is located six miles from the village by a pleasant drive. The lake is about five miles long. On a projecting promontory at the lower end stands **St. Catherine's Hotel,** a summer house pleasantly situated. The *St. Catherine, Haystack,* and *Moosehorn,* mountains to whose lofty peaks pilgrimages are often made, look down upon its quiet waters. **Lake Bomaseen** near by, although less in magnitude, is a place of much attraction. **Carter's Falls,** the *Bowl,* the *Gorge,* and **Middletown Springs** are also objects of interest to visitors. Rensselaer and Saratoga Railroad.

The **Bread-Loaf Mountain** it is not proposed to describe, nor the *Bread-Loaf Inn* where visitors rest, and are at home; but the quaint name is given to catch the fancy of tourists who would leave the beaten tracks of travel for a rustic jaunt among the emerald mountains of Vermont, — a real stage-coach ride of the olden time, up hills that are long and steep, past gorges that are rugged and deep, for a quiet rest beyond. Leave the Central Vermont Railroad at Middlebury, and the stage will take you eleven miles to Ripton, and set you down at the Bread-loaf Inn. Good trout-fishing in the neighborhood.

Lake Dunmore, Vt., derives its name from this historical incident. It is said, that, about the year 1770, Lord Dunmore visited this region, and, becoming enamoured of the beauties of this lake, waded into its crystal waters, and, pouring wine upon it, said, " Ever after this body of water shall be called *Lake Dunmore* in honor of the Earl of Dunmore." Although this lake is but a half-dozen miles in length, its romantic situation, surrounded by high hills, the great depth and purity of its waters filled with gamey fish, and the many pleasant places in the vicinity, invest it with a rare interest to the guests of the Lake Dunmore House. Go to Salisbury, Vt., *via* Vermont Central Railroad, thence four miles by stage.

ILLUSTRATED PLEASURE ROUTE No. 11.

Boston to Providence, Stonington, New London, Shelter Island, New Haven, and New York City.

SHORE LINE RAILROAD.

Engraved expressly for Bachelder's "Popular Resorts, and How to Reach Them."
PROVIDENCE DEPOT, COLUMBUS AVENUE, BOSTON.

The **Shore Line Railroad** is a consolidated route of several companies, — the Boston and Providence, 44 miles; Providence and Stonington extending to New London, 62 miles; New York, New Haven, and Hartford, 122 miles: making the whole distance by rail, 232 miles. Upon the whole, the Shore Line is not merely the shortest to New York from Boston: it is the pleasantest, and in that respect the best.

We wish to make clear notes of this important route. Before speaking briefly of the chief places along the way, we specially impress upon the reader to secure a careful look over the newest and very latest wonder of Boston's century, — the famous castle, or depot, just finished by the Boston and Providence Railroad Company in Boston. In 1834 this road "opened shop" in a depot of great elegance and imposing proportions — for that day; so pretentious, in fact, that it held up its respectable head in a comely and quite fashionable way, until January, 1875, when its president, directors, and company sat down in the present regal edifice. Two stone tablets at the entrance symbolize the whole significant story in a quiet way. The one records the names of the president, directors, superintendent, treasurer, and architects of 1834; and the other gives the names of like officials, under whose authority and care the company replaced the new for the old, forty years afterwards. The building is an eloquent witness; but the tablets reveal the secret of its birth, — the brains and the loins.

It would be out of place here to tell how many gas-burners, how many miles of piping, how many rooms and for what, and all those minuter statistics which show the vastness of the whole work. The exterior, a perspective view of which is given, is an imposing, immense, and graceful composition, mainly of brick. Bricks are durable, but not elegant nor artistic, except in their arrangement; and yet the outside gives no conception of the splendor within. The approach and entrance is happy in its conception and elegant in execution. The impression made by the grand hall for passengers is that of cathedral opulence and sumptuousness. There is no gingerbread effort at effect; but there is a profound sense of massive grace, of princely cost and lavishment, and of architectural refinement, which is the type of adaptation to all time.

A building more than the seventh of a mile in extent must arrest contemplation; but to stand in a hall rich in tasteful outlines beyond power of words to fairly describe, expanded to one hundred and eighty feet, forty-four wide, and eighty feet to the ceiling, is a new thing under depot suns. Half a dozen country stations with all their pine devices, in wriggling sinuosities and affected grace, could be set in this immense case, and have comfortable elbow-room besides. One of the perfections of good taste in this structure is the space allotted for every room, not forgetting the humble and the ignoble.

Not only every want and demand of the public, but every arm of the working force, is provided for in the same sumptuous and luxurious manner, even to baths for the conductors. In fact, the new depot of the Boston and Providence Road is a marvel of beauty, utility, and modern progress. Others cover more earth-space, but this is *the* depot for the million. Its influence is not lost in its immediate atmosphere. Not only will every engineer, every conductor, every employee, feel a sort of proud identity with this palatial property; but, better still, the thousands who travel the road will feel the exulting and flattering consciousness, that all this unstinted bountifulness was intended for "us." It is "our" new depot to them; and the idea is full of shekels. "There are millions in it."

On the route by the Shore Line to New York, from Boston, objects of rare interest continually occur; as, indeed, they must in a region so alive with people and their industries. All along the course to Providence, thriving towns and pretty villages are found, with the gently rolling intervals of country farms and cottages. The cheerful co-operation of the railroad company in advancing facilities and encouraging all objects of enterprise has stimulated a healthful suburban growth. The city of Providence is of growing renown, as the focus of an immense summer travel, *en route* for Newport and the multitude of summer resorts, located upon the delightful shores of *Narraganset Bay*. The little thread of water, which runs from the bay up to Providence, is constantly enlivened by magnificent excursion steamers, puffy little "tugs," and graceful pleasure-boats, in "the season." Fresh breezes and fresh clams are the staple joys of all who go to the famous resort of **Rocky Point**. The city is a very pleasant place with numerous attractive local curiosities, and suburban drives. It is of goodly size, about a hundred thousand inhabitants. At *Kingston* tourists leave by stage for **Narraganset Pier**, a summer watering-place of growing popularity. Of its twenty hotels, all are well patronized in summer. It should be better known.

Stonington is a remarkably neat, pretty, and solemn place. It is a favorite resort in the summer for a select class, and has an excellent and well-kept hotel. **Watch Hill**, but three miles across the bay from Stonington, is a delightful seaside location, being directly upon the coast, and possessing all the characteristics peculiar to the ocean's margin. It is readily reached and largely visited from Stonington.

Stonington is also important as the eastern terminus of the *Steamboat Line* from New York, a description of which supplements this article. At this point the route strikes *Long Island Sound*, which it follows closely, by which the tourist is cheered by many fine marine views, and invigorated by cool salt-water breezes.

GRAND CENTRAL DEPOT,
Forty-second Street and Fourth Avenue, New York City.

New London is another conspicuous and historic place, following Stonington, crossing the Thames River by ferry. A few days' tarry at the *Pequot House*, at the mouth of the river, will incline the guest to repeat his visit. This is a first-class modern hotel, and is chiefly occupied by wealthy New York families in "the season."

Among the notable sights is Fort Trumbull, one of our good Uncle Samuel's peculiar sea-coast, burglar-proof safes, in which he keeps his "mad" stored up. It looks as though it was a very solemn and peaceable place, however; but looks are deceptive, you know.

The temptations to visit Neptune's domain are unusually great, and afford cheery sails and "lots" of plunder. This is the point from which to leave for *Shelter Island Park*, located on Shelter Island, near Greenport, L.I., across the Long Island Sound, directly opposite from New London, from which place a steamer runs regularly.

New Haven is reached through quietly diversified, but pleasant landscapes. The city itself is justly inclined to be a whit aristocratic, as it contains *Yale College*. It is called "The City of Elms," and holds many valuable and agreeable objects for the curious. It has a sweet, picturesque fringe of country scene, and will repay the time of a few days' tarry; for which good hotel accommodations will be found. Fine steamers run from New Haven to New York twice daily, and connect with other points also. There are steamers likewise from Bridgeport to New York. This is an enterprising, thrifty place beyond New Haven.

Pursuing the route, there are entered and passed successively Southport, Westport, Norwalk, Darien, Noraton, Stamford, Greenwich, Port Chester, New Rochelle, and other places numerous but of less note. The journey ends in the far-famed *Grand Central Depot*, corner of Fourth Avenue and Forty-Second Street, in the great city and pandemonium of New York. This renowned depot, the centre of an amount of travel almost defying computation, we present in picture form; and it is a worthy close of a trip commenced from the new Boston and Providence Depot.

The peculiar advantage this route affords to *Western travellers* lies in the fact that the western trains of the New York Central and Hudson River Railroad also have their termination in the Grand Central; so that no time is lost, nor inconvenience felt, by change of cars. Still further liberality is found by Southern travellers, that their transfer is by coach and free of charge, if they have procured through tickets, — a fact of which the knowing ones will take heed. The Fourth Avenue horsecars start from this depot, by which, with connecting roads, tourists may visit any portion of the city. The system of baggage delivery prevailing here is too well understood to require description.

EXCURSION THROUGH LONG ISLAND SOUND,

BY THE STONINGTON LINE.

NOTHING conduces so much to the pleasure of travel as a feeling of security. Whether flashing through the valleys of a beautiful landscape, around the hills, along the streams, or across the broad prairies; whether skimming the waters of some placid lake, stemming the current of a mighty river, or ploughing old "ocean's billows,"—the pleasure of the excursion will be in direct ratio to the confidence of the excursionist in the character and reliability of the route. In this particular THE STONINGTON STEAMBOAT LINE stands at the head of steam travel in America.

These boats possess the advantage of having been substantially built for outside service. They lay low in the water, presenting less surface to the winds, and in storms ride the waves "like a thing of life." They are unsurpassed for speed, comfort, and safety; and, whatever the state of the weather, *always make the trip, and are sure of connections.*

The change from the busy whirl and heated streets of a crowded city to the open harbor, where the sea-breeze sweeps unobstructed from shore to shore, is a source of great relief; and the sail from New York, through the harbor, up the East River, through Hell Gate, and down Long Island Sound, is one of the most delightful on the coast. The ferry-boats fly hither and thither like things of life. The gayly-dressed ships, bearing the fruit and merchandise of foreign climes; the forest of masts, with their streaming pennants, which for miles line the wharves along which we sail; the magnificent suburban residences and fine public buildings, with cultivated grounds, which adorn the banks; and the receding city clothed in the rich, warm glow of a beautiful sunset, — combine to make this sail one of the enjoyable episodes of a pleasant tour. THIS IS THE GREAT INSIDE LINE, leaving New York every afternoon (Sundays excepted) from Pier 33, North River, at 5, P.M., in summer, and 4 in winter, and continuing to Boston *via* Stonington, entirely avoiding Point Judith, a dangerous promontory, against which, during storms, the waves dash with fearful violence, making the passage, if not always dangerous, at least unpleasant to persons unaccustomed to sea-life.

The **Providence Railroad**, by which passengers from the boat continue to Boston, is one of the best appointed in the country. Its *Chair Cars* are a great luxury, and add much to the comfort of tourists.

THE STONINGTON LINE POSSESSES ANOTHER IMPORTANT ADVANTAGE.

Should any detention of the cars, or the probability of a rough or foggy night on the Sound, render such a course desirable, passengers from Boston for New York can change cars before reaching the boat, and continue on the SHORE LINE by rail, thus insuring Southern or Western connections in New York.

The Stonington Line presents unrivalled claims in the seaworthy character of its boats, an advantage fully tested during the unprecedented cold winter of 1874–5.

The "Stonington" and "Narragansett" were staunchly built for outside service: they lie low in the water, and in storms ride the waves "like a thing of life."

The "Rhode Island" is a splendid boat in model, architecture, and finish; it is furnished magnificently, and is claimed to be the fastest boat on the Sound. The spacious dining-room on the upper deck, the elegant state-rooms, smoking-room, with abundant facilities for promenade both inside and out, are among the luxuries this boat affords.

The Stonington Line is the direct route of approach from New York and the South to the summer resorts of *Stonington*, *Watch Hill*, and *Narragansett Pier*: the latter a place of growing popularity, is patronized largely by New Yorkers. Its isolated position is a disadvantage; yet its twenty or more hotels are always well patronized in "the season."

Passengers by this line reach Boston *via* the Boston and Providence Railroad in season for breakfast at the magnificent restaurant of the Providence Depot, before continuing by the morning eastern or northern trains.

MIDDLE STATES.

New York City — the great heart of the nation, whose throbs vibrate along the rails and magnetic wires which ramify into the remote recesses of the country — will repay a visit from tourists.

But before entering upon the wonderful surroundings, and "how to reach them," of that huge and renowned metropolis of a continent, it may as well be said, that it would be almost impossible to pass through it without travelling into a portion at least of Broadway, — the mighty artery through which surges the dense flood of life and activity of that vast Pandemonium of civilization. And to gaze upon the masses of vehicles of every kind, — stages, carts, drays, carriages, handcarts, running and racing, men and women, dogs, and every possible animated and moving thing, — is a scene even more stirring and exhilarating than are Niagara Falls, or the panoramas of the grandest mountains. In fact, Broadway, in full activity, is "a sight," — and well worth a long journey to see, if one saw nothing else. Indeed, it is a bewildering wonder; and while a New-Yorker moves amid all the seeming perils and inexpressible confusion, —

"Calm as a summer's morning," —

the visitor from a moderate city, or quiet town, is all aglow with excitement over the strange scenes. By all means see Broadway once.

As might be expected in a city so immense, amid industries and wealth so vast, stupendous works of various kinds are required and in constant development. Those already completed, in progress, and in contemplation, can hardly be enumerated. The great bridge across the East River, the tunnel beneath the North River, the gigantic railroad that cuts under ground, the startling railroad that travels in the air, the Croton-water masonry, the Central Park, the Grand Boulevard, and numerous other marvels of progress and cost, show the expansive energies of a million people, bent on improvement and wealth.

But New York is not a handsome city inside. Like all "huddles" of men in huge congregation, dinginess and squalor are set beside the brilliants of splendor and wealth. Every phase of human existence can be found in New York, — from the filth and brutality that would make a Hottentot blush, to the magnificence and luxury which might excite royal envy. Edifices, public and private, upon which money and skill have been unstintedly lavished, are numerous, of course; but the shabby, rickety, tumble-downety shanties are far more abundant. New York is in its full robust vigor, in fact; but many portions of it have passed it, and have a sadly old and jaded general look. A large number of pleasure routes radiate from the city to the numerous popular resorts for which the Empire State is famous.

NEW YORK CITY AND CATSKILLS.

New York City does not come within the design of this work, as a place of "popular resort." It is the great cities whose tens of thousands swarm to distant places, in pursuit of rest and recreation, when the sun pours down its summer heats. Cities have their peculiar points of striking interest, distinctive of art and wealth, refinement and cultivation. These are about as well studied, in the main, at one season as at another. Nature has there been subdued; and more formal things usurp her claims. But all our notable cities have their fringes of exquisite charms, replete with luxuries and delicacies, to which the multitude make frequent resort; and from these prolific centres the pilgrims in pursuit of ease or pastime make their summer journeys. The vast suburban regions around New York present an infinite variety of nooks and resting-places free from heat and glare and city turmoil. To enumerate these in detail would be useless. If we can glance at the leading lines of inviting travel, and places for repose, the balance can all be taken in while thus upon the wing.

UP THE HUDSON.

Taking royal precedence of all rivals, commencing at the city of New York, is the **Hudson River**, — or, rather, its grand and glorious shore views. Novelty, in describing this renowned river, has long since passed out of the possible. Its panegyrists embrace the ablest pens and the most gifted minds, not to speak of the hosts who have tried and failed. Indeed, the scenery that paints the margin of the Hudson, and as far into the remote as vision can reach, simply defies the power of descriptive delineation in printer's ink. The delicate and appreciative colors of the true artist alone can exhibit the tenderness of the tints and shades; the gently serpentine lines; the valleys and verdure; the modest undulation; the sharp and rugged ascent ; the grand and majestic mountain curves and piercing summits, with their soft haze, virgin blues, and rich, deep purples; and all these repeated, like a dream echo, in the water mirror between. It is genius only that can attempt to convey some grateful idea of how the scenery of the Hudson River fascinates and delights all minds and all grades of people, when viewed in the full glory of a robust summer. To enjoy the river to advantage, one should make the day trip, by steamer for Albany. What will be seen must be left for the reader to learn by study, as he winds along the sinuous route of the river. It will richly repay for the time and cost of the trip, as a rare painting by nature, graced by many a gem of architecture and art.

It should be borne in mind, that along the Hudson River, and at points not remote from its waters, are numerous memorable localities, where some of the sharpest conflicts and most momentous events of the Revolutionary war occurred. The holding of New York City by the British; their efforts to extend their occupation, and that of the patriots to hem them into as narrow a space as possible, — these, with the distractions which attend all like scenes, stamped the still living impress of the struggle upon many a field of strategy, skirmish, and battle. Of these, Forts Washington and Lee (both close to the city) are notable; also Fort Tryon and King's Bridge. All these spots are mentioned in histories of the Revolution, especially the desperate battle at King's Bridge, in 1777.

Yonkers is as familiar to a New Yorker as the Central Park.

The lover of old stories will find rare studies of old things around Piermont, N.Y., more especially the jail in which Major André was confined, and the spot where he was executed. These are at the ancient town of Tappan, near Piermont.

Washington Irving's home, "Sunnyside," is plainly seen on the right as you ascend the River Hudson.

Tarrytown, N.Y., where Major André was arrested, is a notable place. Cooper's graphic descriptions of the "Skinners" and "Cowboys" are laid in this region; and Irving's "Sleepy Hollow" is also close by.

Sing Sing, N.Y., is chiefly noted for its great prison and the Croton Aqueduct.

Croton Point, N.Y., holds the great lake and the vast reservoirs which supply the city of New York with water. Some of the grandest triumphs of modern engineering skill are here to be seen.

At Haverstraw, N.Y., Arnold and André met to arrange for the surrender of West Point. It is about forty miles up the river.

The famous "Stony Point," the scene of "Mad" Anthony Wayne's gallant exploit, lies just above Haverstraw, in New York State.

Peekskill, N.Y., has several Revolutionary reminiscences in its midst. On the opposite side of the river is the place where Capt. Kidd is said to have buried the treasures so much sought for, but not yet found.

Ascending Hudson River, and once past Peekskill, the grand diversities of "the Highlands" open to view, and continue to excite wonder and admiration, beyond the power of language adequately to express. These commence about fifty miles up the Hudson, and are probably unsurpassed for romantic scenery by any river travel in the Old World or in the New. To mention even the more notable, much less all the familiar features along and near this river, or to attempt a detailed description of them, would demand too much space.

West Point, the most renowned fortification on this continent, stands at the entrance to the Highland scenery of Hudson River, N.Y., and is a specially conspicuous object of interest to strangers.

Back from the Hudson River, N.Y., some dozen miles, rise the celebrated **Catskill Range, or Kaats-Kills of New York,** said to have been named by the Dutch, on account of the catamounts found there. By the Indians they were called Ontioras, meaning "a cloud-like appearance." These mountains are a part of the great Appalachian chain, extending from Canada East to the Gulf of Mexico. Their especial point, however, is the range following the course of the Hudson River for twenty or thirty miles, — lying twelve miles west, separated by the richly productive Catskill Valley. It is at this part of the Hudson that the landscape is the most charming; and tourists always meet with glad surprise this, the objective point of the "Beautiful Hudson."

The Catskill Mountains, renowned in story and in song, have long been famous as summer resorts. Thousands of those eager to escape the heat and discomforts of great cities annually visit the haunts of "Rip Van Winkle," which Washington Irving in one of his charming legends so gracefully portrays, where amid grand views and picturesque scenery, the summer is passed pleasantly and rapidly away.

The many mountain streams filled with trout, the wild unbroken forests abounding with game, the cool temperature, and pure waters, are among the many inducements offered to the tourist, the sportsman, and the invalid.

The advent of the *New York, Kingston,* and *Syracuse Railroad* through this region, opens up a new route which renders the trip both easy and pleasant, alike to the aged and infirm, and has greatly increased the popularity of these retreats. The little travel, time, and expense required in reaching these regions excel all other mountain resorts.

The high prices exacted for board, the reign of fashion, and whirl of excitement, at the fashionable watering-places, deter many from seeking that recreation which their health requires. Throughout this region, in isolated positions and in considerable clusters, are large hotels, and attractive boarding-houses, where, at prices ranging from six to ten dollars per week, ample accommodation, a good table, and kind attention await the guest.

There are two distinct routes by which tourists approach the Catskills: one from Kingston *viâ* the New York, Kingston, and Syracuse Railroad; and by the old popular line by stage from Catskill Station: both of which are described. Visitors from New-York City can go by either of the elegant and commodious steamers "Thomas Cornell," or "James W. Baldwin," from the foot of Harrison Street; the fast and famous steamer "Mary Powell," from the foot of Vestry Street; the Albany *Day Boats* "Drew" or "Vibbard;" the New York Central and Hudson River Railroad, and the Erie Railway.

APPROACH FROM KINGSTON STATION.

At **Kingston**, all of the above-named steamers and trains connect daily with the New York, Kingston, and Syracuse Railroad, to the mountain regions of Ulster, Delaware, and Greene Counties.

This company has recently purchased handsome coaches, which are under the charge of attentive officials; and the traveller will find the trip to the mountains easy and pleasant. As the train winds its way around and up the mountains, gorgeous scenery surrounds him on every hand. From the car-window the tourist looks out upon the grand, majestic Catskills, each peak rich in romantic legends. Drinking in the pure cool breezes of the mountains, inspired by the picturesque grandeur of the scene, he is sure to remember it as one of the finest views that has greeted him in any quarter of the globe.

A ride of nine miles brings the tourist to **West Hurley.** This is the point of debarkation for passengers bound for the *Overlook Mountain House.*

It is situated on **Mount Overlook**, the most lofty of the summit of the Catskill range. Its height is thirty-eight hundred feet above the level of the sea. The scenery from its peak is of the boldest and most romantic description. The hotel has a capacity to comfortably accommodate five hundred guests. The telegraph and postal facilities are ample. The temperature is remarkably cool, the thermometer seldom reaching higher than seventy-eight degrees.

Eighteen miles farther up the Shandaken Valley brings us to **Phoenicia.** Here passengers destined for *Hunter*, Greene County, will find stages in waiting. The scenery from Phoenicia to Hunter is perhaps the most attractive of any among the Catskills. The drive through the famous *Stony Clove*, and a visit to the *Kauterskill Falls* and *Plattekill Ravine*, are of themselves sufficient to attract many to this charming resort.

Still travelling through a beautiful valley, replete with the most romantic scenery, the tourist arrives at **Shandaken.** The scenery here is the annual study of a large number of artists: it is picturesque and exceedingly beautiful. This is also the point of debarkation for tourists crossing the "Notch," and visiting *Westkill* and *Lexington.*

The Notch is one of the most curious features of the mountain. The public road crosses the mountain at this point, through a narrow defile with abrupt precipices about twelve hundred feet high on either side, which appear to close in at the top. In summer it affords an impenetrable shade, where snow and ice remain nearly the whole year through in the clefts of the rock.

Big Indian is thirty-six miles from Kingston at the foot of the grade ascending *Pine Hill*, and about five miles from the summit grade of the road, and is one of the most favorite resorts for trout-fishing. The scenery is wild, and presents some of the most fascinating mountain views.

Again taking the train, while we slowly wind and twine our way around and up *Pine Hill*, we gaze with silent admiration and awe upon the magnificence and sublimity of the scene. It is broad and grand, and beyond the power of reproductive art.

Having crossed Pine Hill at an elevation of nearly two thousand feet above the level of the sea, we soon glide into *Margaretsville*, Delaware County. This inviting little village is delightfully situated on the banks of the Delaware River. We are now in the vicinity of the famous fishing and hunting grounds of the *Beaverkill* and *Millbrook*. There are two good hotels located here.

Jumping aboard the train again we soon find ourselves at *Roxbury*. The tourist entering this village is at once impressed with the neatness of its appearance, and the beauty of its location.

At *Moresville* passengers for Prattsville, Ashland, Windham, and Hensonville in Greene County, will find splendid four-horse coaches in waiting.

Prattsville. — This is an attractive little village cosily nestled among the Catskills, and has a population of seven hundred inhabitants. Here are located a number of fine boarding-houses.

Windham. This is the headquarters for city boarders in Greene County. Of all the towns upon our route, none are more attractive to the pleasure-seeker; none present more pleasant social aspects, or equal it in its picturesque scenery, and grand surroundings.

Hensonville is a beautiful mountain village, situated about three miles from Windham, and contains a number of first-class boarding-houses, and is each year increasing in popularity.

Stamford, Delaware County, is the present terminus of the railroad. This charming little village, situated over eighteen hundred feet above the level of the sea, will long detain the visitor by the pleasant strolls which it invites among its shady streets, bordered by cosey residences and elegant pleasure-grounds; by the many drives which are afforded in its environs through inviting groves, into beautiful villages, and over rippling brooks. Those desirous of passing the summer among the mountains can certainly find no pleasanter accommodations nor more polite and kind attentions, than those we guarantee them to receive from the hands of the inhabitants of this charming village.

APPROACH FROM CATSKILL STATION.

Catskill Station is most accessible by way of this river, either by rail or steamboat: several trains and boats run daily, south from Albany, and north from New York City, connecting with the ferry which crosses the river to Catskill proper, including the world-renowned day-line steamers, "C. VIBBARD," and "DANIEL DREW," which leave New York every morning. A daily line of stages conveys passengers from the village across the valley to the mountains about twelve miles off.

If the visitor is in search of mountain scenery alone, he will leave the fashionable hotels of the river and village neighborhood, and proceed at once by the stage-road to the terminus of the lovely valley of Catskill Creek, where he will find himself at the seat of the hill and mountain region.

About six miles on from the village, he will pass the ancient Dutch hamlet of *Kiskatom;* and, along a mile or two, "*Sleepy Hollow*" of Irving's legend is seen. It is rather paradoxical to breathe here, not a slumbering, dream-like air, but, instead, a literal "Rip Van Winkle" in shape of a wide-awake hotel. Thrift and enterprise have entered into that "Hollow" with an earnestness that would bring dismay to the charming writer who aimed to immortalize the spot as one of restful, unchanging scene.

About two miles from the Hollow, the road turns, and ascends *Pine Orchard Mount;* and here the view includes an area of ten thousand square miles: the eye can reach four States. To the west is a varied mountain view; to the east, a wide half-circle of etherealized blue landscape. "On the horizon, the Hudson Highlands, the Berkshire and Green Mountains, unite their chains, forming a continuous line of misty blue. The Hudson, its broad valley studded with white villages, is stretched below for many leagues." This view has been enthusiastically described by some of our best American authors, and Harriet Martineau was more moved by it than by Niagara itself. The mirage, the sunrise over the Taghkanics, and the raging of a thunder-storm, are objects of especial attention here.

North Mountain is reached by a path leading from the hotel of Pine Orchard Mount which passes the *Bear's Den*. On the way to *Pudding-Stone Hall* and to the *Fairy Spring*, the *South Mountain* is ascended: it is a pleasant path, and the view from the summit extends over a vast space, taking in certain peaks of New Jersey. The two *Cauterskill Lakes*, about a mile and a half off, abound in fish.

An old road, exceedingly rural, leads from the broad rock platform of Pine Orchard to *Moses' Rock*, and, for a mile or two on, there spreads a deep and well-wooded ravine. Just below are *Cautersville Falls*, interesting as the outlet of the lakes. These falls spring over the rocky cliffs in two jets; the first one hundred and seventy-five feet, the second eighty feet high. A curious performance can be here carried on by accomplished guides, who will, by aid of a dam, and for a small fee, effect a "freak of nature." The natural flow of water is not at all times, especially in summer, sufficient to display the desired effect of the falls; and, in order to satisfy sight-seers, the water is turned on from a dam, and the object is furnished for value received.

Bastion Falls are a quarter of a mile down the ravine, on the way to *Cautersville Clove*.

This Clove is the favorite resort of artists; and the section leads upward with gradual rise, and west to the lofty plateau of *South Mountain*, passing pretty dots of brook scenery. A ravine leading from this summit contains the famous *High Rocks*, and also the *Fawn's Leap Falls*; and farther along, near the outlet, is the entrance to *Hain's Falls*, an imposing cascade one hundred and fifty feet high, with one or two less notable ones above and below. The *Washington Profile Rock* is near the bridge, crossing on to the Clove road. Many graceful cascades occur in the Clove stream, which follows along the slopes of *High Peak*; the road passing on with it diverges at *Hunter*, running to the south-west, and down to *Esopus Valley* near *Overlook Mountain*, and returning through Stony Clove. This stream is excellent for trout-fishing.

Hunter's Glen is narrow, and wildly grand. It was originally settled by "Cow Boys," a band of border banditti. Near it, is the sharp ragged peak, the *Colonel's Chair*, and also *Hunter Mountain*.

Two miles from Hunter Hamlet a portion of Stony Clove gorge lies continuously sunless, holding to its bosom ice-depths throughout the year, — a fitting place for the Devil's Tombstone, which is in this gloomy ravine. At this place we are twenty miles from Catskill village. Farther west are the lonely glens of Lexington; and on, a distance of thirty-six miles west of Catskill, are the far-viewing *Pratt's Rocks*. Looking north towards Hudson, *Mount Merino* is seen overlooking *Matteawans*, *Catskills*, *Taghkanicks*, the *Green Mountains*, the *Luzerne Mountains* at Lake George, and many miles along the Hudson.

About six miles south of the Canterskill Clove, **Plattekill Clove** is entered from Saugerties Plains by a road running along the foot of the mountains or from the Tannersville plateau. It is an exceedingly weird, deep-descending gorge, incased in massive cliffs, and is traversed by a rambling brook and rough path. *Black Chasm Falls* are in this Clove: they have an altitude of three hundred feet. On towards the north, *High*

Peak and *Round Top Mountains* are plainly visible. *High Peak*, the most prominent of all this region, is often ascended by venturesome ladies. The path is rough and winding, and starts from near *Hain's Falls*. It is thirty-eight hundred and four feet high, and is conceded to give the most extensive view of this region. Next in consequence is the symmetrical *Round Top*, thirty-seven hundred and eighteen feet high. These two mountains are isolated from their sisters of the range, being separated by the deep ravines of *Plattekill* and *Cauterskill Clove*. *Blackhead*, near by, and six miles north of Catskill Mountain, is a steep and cone-like peak.

There are remaining, in adjoining counties, several hundred square miles of merely officially explored territory. The land is scarcely inhabited, and is a region of profound wildness. It is made up of tall, savage-looking mountains, covered with rank forest growth, intersected by sparkling trout-streams, forming themselves into occasional cascades, presenting to the adventurer rare scenes of interest.

Three or four days will suffice to see the Catskills *in a general detail*, a week or more, however, ought to be given to perfectly satisfy the tourist. They are mountains of more than ordinary capacity to interest. The artist Thomas Cole, of "Course of Empire" fame, made this region his home with an enthusiasm equalling the well-known Niagara hermit; in fact, his well-earned and favorable reputation dated from his painting of the Cauterskill Falls. It is, perhaps, unnecessary to add that the visitor will find here good accommodations. It is hoped that a future edition will present this region, so replete with scenic beauty, fully illustrated to the public.

ILLUSTRATED PLEASURE ROUTE No. 14.

New York and Philadelphia to Bethlehem, Mauch Chunk, Wilkes Barre, Scranton, Richfield and Sharon Springs, Howe's Cave, Saratoga, Lakes George and Champlain, the Adirondacks, Montreal and Quebec. Also a New Route from New York to Long Branch and North Mountain, Pa.

CENTRAL RAILROAD OF NEW JERSEY, NORTH PENNSYLVANIA, ALBANY AND SUSQUEHANNA, RENSSELAER AND SARATOGA, AND NEW YORK AND CANADA RAILROADS.

THE depot of the **Central Railroad of New Jersey**, in Jersey City, is reached by the railroad company's splendid ferry-boat from the foot of Liberty Street, New York, from whence cars continue through Northern New Jersey to Easton and Bethlehem, Pennsylvania. At the latter place connection is made with the North Pennsylvania Railroad, leading from Philadelphia to Bethlehem, whence the route continues over the rails of the Lehigh and Susquehanna Railroad (a leased road of the Central Railroad of New Jersey) up the valley of the Lehigh, past Scranton, to Green Ridge, connecting with the Delaware and Hudson Canal companies' railroads, to Cooperstown, Sharon Springs, Howe's Cave, Albany, Saratoga, Lakes George and Champlain, Montreal, Quebec, and the Adirondacks.

NORTH PENNSYLVANIA RAILROAD.

Although Philadelphia has been denominated the "City of Homes,"— and justly so, for in no place of its magnitude in the world is the general population "housed" in the same independent and comfortable manner, — yet, as the summer solstice approaches, there is a certain portion of the community, that, tiring of the heated streets and sultry atmosphere, are willing to leave the clustering comforts by which they are surrounded, and are anxious to escape into the open country, to revel among the verdant fields and leafy groves, to listen to the murmuring rill, or the deep diapason of the ocean's roar, to climb the mountain, or thread the valley, and at the same time to drink in deep draughts of the pure, life-giving air that will invigorate them for their returning duties in the metropolis, whether their accustomed routine be business or pleasure. One of the favorite routes leading from the city, and largely patronized by the pleasure-seekers, is the North Pennsylvania Railroad, running from Philadelphia to Bethlehem (with several lateral branches), where it unites with the Lehigh and Susquehanna division of the Central Railroad of New Jersey, and with the Lehigh Valley Railroad. From an imposing and well-ordered depot at the corner of Berks and American

Streets, easily accessible by the lines of street-cars from all points of the city, numerous daily trains, at short intervals, are receiving the thronging passengers, and distributing them upon their several errands ; whether their destination be the suburban homes of which so many grace the line of the road, or the more lengthened trip to mountain, glen, or lake that this route offers on such favorable terms, with such excellent accommodations.

The line of the road is singularly beautiful in a quiet and rural point of view. Before leaving the city limits, it strikes through large plantations owned by old and wealthy families, who, keeping their possessions intact with the proper admixture of glebe and woodland, contribute more to the enjoyment of the cultivated eye in viewing the beautiful results, than would extended rows of brick and mortar. Progressing up the road, the country opens out on either side in exceeding beauty. Valleys stretching off for miles, dotted with hamlets, sparkling with streams, and showing evident marks of thrift and cultivation, greet the eye, while on many an eminence may be seen handsome and substantial mansions, surrounded by grounds decorated in all the perfection of landscape-gardening.

Nor is the useful entirely subordinate to the beautiful. Several belts of hematitic iron-ore cross the road in different localities, at each of which may be seen the lofty stacks and puffing engines of the blast-furnace, with the accompanying aggregation of dwellings, and other marks of this industry upon which Pennsylvania builds so firmly the edifice of her greatness. At Bethlehem the North Pennsylvania makes connection with two diverging roads,—the Central Railroad of New Jersey, leading eastward to Easton, and through New Jersey to New York City, and northward through the Lehigh and Wyoming Valleys to Scranton, and so on *via* the Albany and Susquehanna Railroad; and also with the Lehigh Valley Railroad, leading eastward to Easton, and northward through the Lehigh, Wyoming, and Susquehanna Valleys, to the New York State line, joining at that point with the Erie Railway, Ithaca and Athens, Southern Central, and other roads *en route* to Watkins Glen, Niagara Falls, the lake system of Central New York, the Great Lakes, and all the desirable watering-places of New York and Canada. The North Pennsylvania route has for some years been a favorite one for pleasure-seekers; and each season increases the number of those availing themselves of its speedy transit, comfortable cars, and admirable accommodations.

There are several branches striking off from the main line; at Lansdale, one running eastward to Doylestown, the county seat of Bucks County, and one running westward to Norristown, the county seat of Montgomery County; one at Abington, running to Hartsville and Hatboro', flourishing villages in Montgomery County, upon all of which are enterprising and growing towns.

PENNSYLVANIA SCENERY — UP THE LEHIGH.

THE remarkable unfolding of the mineral resources of Pennsylvania during the last few years has developed some of the finest scenery on the continent. Deep gorges, bold precipices, and wild ravines, heretofore untrodden by human foot, now sparkle with the light of civilization. The screaming locomotive, guided by science, darts into the recesses of the mountains. Forests are levelled, valleys cleared, houses erected, cities reared, mines opened; and the very hills pour forth their hidden treasures. This industrial research has opened up a new field for pleasure seekers. Probably no other locality on the continent has received a like increase of visitors. A few years ago the extensive coal region of Pennsylvania was comparatively unknown to the tourist: now thousands visit it annually, and return filled with admiration of the wild beauties it contains.

LEHIGH VALLEY, MAUCH CHUNK, PENN.
Looking South from Mt. Pisgah.

ROUTE OF APPROACH.

The direct route of approach to the coal regions of Pennsylvania from New York, New England, and the Provinces is by the **Central Railroad of New Jersey, its Branches and Connections**, and from Philadelphia by the **North Pennsylvania Railroad**.

This also is the most direct and the shortest route from New York to Easton, Allentown, Wilkes Barre, Reading, Harrisburg, Williamsport, the Oil Regions, Pittsburg, and the West, and is one of the very pleasantest to North Mountain and Watkins Glen (elsewhere described), and when connected will embrace one of the finest and most varied pleasure trips on the continent. It has also been opened as a through route from New York and Philadelphia to Saratoga, via Mauch Chunk, Wilkes Barre, Scranton, &c. (see description). We leave New York from the foot of Liberty Street, by the Central New Jersey Railroad Company's splendid ferry-boats to Jersey City, from which point our route by rail commences. The road leads at first in a general westerly direction, through a fine agricultural region, interspersed with thriving villages and elegant suburban residences. It is also a field replete with historic memories. *Washington's Rock*, the lofty crag from which that revered general was wont to study the position and note the movements of a foreign foe, is plainly visible from the cars.

At Hampton Junction the "Delaware, Lackawanna, and Western Railroad" connects for Delaware Water Gap, Scranton, Great Bend, and Binghamton.

At Phillipsburg, a picturesque town built on a bold bluff on the left bank of the Delaware River, opposite Easton, which it overlooks, the "Central Railroad" connects with the "Morris and Essex" and the "Belvidere" Railroads.

Easton is delightfully located at the confluence of the Lehigh and Delaware Rivers, the former leaping over a dam of twenty-one feet at this place. The town is approached by a magnificent bridge, one thousand feet long, and twenty-two feet high. It is constructed of wrought iron, resting on heavy cut-stone piers, and, including the rock-cuts in the vicinity, cost $650,000. Beneath this pass diagonally the Canal, "Belvidere Railroad," and foot-bridge; and under all rush the waters of the wild mountain torrent in its race to the sea.

This bridge connects the "Central New Jersey Railroad" with the "Lehigh and Susquehanna" Division on the north bank of the Lehigh, and with the "Lehigh Valley Railroad" on the south bank. By the latter route we continue to Bethlehem and Allentown. The picturesque beauty of the scenery increases from Easton, the cars following the graceful curves of the river, which is fringed and shaded by beautiful trees, while bold hills, clothed with luxuriant foliage, compose the background.

At Bethlehem the "North Pennsylvania Railroad," from Philadelphia, intersects with the "Lehigh Valley" and the "Lehigh and Susquehanna" Roads, contributing its quota of tourists from Philadelphia and the South. The "Lehigh and Lackawanna" Branch to Chapman's also leads from this point.

At Allentown the course of the river is from the north-west, up which the "Lehigh Valley" and "Lehigh and Susquehanna" Railroads extend, while connection is also made with the "Allentown Line;" which comprises the "East Pennsylvania Railroad," thirty-six miles from Allentown to Reading, and the "Lebanon Valley Railroad," fifty-four miles farther, to Harrisburg. The general course of this route is westerly. The scenery is unusually fine; and, differing entirely in character from the "New York and Allentown" section, it adds to the variety and pleasure of the tourist. At Harrisburg connection is made with trains on the "Pennsylvania and Northern Central Railroad," affording ample facilities to go North, South, or West.

THE LEHIGH VALLEY.— RESUMING FROM EASTON.

COAL VEIN

The "Lehigh and Susquehanna" Division of the "Central Railroad" connects at Easton, and, following the tortuous course of the Lehigh, winds its picturesque way through the mountains to the Susquehanna at Wilkes Barre, up which it follows to Pittston, and thence on the east bank of the Lackawanna to Scranton.

This is a main line, into which lead, from every direction, branches filled with trains burthened with the rich mineral products of this remarkable region. Coal is not the only product: iron, slate, &c., are manufactured in great abundance.

Either of these is found in quantities sufficient to insure the wealth and prosperity of any section of the country. These industrial pursuits form an interesting source of information, as well as amusement, to the tourist. He often gazes in amazement upon the curious mechanism and ponderous implements employed. The enormous expenditures which have been made to develop and frequently to prepare to develop these enterprises, are a source of wonder. And when we realize that these features are but adjuncts to one of the finest combinations of natural scenery in America, we can better understand its growing popularity.

LEHIGH GAP,
(Looking Down).
Central Railroad of New Jersey

" Soon after leaving Bethlehem, the mountains approach the bed of the stream, and at 'The Gap' fling themselves directly in its path, leaving no resource but to go through them, which it has accordingly done, cleaving the mountains from summit to base in its efforts to escape.

"It is not until the vicinity of Mauch Chunk is reached that the peculiar features of Lehigh Valley appear in perfection."

MAUCH CHUNK, PENN.
Mt. Pisgah and "Switch-back" Railway.

This wild, picturesque, and popular region is reached from New York and Philadelphia *viâ* Central New Jersey, North Pennsylvania, and connecting railroads.

The arrival of the morning trains at Mauch Chunk from New York and Philadelphia is at the hour of noon; and a hot dinner at the Mansion House is waiting to be served. This, to the frequenter of the Lehigh Valley, is the only announcement necessary; but to the stranger I will add that "The Mansion" has no superior in this region. It needs but one visit to insure a second. The cars of the "Central Railroad of New Jersey" stop at the door; and its location on the banks of the Lehigh, overhung by rugged mountains, all

MANSION HOUSE, MAUCH CHUNK, PENN.
Central Railroad of New Jersey.

clothed with the fragrant rhodendron, is picturesque to the last degree.

The visitor to Mauch Chunk is advised to go without any pre-arranged plans. It is not a place to "do" by programme, as many tourists travel. It contains too much, has too many features of interest, so startling in their character, so grand in conception, and so beautiful in detail, that any previous plan of operations must in execution fall to nought. It is better to go untrammelled.

After finding yourself comfortably domiciled, go first to the veranda on the front of the house, and leisurely study the scene, an engraving of which is herewith submitted. It is truly a wonderful view, pleasing in art, yet far more so in nature. A glimpse of the entrance to the town shows through the narrow street to the left. Splendid residences cling

Engraved expressly for "Bachelder's Popular Resorts, and How to Reach Them."
VIEW FROM THE MANSION HOUSE, MAUCH CHUNK, PA.

1. East Mauch Chunk
2. Bear Mountain.
3. Lehigh Valley R.R.
4. Lehigh & Susquehanna R.R.

to the hillside beyond, over which a few marble monuments indicate the village cemetery. Beyond this we take the cars for the "Switch-back" Railway. The "dam" in the left middle ground throws the water into

THE FLAGSTAFF.

the canal, whose boats, loaded with the "black diamonds" of this region, we have seen by the wayside. Immediately before the door is the platform of the "Central Railroad of New Jersey," where passengers are left and received from every train. The light iron bridge leads to the depot of the "Lehigh Valley Railroad," on the opposite side of the river. *Bear Mountain* is the central feature of the landscape.

But the mountain on the right receives the greatest homage from visitors. From the "Flagstaff" on its summit you get the view suggested by the above engraving, though vastly superior. It is too extensive, too grand, to receive justice from the artist's pencil. The topography of the whole country is spread out before you. It seems a moving diorama, through which you trace the serpentine windings of the Lehigh Valley, with its river, its railroads, and canals.

An excursion over the "Switch-back" Railroad will also be in order. Strictly speaking, the "Switch-back" has ceased to exist, and a *gravity* road has taken its place; but the name remains. The first improvement in the "Switch-back" Railroad — for conveying coal from the mines about Summit Hill, ten miles distant, to the boats of the "Coal Navigation Company" at Mauch Chunk — was by employing *gravity* one way, the grade being sufficient to insure this. Mules were taken down on the train to draw the cars back. This was subsequently improved by the construction of planes over the intervening elevations. Mt. Pisgah and Mt. Jefferson, up which the cars were drawn by stationary engines on their summits; the altitude thus gained being sufficient to turn the grade to *Summit Hill*, to which place the cars returned by their own gravity. This means of transportation answered well its purpose until the great demand for the anthracite coal of the "Lehigh" warranted the construction of a steam railroad, — the cutting of solid rocks asunder, and piercing the mountain barriers with tunnels. The success of this last enterprise relieved the "Switch-back," or, more properly, "Gravity" Road; but the opening of this region at this time as a "popular resort" suggested the use of the "Switch-back"

MT. PISGAH PLANE.
"Switch-back" Railroad, Mauch Chunk, Penn.

Engraved expressly for "Bartholf's Paquebot Houses, and How to Reach Them."

ONOKO STATION, MAUCH CHUNK, PA.

1. Moyer's Rock
2. Central R.R. of New Jersey
3. Lehigh Valley R.R.
4. Glen Onoko

as a pleasure route for excursionists. Passenger cars have been substituted; and the same powerful machinery used for coal-cars is now applied to the light pleasure traffic. Carriages from the depots and the hotels take passengers to the base of Mt. Pisgah, though it is but a short distance for those who prefer to walk; and the fine scenery will repay the effort. The *plane* of Mt. Pisgah rises one foot in three for 2,322 feet. The cars are drawn up by a stationary engine on the top, connecting with an iron band six and one-half inches wide, which runs over a drum eighteen feet in diameter. The passenger car is followed by a *safety car*, supplied with a long iron bar following in a "ratchet," which, in case of breakage of engine or bands, securely holds the cars against accident; and its efficiency may be judged by the fact that there has never yet been an accident. On reaching the summit of Mt. Pisgah, the car starts by its own gravity down the opposite grade. Its course is gradual, following the tortuous sinuosities of the surface; now glancing under the shade of broad-spreading trees, for a moment refreshing all with their cooling shade; anon skirting the brink of a beetling crag, unfolding glimpses of the changing scenes below. Now we glide along the mountain side, and skim through the valleys, clearing at a bound the noisy streams which foam and boil far down among the rocks.

The ride is exhilarating beyond description. Without motive power, we seem to *fly through the air*. The winter coasting which delighted our childhood days tames in comparison.

The car is under the complete control of the brakemen. It would acquire a speed of forty-five miles an hour, but is kept at eighteen. Six miles our downward course is held, to the base of *Mt. Jefferson*, up which we are drawn as before, and again descend a single mile to *Summit Hill*, where a half-hour's stay is made. This is a mining hamlet, whose chief attraction to the tourist is the "Burning Mine," which has been on fire since 1832. The homeward ride is pleasant; we have no more *planes* to rise; our altitude is sufficient to give the grade, down which we glide nine miles to the point of starting. The pleasure of the party increases; familiarity with the scene has banished the fear of fancied dangers; and all return feeling that they have received an unusual amount of satisfaction for a dollar; and not unfrequently repeat it the next day.

It would seem that enough objects of interest about Mauch Chunk have already been described to insure its popularity; but the most beautiful feature remains.—**Glen Onoko.** Two miles above the village this fascinating spot is located. Cars by the "Central Railroad of New Jersey," and "Lehigh Valley Railroad," make several trips daily. It consists of a depression in the mountain, from which a fiery stream springs a thousand feet by successive leaps to the valley below, forming among the rocks and precipices a rare combination of waterfalls and cascades, which are clothed with deep evergreen foliage, and ornamented by the bright flowers of the rhododendron. At much time and expense a good path, stairways, and rustic bridges have been constructed, to facilitate the visitor. The accompanying "cut" of Onoko Station will convey a good idea of the locality, which is known to boatmen as the "Turn-Hole," from the "eddy" in the river formed by the current. The bluff on the left, through which the "Lehigh and Susquehanna" Division of the "Central Railroad of New Jersey" passes by tunnel, is properly known as **Moyer's Rock,** and possesses a traditional interest. The story is told in this wise: During the early settlement of the country, a noted hunter and Indian-fighter, living in Mahoning Valley, four miles south, who had hitherto eluded all attempts at capture, was surrounded, taken prisoner, and disarmed, by five Indian warriors, and left on the summit of this rock for security, guarded by two of their number, while the others hunted for game. Moyer was sorely perplexed. To fight alone two armed Indians was not to be thought of; and long he pondered. Suddenly starting, he listened intently, then relaxed into his former quiet. The Indians watched him unmoved. Again he started; and, creeping to the very brink, throwing into his countenance all the interest he could command, he gazed intently down. The ruse succeeded: overcome by curiosity, the Indians unguardedly moved to his side, and sought to discover the source of interest; when, with the spring of the tiger, he seized and dashed them to the rocks below.

The visitor to Glen Onoko should be well shod and suitably clothed, the refreshing coolness of the atmosphere rendering an extra "wrap" acceptable. The successive cascades, waterfalls, and other objects of interest, at Glen Onoko have each received appropriate names, and are worthy an individual description; but there are so many other interesting features of this picturesque region which demand a passing notice, that we must leave details to local guides.

Mauch Chunk is not, as many suppose, a mining town, but is, rather, the great coal-depot or shipping-mart of the Lehigh Valley. The production of coal is a subject of growing interest to the people of America; and, although it is not within the province of this volume to enter into a detailed description of the manner of working a coal-mine, yet a few lines for the benefit of those who would like to investigate the subject while in this region will be in place. The coal is found in veins of various thickness, and differently situated,—sometimes level, sometimes curved, often at an angle, and occasionally cropping out at the surface, from which the entrance is made. The experienced geologist can predict with approximate correctness the location of a vein of coal, and estimate the thickness of the overlying strata which must be pierced to reach it.

Sometimes these tunnels enter at the upturned edge of the vein, and

COAL BREAKER.

MINE

descend with its inclination, and are termed *slopes*. These apertures are generally about eleven feet wide by seven feet high, and contain two railways,—one for the descending and one for the ascending cars,—and a "pump way" (for the mine must be continually cleared by the most powerful pumps), and a travelling or "man way." The slopes vary in length, frequently descending to great depths, passing at times under towns and rivers. The longest slopes in the anthracite regions are at New Philadelphia, or Lewis Vein, 2,700 feet; and at Diamond Vein, which is from 2,800 to 3,000 feet: these are on an incline of about 45°. "Gangways" are turned off to the right and left; and in working the coal a "pillar" is left every few feet which sustains the overlying strata of rock, and prevents it falling in.

Within a distance of from two to five miles from the town of Wilkes Barre, through which this route leads us, there are worked over forty mines, producing in some cases 1,500 tons of coal daily from a single mine.

The "Nesquehoning Valley Branch Railroad" leads from Mauch Chunk, and, connecting with the "Catawissa Railroad," extends to Williamsport, a distance of ninety-two miles.

This route is noted for the wildness of its scenery, its deep ravines, and high bridges, and must eventually become very popular with the pleasure-seeker.

THE LEHIGH.
Looking North from Mount Pisgah, Mauch Chunk.
Central Railroad of New Jersey.

From Mauch Chunk northward the Lehigh Valley is little better than a cañon enclosed between high mountain walls, at whose base the narrow stream tumbles and foams; its waters now displaying the rich amber hue which they have distilled from the roots and plants in the swamps around their source, now white from their encounter with rock or fall. High rocks hang directly overhead, and threaten to fall at any moment upon the trains which constantly roll beneath; branches wave, and flowers blossom on the hillside, so close to the railroad track that the passengers can almost reach them without leaving their seats. Here and there a miniature waterfall springs from the mountain top, and glances, a ribbon of foam and spray, to the river at its foot; and at frequent intervals ravines cut in the mountain side present a confusion of rocks and wood and water to the eye of the traveller as he flashes by. Traced back a little from their mouths, these glens often show a wealth of beauty, a succession of snowy cascades, transparent pools, and romantic nooks, which are an ever fresh surprise to the explorer.

At Penn Haven, seven miles above Mauch Chunk, the "Lehigh Valley Railroad" connects with the "Mahanoy, Beaver Meadow, and Hazelton" Branches. The "Lehigh Valley" here crosses the river, and runs on the east bank to White Haven.

Fifty years ago this whole valley was a wilderness, with one narrow wagon-road crawling at the base of the hills beside a mountain torrent which defied all attempts to navigate it. Now the mountain walls make room for two railroads and a canal; but the tawny waters of the stream are nearly as free as ever. Here and there, indeed, a curb restrains them; and once an elaborate system of dams and locks tamed the wild river, and made it from Mauch Chunk to White Haven a succession of deep and tranquil pools. "But one day in 1862 the waters rose in their might. Every dam was broken, every restraint swept away; and from White Haven to Mauch Chunk the stream ran free once more. The memory of that fearful day is still fresh in the minds of the dwellers of the valley; and the bed of the torrent is still strewn with the wrecks that went down before its wrath.". . .

Nescopec Junction is a place of little importance; but the "Nescopec Branch Railroad" leads nine miles into a valley filled with wild and picturesque scenery.

This whole region is strange to the visitor. The valleys are deep, the precipices are bold and high, and the mountains steep. Even the waters rush with greater violence than in tamer countries. But the public will soon understand this scenery better. The artists, the pioneers of pleasure travel, have already heard of it, and each year visit it in increasing numbers. Soon the tide will set up this valley, hotels will be in demand to meet it, and the *press* will herald its praises.

Persons residing in our large cities hardly realize how quickly and for how small a sum these romantic places can be enjoyed. The morning train from New York or Philadelphia takes you to Mauch Chunk in season for dinner,— dinner steaming hot at the Mansion House. The "Switch-back" and Glen Onoko can be visited in season to return at night.

The subjoined description of the Nescopec region is from "Lippincott's Magazine:" —

"We walked about a half-mile along a wood-road, struck into a foot-

PROSPECT ROCK.
Nescopec Valley.

path, and followed it a hundred yards or so, and without warning walked out on a flat rock, from which we could at first see nothing but fog, up, down, or around. It was a misty morning; but we made out to understand that we were on the verge of a precipice, which fell sheer down into a tremendous abyss; and when the fog lifted we looked out upon miles and miles of valleys, partly cleared, but principally covered with primeval forests. We were on **Prospect Rock.**

"Presently our guide took us by a roundabout way to Cloud Point. This is a commanding projection on the other side of the glen; and here a still wider view — another, yet the same — lay before us. There is something indescribably grand in the solitude of this scene, — forests

of giant trees lifting high their heads, in places, where growths for thousands of years have stood before, through which peer rough-visaged rocks which the hand of Time has failed to smooth. We gazed with delight on the beautiful landscape, then descended into Glen Thomas, a gem of scenic loveliness; fresh in

CLOUD POINT, UPPER LEHIGH.

its pristine beauty and granduer.

"Our visit was made on the first of May. We found here miniature glaciers, formed by the water falling over the rocks, the ice three feet and more in thickness; while not a hundred yards away May-flowers were blooming in fragrant abundance. This region is filled with an untold wealth for the artist and lover of nature." And the time is not far distant when the travelling public, wearied by oft-repeated visits to old resorts, will demand the opening of these fresh and charming scenes.

From White Haven to the "Summit," on the main line, the landscape is more tame: the soil is poor; and the trees present that stunted appearance usual at high latitudes. But this brief respite tends to make the startling scenery through which the road soon passes even more effective. Having passed the crest of Wilkes Barre Mountain, the train glides rapidly down the opposite grade, and soon enters that wonderful gorge known as **Solomon's Gap**, the scene of the annexed engraving. This is the head of a system of *planes* by which loaded cars from the coal-fields below are raised by the Company of the Central Railroad of New Jersey.

We get here the first glimpse of "Wyoming Valley," which we are approaching at right angles. The Susquehanna can be seen in the valley, beyond which ranges of mountains rise in the blue distance. In altitude we are far, far above the Wyoming Valley; and the construction of the road by which it was reached was a rare feat of engineering skill. It is but three miles in an air-line to the small village of *Ashley*, seen below; yet, to overcome the grade, for eighteen miles the cars glance along the mountain sides, following in its zigzag course its varied irregularities.

GLEN THOMAS.

Engraved expressly for Batchelder's "Popular Resorts, and How to Reach Them."

1. Lehigh Valley Railroad.
2. Susquehanna.

SOLOMON'S GAP

3. Ashley.
4. Stationary Engine and Railroad "Plane."

We enter the gorge, and turn to the right, while across the valley can be seen the line of the "Lehigh Valley" Road, which, having kept us company from Bethlehem, now turns around the point to the left, to meet us twenty minutes later at the town below. The view from a half-mile below Solomon's Gap is remarkable. (See engraving.) Its composition varies so decidedly from any witnessed in the Lehigh Valley, that it always awakens feelings of surprise and awe. Here, surrounded by scenes of the wildest grandeur, the beautiful Wyoming Valley bursts like a flood of light suddenly upon you. The train glides smoothly on, the scene unfolds, and we are soon at *Ashley*, near the foot of the mountain.

From Ashley, ninety miles from Easton, the "Nanticote Branch Railroad" extends twelve miles to Nanticote, on the Susquehanna River. Wilkes Barre, ninety-nine miles from Easton, is located in Luzerne County, in the Valley of Wyoming, on the north branch of the Susquehanna. At this place visitors to the "North Mountain House" change to the "Lackawanna and Bloomsburg" Road; but it will be better to spend the night at Wilkes Barre. Fortunately they will find at the "Wyoming Valley Hotel" a house replete with every thing necessary for the comfort of guests. It is pleasantly located on the banks of the Susquehanna, of which it commands some charming views.

The town of Wilkes Barre possesses historical associations of rare interest: its tragic deeds have oft been the theme of the historian's pen and the poet's muse. It is also a well-built town, and its surroundings are pleasant; and it will, withal, prove an interesting place of sojourn for tourists.

From Pittston, nine miles above Wilkes Barre, the road leaves the Susquehanna, and follows the course of the Lackawanna twelve miles, through Scranton to Green Ridge, where it connects with the "Delaware and Hudson Railroad" for Cooperstown, Sharon, and Saratoga Springs, Albany, Lake George, Lake Champlain, Montreal, &c.

"From New York and Philadelphia, the tourist to Saratoga, Watkins Glen, Niagara Falls, and the West is, by this route, transported through a wild and picturesque region, comparatively unknown to tourists. Much of the scenery is unlike that of any other section of the country; and, if only to gain a knowledge of the operation of the mammoth collieries of Pennsylvania, which have been scarcely alluded to in this article, it will amply repay an excursion on the 'Central Railroad of New Jersey,' its connections and branches."

If the tourist contemplates a through trip from Philadelphia or New York by the route described, Scranton will be found the natural place to spend the night; and for this purpose, or longer, good accommodations will be found at the Lackawanna Valley or "Wyoming" Hotels.

LACKAWANNA VALLEY HOUSE.

Three railway lines centre at Scranton,— the Lehigh and Susquehanna, over which we have come; the Delaware, Lackawanna, and Western; and the Lackawanna and Bloomsburg. In addition is the Pennsylvania Coal Company's Gravity Railroad, of which a fuller description will be given.

A half-hour's drive into the suburbs along a romantic and picturesque road leads to a deep cañon, which the writer has christened **Scranton Gorge,** through which rushes a wild, turbulent stream, hemmed in by towering trees, and bounded by ramparts of stone, popularly known as *Roaring Brook*. As its name suggests, and from the description already given, the reader will see that this is one of those romantic dells always pleasing to the lover of Nature in her wildest moods. The stream, which is of the purest water, comes frolicking down the mountain side, now leaping some slight obstruction or miniature cascade; now, overhung by rock or vine, it moves lazily along, till at last the brink is reached, and, with a startled leap, down it springs, a beautiful cascade, into the shadowy depths of Scranton Gorge. To this an Indian name is given: they call it **Nayaug Falls.**

Engraved expressly for Bachelder's "Popular Resorts, and How to Reach Them."
NAYAUG FALLS, SCRANTON GORGE.
Near Scranton, Penn.

But Scranton's great attraction is the "Switchback," over *Moosic Mountain* and the Highlands beyond, *via* the Pennsylvania Coal Company's road, from Scranton to Hawley on the Delaware, thirty-five miles distant. This is the longest gravity road in the world. As a pleasure route it is comparatively new, and is so entirely unlike ordinary pleasure routes that it is sure to become one of the attractions of this region. Preceding its description, the reader should have a brief history of this remarkable road, which, although in character not of unusual construction in the mining regions, is elsewhere of very uncommon occurrence.

Although it has for many years formed an important link in the system of coal transportation in Pennsylvania, its history abroad is very limited. Strangers visiting Scranton manifested such interest in it, that the management determined to shorten the route, and place excursion cars on the tracks to accommodate them.

Engraved expressly for Bachellier's " Popular Resorts, and How to Reach Them."

SWITCHBACK RAILROAD, MOOSIC HIGHLANDS. Scranton, Penn.

To increase the facilities for coal-transportation to market, it became necessary to construct a railroad from Scranton thirty miles across a mountainous country to the village of Hawley. To *grade* such a road for locomotive use was found simply impracticable; but the difficulty was overcome by erecting stationary engines on the summits of the mountains; from these a broad iron band extends down, and attaches to the car; by this you are drawn to the mountain top, from which the road again descends on the opposite slope, not direct, but following at an angle along the side, falling at easy grade, governed safely by the brake.

The descent on the opposite side is frequently by a circuitous route of many miles. When fully made, and another mountain reached, the cars are drawn up this as before, and again descend. And so by nineteen planes the route is passed for thirty-five miles and back. To the uninitiated the first thought will be danger. The best answer is, that in many years of constant use for passenger travel (for since its construction it has been used as such by the local inhabitants), not an accident has occurred. The same machinery that lifts the light cars with living freight constantly draws six loaded coal cars of more than ten times the weight.

Engraved expressly for Bachelder's "Popular Resorts, and How to Reach Them."
JONES LAKE, SWITCHBACK RAILROAD.

The most attractive section of this route is between Plane No. 6 at Dunmore, at which station tourists usually embark, and No. 19 near that charming little sheet of water, **Jones Lake**. Plane No. 19 is only fourteen miles distant, and is on the return route from Hawley; but it is reached by a connecting track, allowing tourists, when they desire, to cross to Jones Lake, and have several hours at this delightful spot while the balance of the train has completed the entire trip to Hawley and back.

Moosic Mountain, or *Moosic Highlands*, as the section crossed by the " Switchback " Railroad is commonly termed, is a spur of the Blue Ridge. The line of the road passes beyond over an undulating country to the Delaware River at *Hawley*, intersecting with cars and canal for the Hudson *viâ* Port Jervis. The outward trip is by the " loaded track," as it is termed, as the coal-cars go by it *loaded*; the " light track " is that by which we return.

Language fails to describe the singular sensation produced by a ride on these cars. If in winter we coast down the slippery hillside, or if in summer we " scud before the wind " in a sail-boat, the mind, from early associations, has been prepared for the sport: but this is a new experience which cannot fail to please.

The " first sensation " experienced by the tourist is while ascending Plane No. 6, from Dunmore. The signal is given, and the cars start with a throb. Up, up they go above the village, above the tree-tops, above the checkered city which now lies at your feet,

WYOMING HOUSE.

above the broad valley, each moment opening wider to view, and through which in tortuous course winds the Lackawanna, above the mountains on your right and left, till the whole landscape gradually unrolls, and like a map lies spread out before you. This sensation is so magical, so exhilarating, that it rises above the ordinary forms of description.

Between Dunmore and *Jones Lake* there are seven inclined planes, each two thousand feet in length, and each having an elevation of two hundred feet. At least *two* of them are usually overcome before the tourist has settled himself down for a full appreciation of the situation, and he is then prepared to enjoy the novelty of the scene.

DELAWARE AND HUDSON CANAL COMPANY'S RAILROADS.

From Scranton the route continues past Green Ridge, the terminus of the Central New Jersey's leased road, the Lehigh and Susquehanna Railroad, and Carbondale, a city of marked enterprise and thrift, to the junction of the Nineveh Branch with the Albany and Susquehanna Railroad. Although the scenery from Philadelphia and New York to Scranton varies continually, and presents new charms with every mile of road, that from Scranton to this place will be found exceedingly attractive. After being whirled along a route hemmed in by scenes of the wildest grandeur, it is a relief to change to pleasing landscape, where the lines melt in harmony, and are clothed with the picturesque.

The train on the "Albany and Susquehanna" with which we intersect is from Binghamton, with passengers from that city, Elmira, Watkins Glen, and the West, *en route* to the same points of interest to which we are bound.

Our route in a north-easterly direction lies along the banks of the Susquehanna, whose placid waters move lazily on in remarkable contrast to the turbulent Lehigh, up which we have so recently passed. We cross and recross the stream as we glide onward through this fruitful agricultural region, — a fine rolling country with broad fields sweeping down to the river-banks, dotted here and there with thrifty farm-houses, and interspersed with flourishing villages.

Cooperstown, on a branch road to the left, will next attract the attention of the tourist.

"This popular resort is the county seat of Otsego County, N.Y., and is situated at the south end of Otsego Lake. It is one of the literary Meccas of America; for here was the home of J. Fenimore Cooper, and in these scenes he wrote those wonderful American stories, which the English-reading world have placed on a level of popularity with the undying fictions of Walter Scott. In his 'Deerslayer,' he describes the lake and surrounding hills. A late guide-book says, 'The same points still exist which "Leather-Stocking" saw. There is the same beauty of verdure along the hills; and the sun still glints as brightly as then the ripples of the clear water.' The whole region is full of interest, because of the creations of Cooper's genius; and his romances have a new zest and beauty when read amid the scenes that inspired them.

The surroundings of the lake are all beautiful; and the entire region is full of interest. In close proximity are several favorite resorts accessible by a small steamer which runs on the lake, which is widely famed for its bass and pickerel fishing."

Two miles out from the village is *Hannah's Hill*, made celebrated by the great novelist Cooper, as a summit furnishing the most bewitching of scenery, the beautiful lakes of his much-loved Cumberland scarcely competing with them. It is also said, that the name "Hannah" was given in compliment to Cooper, it being the name of his favorite daughter.

From this elevation, only two miles eastward from the village, is seen *Mount Vision* overlooking the enchanting *Otsego Lake*, of which Cooper sang in charming prose verse, " A broad sheet of water, so limpid and placid that it resembled a bed of the pure mountain atmosphere compressed into a setting of hills and woods. Nothing is wanted but ruined castles and recollections, to raise it to the level of the Rhine."

Near Hannah's Hill is *Leather-Stocking Cave*, only a mile and a half from the village; and *Leather-Stocking Falls*, or *The Panther's Leap*, is at the top of a wild gorge near by, and at the head of Otsego Lake. *Council Rock* is a mile or two on, a round-topped surface, five feet high, where the Indian tribes of "long ago" were given to meet, and form their treaties.

The *Cooper House* furnishes first class accommodations to tourists who may visit this beautiful village and its many points of interest in the neighborhood. "The Susquehanna River takes its rise in Otsego Lake, and after winding through forest defiles, across broad meadow lands, past rural hamlets and pretentious cities, for nearly four hundred miles in a southerly direction, finally rushes through the outstretched arms of Chesapeake Bay into the welcome bosom of the Atlantic Ocean." *Cooper's Monument* forms a lasting memorial of a great man, and is a silent reminder that the visitor treads historic ground.

Richfield Springs. — "These springs, long and favorably known, are in Otsego County, N.Y., in the vicinity of Cooperstown, and seven miles distant from Otsego Lake, which is one of the sources of the Susquehanna. The great river of Pennsylvania here extends his arms, and intwines his fingers with the tributaries of the Mohawk, as if to divert that gentle river from its allegiance to the Hudson. The village of Richfield Springs is situated on a narrow plain near the head of **Schuyler's Lake**, which is five miles in length, and a mile and a quarter at its greatest breadth. This little lake is surrounded with high hills on every side except the northward; and, being but a mile from the springs, forms the principal attraction for visitors. According to tradition, the waters of these springs were sought for their medicinal virtues, by the Indians, long before the advent of the white men. A healing prophet of the Iroquois dwelt on an island in the midst of the lake; and the suffering came to him, to be cured by the waters he secured at night and conveyed secretly to his retreat. But the Great Spirit became angered at his pride, and sunk him and his island beneath the deep waters."

These springs may also be reached by rail *viâ* Delaware, Lackawanna, and Western Railroad.

Returning to the main line of the "Albany and Susquehanna," the road continues to follow the same interesting valley, with water views, hills, and rich cultivated scenery on either side. The country has an inviting look. At *Cobles Kill Junction* the "Cherry Valley Branch" train is in waiting to take us to **Sharon Springs**, fourteen miles to the left.

This famous summer resort is in Scoharie County, and has since 1830 maintained its full share of popular favor. The village of Sharon is a mile or two from the station, with coaches in constant attendance; and during the "season" the hurry and excitement of getting off is just sufficient to spice the occasion, and gives us a reminder of the days when this popular place was reached entirely by stages, as are most of the Virginia springs at this time. Sharon seems to be located in and around a valley scooped out of the northern face of a ridge of land. Although in a valley you are still on a hill, as can be best seen from the broad veranda of the *Pavilion Hotel*, which, facing the north, commands one of the most extensive and satisfactory views known in the State, — the rich valley of the Mohawk bounded by mountain chains in blue, rising in the north, extending even to the Adirondacks, and the Green Mountains of Vermont.

The immediate location of the springs is in a ravine hemmed in by vine-clad precipices hundreds of feet high. Among stately groves of primitive trees, serpentine walks and rustic seats add to the cultivated charms of the scene; but the great attraction and wonder is the springs, five in number, chalybeate, white sulphur, blue sulphur, magnesia, and pure water, all of which gush from the ground within a few feet of each other, and form the great attraction which annually fills this romantic village.

Having returned to the Albany and Susquehanna main line, we are in close proximity to one of the most noble natural curiosities in the land, — *Howe's Cave;* and yet so little is known of it abroad that the tourist may find it difficult, even a few miles away, to gain any reliable information about this wonderful freak of nature.

Howe's Cave is within a gunshot of the station of the same name, and is said to be the second in size in the United States. It is estimated to be eight miles long, with many portions yet unexplored. The author, in company with a party of ladies, visited the first two miles of it in 1874 with the most satisfactory results. Being somewhat pressed for time, and not having taken this "stop" into account, the thought occurred that this cavern could be visited as well at night as by day; and it was arranged to spend the night at the *Cave House*, which, by the way, we found one of the most homelike places on our tour.

CAVE HOUSE, HOWE'S CAVE.

The house is built of stone, finished in hard wood, with accommodations and appointments so comfortable and chaste, and grounds so well kept, that it seems more like a private villa than a public hotel. The size is its only drawback, and it will doubtless be enlarged to meet the increasing demands of travel. It will always be safe, however, to stop over a train, which will allow time for dinner, and to examine the cave, and by previous arrangement longer stop could be made if desired.

We visited it in the evening; the passage for the first three-fourths of a mile was lighted by gas (I think the only cave that is so lighted). And here it may be said, that to this point any person, lady or gentleman, can go, who is physically able to make the same distance on a country road artificially lighted. Beyond that it is more difficult, though the ladies of our party found no trouble. Generally, it is comparatively dry walking, in places wet, and inclined to be slippery on the moistened clay, for which reason a heavy leather boot is preferable to rubber.

The entrance is not from, but in close proximity to, the house, and is by a descending stairway to the "Reception-Room," an apartment of considerable dimensions. It possesses an even temperature of sixty degrees the entire year, rendering it comparatively warm in winter, and delightfully cool in summer. There is good air, with free circulation.

The floorway to the lake, about three-fourths of a mile, has been improved, and is comparatively level; and along this section, also, several artificial cuttings have greatly facilitated the passage. The cave is nowhere for any considerable distance of the same dimensions. It lowers in places until you are glad you are no taller than you are, then widens and increases in height until the top is lost in obscurity, and can only be measured by a string attached to a lighted hot-air balloon.

Among the names given to different localities are "The Reception Room," "Washington Hall," "Bridal Chamber," "Chapel," "Harlequin Tunnel," "Cataract Hall," "Ghost-Room, or Haunted Castle," "Music Hall," "Stygian or Crystal Lake," across which you are ferried. Beyond the lake are "Plymouth Rock," "Devil's Gateway," "Museum," "Geological Rooms," "Uncle Tom's Cabin," "Giant's Study," "Pirate's Cave," "Rocky Mountains," "Valley of Jehoshaphat," "Winding Way," and "Rotunda." These are among the names given to different sections by Mr. Howe the discoverer, or by visitors. Stalactites and stalagmites are everywhere to be seen. A stream of running water comes from the cave. This is influenced by the seasons, and is subject to rapid rises and heavy floods, sweeping with it every movable thing left therein. Indeed, this phenomenon has been utilized to clear the cavern of surplus earth and other *débris*. There are several side passages yet unexplored, in which the sound of running water can be heard. Howe's Cave possesses the advantage of easy access, only thirty-nine miles from the city of Albany, on the line of a prominent railroad, and within a gunshot of one of its stations; and, with good hotel accommodations, it is not surprising that it is now receiving so much attention. There are other and smaller caves in this vicinity, but this was only recently discovered; and, for considerable time, Mr. Howe's familiarity with such scenes caused him to forget that what might be so simple a thing to him would be a curiosity of great interest to others. And it is only of recent date that preparations have been made to accommodate all who come. Tourists will also be glad to learn that this cavern is located in the midst of beautiful scenery, sufficient of itself to attract attention. The reader will have seen that this entire region, since we struck the "Albany and Susquehanna," is admirably adapted to summer recreations. This may be said particularly of **Schoharie.**

This place is the county seat of Schoharie County, N.Y., and is located thirty-three miles from Albany, in one of the loveliest and most picturesque valleys of the Empire State. Its splendid drives along the banks of the Schoharie River, and its mountain scenery, is unsurpassed. From the top of the Sager-Warner Mountain, four miles east of the village, the tourist has spread out to view an extent of territory as far as the eye can reach. To the south he sees the Catskill Mountains, to the north the Adirondacks, to the east the Green Mountains of Vermont, and to the west a vast expanse of hill and dale. Three miles to the west of Schoharie is located the famous Howe's Cave. Sharon Springs is but twenty miles from Schoharie, and Saratoga Springs but twenty-six miles. A trip can be made by rail to either of these places, and return the same day. Two good hotels in the village are specially adapted for summer visitors. The Parrott House accommodates about a hundred and twenty guests, and Wood's Hotel seventy-five. Three express trains per day each way connect Schoharie with the city of Albany. With a population of eighteen hundred, four churches, an academy, and choice society, it is not surprising that the place is filled by visitors every summer, especially when the cost of living is less than half what it would be in the city.

Continuing east from Howe's Cave or Schoharie, the tourist can go direct to Albany, connecting with the Hudson River boats or cars for New York, the Boston and Albany Railroad for Boston, or continue up the river to Troy or Saratoga. The scenery by this line is a continuation of the same pleasing succession of landscapes along which we have passed, station after station gliding by without special attraction. Near New Scotland is located *Lawson's Lake*, which will command the attention of the curious. The outlet of this sheet of water sinks, and passes for a mile and a half through a subterranean passage, or cavern. In formation it possesses the characteristics of all the caves for which this region is noted. Stalactites and stalagmites abound. There are several caverns in this neighborhood, varying in size, one of which was formerly occupied by a band of smugglers. The more direct route to Saratoga is to go from Howe's Cave *via* Schenectady. This "branch" diverges to the left from *Quaker Street*, leading north-east down into the valley of the Mohawk. At Schenectady the route crosses the line of the Central New York Railroad, continuing to Ballston and Saratoga.

Saratoga Springs, N.Y. — Elsewhere, very brief allusion is made to Saratoga and to Niagara Falls. These famous places are so well known and generally understood, that any special account of their peculiarities would seem to be unnecessary and superfluous. Possibly, however, more particular notice may be desirable by some who read this work, to aid in deciding "Where am I to go?" in vacation time, and who wish to consider the whole field. Saratoga Springs may be visited from New York City, either by the Hudson River to Albany, or by the New York Central Railroad; and a new and very desirable route, from the romantic scenery it passes, is *via* Central Railroad of New Jersey, through Mauch Chunk, Wilkesbarre, Scranton, &c. (see description of Central Railroad of New Jersey); or from Boston, — circuitous but diversified and charming routes, — by various railroads; those from the Fitchburg, the Boston and Albany, and the Lowell passing through every description of inhabited, rural, and mountainous regions, and therefore to be preferred. Taking either of these initial points as the starting-place, ample novelties will invite one's leisure throughout the distances travelled. The chief places of special interest found by the New York line of travel have already been noted. The sweep around the country required by the Boston start is rich in natural and artificial wonders. Commencing at either of the named Boston stations, the first hour passes in the midst of delightful towns and villages, which are mere tributaries of Boston, and are sustained by, and aid to sustain, the great "Hub." Here are residences of perfect taste, and surrounded by rural charms, filling the minds of visitors with continued pleasure. Some of the places through which the lines of travel pass are renowned in Colonial and Revolutionary histories. Those routes which converge at Fitchburg diverge again towards Saratoga Springs, Niagara Falls, the Adirondacks, Lake George, the White Mountains of New Hampshire, and the healthful quietudes of the Green Mountains of Vermont. Whichever course is preferred, the enthusiasm of the refined traveller constantly warms and renews as the glory and splendor of summer verdure, of hills, valleys, meadows, purling streams, and cosey homes, — all speeding by like the flight of birds, — break upon the vision in ever-varying novelty and freshness. It may well be remarked here, that no veteran of the road ever prepares for a pleasure jaunt without first procuring tables of railroad and water lines of communication, and thoroughly mastering his course of march and how he will proceed. He then secures his through ticket, and is prepared to enjoy his pleasure campaign, without the flutter and annoyance of constant doubt as to whither he is moving, and where any change of base should be made. Ladies, especially, ought to ponder this hint.

Should your course from Boston be towards Niagara Falls or the Adirondacks or Quebec, your departure is made from Fitchburg by a different line from the one to be chosen if the aim be for the mountain regions of New Hampshire or Vermont. This the intelligent reader will readily understand. Of course, should you curve around towards the populous State of New York, the chief features of the country will be studded by characteristics of man's busy industry and aggregation into communities. But in the sparsely settled States of New Hampshire and Vermont, Nature still reigns in undisturbed stillness, and in the full beauty and bloom of her pristine charms.

Having decided, then, by what ways you will approach, say, Saratoga Springs, and having reached that fashionable Mecca at last, what are you to do? What is is there to be "done"? Simply nothing, or nearly that, except to drink water from one or all of the thirty odd medicinal springs of the place, and be fashionable, according to the sickly sentimentality of that health restoring and destroying spot.

"Like Newport by the sea, Saratoga is often called the Queen of American watering-places; and this dual sovereignty is generally acknowledged. The hotel system of Saratoga is unrivalled elsewhere in the world; and, although equal to the accommodation of eighteen thousand guests, it is taxed to its utmost capacity in the month of August (the season opens early in June). Broadway is the main street, and extends for several miles, with the chief hotels near its centre, and a succession of costly villas beyond. The village is at its brightest in August, when it is thronged with visitors, and thousands of private and public carriages join in the parade of fashion on Broadway and the boulevard. During the 'height of the season,' the crowds to be seen in all public places, the brilliant balls at the grand hotels, the music of excellent bands, and the many other excitements always prevailing,— make up a scene probably unequalled in the world."

The whole sum of natural scenery, worthy of a walk or ride, afforded by the Saratoga Springs area, is surpassed by almost any rural resort of our land. If the springs were to dry up, the birds of fashionable plumage would flit forever, and the whole of that now populous and prosperous resort would "dry up" as well.

While approaching Saratoga, and within an easy radius of that place, the lover of old stories and romantic adventures may find abundant food for contemplation in hunting up the many historic fields of wilderness campaigns, renowned in the quaint old primitive days. Encounters between our Colonial ancestors and the French and Indians,— extending from Canada, over wilderness and lake, on to Saratoga itself,— with their startling and bloody incidents, fill the mind with a strange fascination. Every schoolboy knows the story, especially the last scene, when the boastful Burgoyne surrendered to the sturdy patriots under Gates.

The opening of that portion of the New York and Canada Railroad which admits of direct rail communication from *Saratoga* to *Ticonderoga via* Whitehall will preclude the necessity of running boats between Whitehall and Ticonderoga as heretofore, which will only go to Ticonderoga; and the branch road between that place and Lake George will do away with the stage-line by which tourists have been transferred from one lake to the other. The fine scenery along this new line of road, the magnificent water views which it affords on the right, as you move north, the rich variety of landscape on the left, with its deep cuts, heavy fills, and rugged mountain scenes, will present entirely new attractions to the traveller. The historic associations of Ticonderoga will command the attention of visitors, who will frequently "lay over" at least one trip to study the relics of the past. From this point they may continue by rail to Lake George, only four miles away, or embark on the magnificent palace-boats for a sail on Lake Champlain.

Lake George. — " Few, if any, among the numerous picturesque lakes in America are more beautiful or more celebrated than this, which lies between the Counties of Washington and Warren, in the State of New York, and is thirty-six miles long, varying in breadth from three-quarters of a mile to four miles, and in many places is four hundred feet in depth. It is in the midst of mountains; and popular belief credits it with islands equal in number to the days of the year. History, as well as tradition, lingers around it, marking many spots with more than ordinary interest. Not the least among these are the ruins of Fort William Henry and Fort George (the former now occupied by a splendid summer hotel). Sir William Johnson, prompted by his loyalty, named it Lake George, after one of the Georges of Great Britain; and this title has been permitted to remain as its designation. A writer, describing the many attractions of the lake, says, 'It has something of interest for every one, — the lover of history, of romance, of beauty, and lovers generally.' "

The sail across this lake is an experience of delightful remembrance. Lake George with its surroundings seems a fairy land of wonderful fascinations. Its points and inlets, its charmingly attractive shores and islands, are admirably adapted to facilitate the pleasures of a camping-out party, where the weary of body and mind, or the despondent and languid invalid, and no less the strong and healthful, will find mind and body invigorated, and the soul elevated, by a sojourn among the picturesque beauties of this lovely lake.

Several of the islands are inhabited, and others contain ornamental structures for summer amusements. On the shores of the lake are several popular resorts and many private villas. After *Fort William Henry*

Hotel, Crosby Side, Recluse Island, &c., are among the more attractive features as you sail down the lake.

The tourist who would continue his visit from Saratoga or Lake George to the White Mountains, *viâ* Burlington, to Montreal, Quebec, or the Adirondacks, will embark on one of the magnificent lake steamers at Ticonderoga.

Lake Champlain. — "This useful as well as beautiful sheet of water lies between the States of New York and Vermont, and extends a short distance into Canada. It is, in extreme length, about one hundred and thirty miles, and varies in width from half a mile to fifteen miles, the water, in places, being near three hundred feet deep. The Vermont shores of the lake are generally fertile and well cultivated; while those of New York are wild, rocky, and barren, rising into vast mountains, and contain rich iron deposits.

"The shores of Lake Champlain are not only interesting in themselves, but they hold many places of celebrity and attractiveness. The ruins of old Fort Ticonderoga stand out upon a high, rocky cliff at the confluence of the outlet of Lake George with Lake Champlain. The remains of the fortress at Crown Point loom up opposite to Chimney Point. The localities where Burgoyne held his famous Indian council, and made his treaty, and where Arnold fought with Carleton, are pointed out. **Plattsburg**, the scene of the battles on the 11th of September, 1814, in which Commodore McDonough gained his signal naval victory, and Gen. Macomb compelled Sir George Prevost to retire into Canada, is the most conspicuous and interesting point on the lake. Numerous natural curiosities exist on its islands and shores; but space will not permit their mention here. Burlington, a beautiful city on the Vermont shore, is well worth the attention of tourists. From Plattsburg the Adirondacks are reached with facility; and it is a starting-point for *Au Sable Chasm*, one of the most remarkable curiosities in the United States."

Those northward bound continue to **Rouse's Point**, from which they can go to Montreal and Quebec, visit the noted springs in Vermont, or turn to the west through a wild region to *Ogdensburg*, and thence by boats to *Alexandria Bay* and the Thousand Islands, and so on to Niagara.

The Adirondacks of New York have sprung into sudden and universal fame and favoritism. The region has all the novelty of a primeval land, diversified by every variety of landscape and unsearched solitudes; and has the freshness and rare American novelty of guides, who alone know the secret of this new paradise.

The atmosphere is remarkably pure, and free from malarious poisons and from chilling damps, so that sudden colds and tormenting fever-heats are scarcely known. At present the Adirondacks may boast solely of its primitive charms; but the region will, it may be feared, be materially altered in this respect ere long, as visitors are annually numbered by thousands.

To ladies claiming invalid propensities, or to those disposed to shrink from rough adventure or the hardships of the explorers, this wild tract is, perhaps, totally uninviting; but for the novelties of camp-life, and utter freedom of conventional rules, it is all that can be desired.

The region is to be found in the northern portion of the State, a wilderness of immense tract, occupying space equal in area to the entire State of Connecticut.

There are several routes by which this wilderness may be entered; the most popular, because it is most flavored of adventure, is by way of Lake Champlain. Leaving the boat at Port Kent, a post-coach conveys the traveller by plank road to Keeseville, from whence he will follow along the *Au Sable River* twelve miles, arriving at *Au Sable Forks;* and at that point, if he has the mountains particularly in view, he will direct his course south, leaving the Saranac Lake region to the west.

On the way from Keeseville, about a mile and a half off, the *Au Sable River* furnishes to the traveller on this route the first natural wonder of the Adirondacks, — a leap of nearly thirty feet into a semicircular basin of great beauty; and this is only the beginning of a series of the wildest of river falls, cascades, and jagged rocks. The walls of the Au Sable vary in rocky height, rising from ninety to a hundred and twenty-five feet. One of its precipices resembles Niagara, leading off the river to a course one hundred and fifty feet below, into scenery of intense wildness, equalling in miniature the picturesqueness of America's greatest fall. It is thus followed up to its source, forming on its way a chain of grandeur and frequent surprises, rivalling any thing east of the Rocky Mountains.

Having reached the *Lakes Upper and Lower Au Sable*, the traveller finds himself completely at the heart of mountain and forest surroundings. Here in this savage gorge, the wildest part of the Adirondacks, the Hudson River takes its rise. The main stream of the Au Sable flows from the north-east, the Hudson south-west, and each from the locally known as *Indian Pass*.

If the visitor prefers a less toilsome journey to this point, starting from the village of *Au Sable Forks*, he can, instead of following up the river on foot or with guide and boat, turn off into a road through the village of *Jay* and on by Wilmington, moving south through the *John Brown Region* on to *White Face* or *Wilmington Notch*. This range, although practically belonging under the Adirondack title, is somewhat distinct, being a sort of "branch range." *Blue Mountain*, *Dix's Peak*, *Nippleton*, *Core Hill*, *Moor Mountain*, *White Face*, and other grand peaks, belong to this group. White Face is the most northern, and, with the exception of *Mount Marcy* its neighbor, the loftiest of this wilderness of crests.

The traveller will prefer here to move by land rather than by water; for although the lakes are numerous, and guides and boats easily obtained, it is among the hills that the chief attractions are found. Having laid aside ordinary travelling costume, and donned backwoods garments, he is ready for his fight.

The monarch of the glen, *Mount Marcy*, or *Tahamus* ("the cloud-splitter"), is 5,467 feet high. The trail to the summit of this mountain is twelve miles, an exceedingly toilsome ascent. *Mount McIntire*, near by, has an elevation almost as great; *Dial*, *McMartin*, and *Colden* are also very lofty; and the ravines, lakes, and waterfalls to be seen from them are inexhaustible.

Many of the wild animals of our northern latitude — the bear, wolf, and wildcat — abound: with several valuable fur animals, — otter, mink, and muskrat. Hunters and trappers are occasionally to be met, although there are no settlements of any account. These men are always ready to act as guides, either to tramp the mountains, or navigate the lakes for pickerel, or the streams for trout. Ten or twelve years ago, moose and deer were plenty, but now are rarely found. At *Mount Seward*, the most inaccessible of all the mountains, the American panther is frequently found.

Having satisfied himself with the mountain research, wearied of the continual hard climbing, and the weird wildness of it all, the traveller may turn with a certain feeling of relief towards the more subdued region in which the lakes abound.

Lake Colden, about six miles west of Mount Marcy, lies almost at the outskirts of the Wilmington Notch: it is a lovely, placid sheet, "perfectly embosomed amid gigantic mountains, and looking, for all the world, like an innocent child sleeping in a robber's embrace."

Beyond Colden, is *Avalanche Lake*, around which stand *Wallface*, *McIntire*, and *McMartin*.

Leaving these behind, the start may be for the valley of the Saranac, by way of *North Elba*, stopping perhaps to visit the *John Brown* Farm, and the bowlder by which the old hero was buried. Stages are in waiting to convey passengers over to *Baker's* and *Blood's* Inns, which are to be seen as the *Lower Saranac* is approached, about twelve miles off. Guides with their boats can be obtained here by those who wish to enter the forest and river streams surrounding the lakes. Excellent fishing is found on *Ray Brook* and other tributaries of the Saranacs.

The favorite trip, next to the easy excursions among the islands and bays of the lakes, is to *Lake Placid.* It is two miles north of the village, and is the most charming of all the Adirondack resorts.

Paradox Pond is near by, and has an inexplicable tidal flow to and from the lake. *Schroon* and *Branch* Lakes, a few miles south, are famous sheets of water, and are often approached by sportsmen.

Round Lake, midway between Upper and Lower Saranac, connecting the two by means of narrow streams branching out from either side, is a little lake gem of unusual beauty. It is round, as its name implies, about four miles in diameter, and surrounded by hills under whose shadows it sleeps in placid loveliness. This lake will well repay the traveller for a week of study. It will not satiate by a never-changing calm, or weary the visitor with continued restlessness; for, like the sensitive-plant, it seems to be a thing of the tropics, and away off in this wild country out of its element. Still and peaceful in the warm sunlight, as if never disturbed, the little islands dotting it with picturesque verdure, it will suddenly, in answer to turbulent wind, ruffle itself, and appear strong and defiant in character, in sublime contrast to its restful mood.

In all this lake country will be found, one after another, in quick succession as it were, almost numberless streams, pools, ponds, and lakes. Going twenty-six miles south-west from Upper Saranac, will be found *Big Tupper Lake*, which, as if Nature were much pleased with its arrangement of the Saranacs, is in imitation thereof; for it is connected to its sister, Little Tupper, by a river stream, and also Round Pond, which, less beautiful than the lake its namesake, was allowed to imitate only in less pretentious title.

The *Tuppers* are but seldom visited, on account of remoteness; but they afford excellent sporting and very lovely scenery.

Lying twenty miles south of *Little Tupper*, through a forest alley holding the *Raquette River*, is *Raquette Lake*, approached through turbulent falls, and exceedingly wild woodland scenery. It is at this point that "Mother Johnson's" is reached, — a low, quaint log house, where the pancakes of Murray fame are served. *Raquette Lake* is twelve miles long, and about five wide. It is seventeen hundred feet above the sea. Its environing forests, graceful islands, and mountain ranges in view, combine to enhance its charms. Fish and game are plentiful here, although few visitors come on account of its poverty of inn and guide accommodations.

About five miles beyond Johnson's the Raquette River widens, flowing north-east into the beautiful *Long Lake*, on whose waters lie the most charming of islands. The largest, *Round Island*, is the most celebrated of any found in Adirondack waters. Hammond says of it, "I wish I owned that island: it would be pleasant to be possessor of so much beauty."

Blue Mountain Lake is twelve miles south. *Utowana* and *Eagle* lying west are connected by inlet with Raquette; and farther west, Fulton Lakes are chained together by a series of ponds. Going north, and all the time following by stream, will be found *North Branch, Big Moose, Beech,* and innumerable smaller lakes and ponds, each partaking of the general lack of civilization, but richly rewarding the hardy adventurer.

It has been generally conceded by painstaking tourists, that, in no territory this side of the Yellowstone, can be realized so completely the spirit and recompense of exploit as among the Adirondacks. It will be a fact rather to deplore than welcome, if, in the great march of civilization, this land, so after God's own making, will be brought under a human plan of architecture and landscape finish. Let it be left to its Maker.

NEW YORK TO LONG BRANCH.

The new short and quick route opened by the Central Railroad Company of New Jersey, June 15, 1874, to Long Branch, is a matter of importance to the citizens of New York, with whom *time* is a great desideratum. Passengers start from the foot of Liberty Street, go *via Elizabeth*, Woodbridge, Perth Amboy (where the line crosses the Raritan River by a bridge a mile long to South Amboy), thence through Red Bank, Oceanport, &c., to **Long Branch,** opening this fashionable watering-place, by an easy, expeditious route, to the panting thousands of the great metropolis. About ten trains are run each way daily, — a cheap and easy release from the narrow streets of the city, and equally narrow pursuits of gain, to the soul-saving worship of the great and good God through the never-quiet, never-ceasing roar of the mighty ocean.

HARVEY'S LAKE, NORTH MOUNTAIN, AND HIGHLAND LAKE.

Returning again on the line of the *Central Railroad of New Jersey*, through the Lehigh Valley, to *Wilkes Barre*, we find several resorts in the vicinity of considerable local interest.

Harvey's Lake is twelve miles north-west of Wilkes Barre. It is a small, deep pool of marvellously clear, cool water. It is approached by a romantic road over the mountains from Wilkes Barre, and is a popular resort for the citizens of that place on account of its fishing and boating facilities. It is a thousand feet above the Susquehanna, and about two hundred feet deep, and nearly circular in shape, forming a clear and beautiful mirror for the overhanging Alleghany Mountains.

North Mountain is another resort in the neighborhood of Wilkes Barre, more distant than Harvey's Lake, yet rapidly increasing in popularity, as will be seen by the following description.

Engraved expressly for "Bachelder's Popular Resorts, and How to Reach Them."

SUMMER LIFE AT NORTH MOUNTAIN HOUSE.

THE NORTH MOUNTAIN HOUSE.

ONE of the most important problems to determine in household matters is, "Where shall we spend the hot months of summer?" A change is required: we must go somewhere. The father has become overtaxed by the cares of business; the mother is wearied by household duties; the children need a respite; the health of all demands this change. "But where can we go?" are the oft-repeated words.

Why, there are places enough,—by the sea, at the springs, or in the mountains. The newspapers teem with notices of them; and books resound with their praises. At Cape May, Atlantic City, Long Branch, or Newport; at Bedford Springs, Saratoga, Watkins Glen, or Niagara; at the White Mountains, Mount Desert, Mauch Chunk, or Cresson; and at hundreds of other fashionable resorts,—houses in abundance are open, servants are ready, and landlords stand smiling at the door to receive you. Their halls dazzle with beauty; their parlors rustle with fashion; their corridors resound with mirth; and their drives are a whirl of excitement. Certainly, with such an array, one need not lack for a choice.

But it is just this rustle of fashion, this whirl of excitement, that deters many of our best citizens from seeking that recreation which their health requires. It was to meet this emergency that the North Mountain House was erected, and a summer home provided where muslin and chintz, common sense and comfort, should supplant the prevailing customs of popular resorts; in a word, where *dress* is not paramount to good taste and social enjoyments. Here are good accommodations for two hundred guests, all amply supplied with the substantial provisions of home, in a climate fresh with invigorating atmosphere, with springs of pure mountain water, where at reasonable rates a man may take his family for the season, and return invigorated and refreshed.

If the reader would locate the North Mountain House, he is referred to that spot on the map of Pennsylvania, between the East and West Branches of the Susquehanna River, where the counties of Sullivan, Luzerne, and Wyoming corner, from which flow the head-waters of Loyal Sock, Muncy, Fishing, Huntington, Kitchen, Bowman, and Mehoopany Creeks. There it stands, on the banks of a charming lake, on the summit of North Mountain, of the Alleghany range, 2,700 feet above tidewater (the highest habitable spot in Pennsylvania), in the centre of an unbroken primeval forest of 25,000 acres, not a house within a half-dozen miles, and "no one to molest or make afraid." The North Mountain House, as a popular resort, is a success. No "rustle of fashion in the parlors;" though its "corridors resound with mirth," and the halls and grounds are radiant with that beauty and alive with that enjoyment which come from good cheer.

There is fishing for those who like it, and hunting in the woods; there is sailing on the lake, and roaming in the groves; there are billiards for rainy days, and croquet for fine; there are scenes for the artist's pencil, and abundant sports for all.

WILD WOODS.

North Mountain is eighteen miles from Shickshinny, on the East Branch of the Susquehanna, through which passes the "Lackawanna and

Bloomsburg" Railroad, connecting north with the "Delaware, Lackawanna and Western," and south with the "Philadelphia and Erie road."

The morning train from Wilkes Barre leaves you at Shickshinny about nine o'clock. A good team can be secured at the hotel; and the drive to North Mountain is delightful. The route is over an excellent upland road, and commands a succession of grand and varied landscape views, in admiration for which the miles grow short, and the distance is the least objectionable feature of the journey. Should necessity require, the trip can be made from the afternoon train; but it is not as pleasant, as we have the evening sun in our eyes, and arrive after nightfall. The road winds among the hills, rising by easy grade to the base of the mountain proper, by which much of its altitude is overcome. The side is steep; but the carriage-way is shaded by forest trees, and is exceedingly picturesque. It is overhung in places by frowning rocks; and the rush of an impetuous stream can be heard in the valley below.

The topography of North Mountain is so different from our general acceptation of the term "mountain," — which usually rises to a summit crest, — that a brief description of its physical characteristics will be in place. Its sides are abrupt, presenting many interesting geological features. The summit surface is generally level, beneath which the outcropping strata of the carboniferous and sub-carboniferous formations are distinctly visible. This forms an extended plateau, broken by gentle undulations, extending thirty miles in length by ten in breadth. It is covered by a dense forest of primeval trees, — oak, hickory, maple, birch, cherry, hemlock, pine, beech, and other varieties usually grown in much higher latitudes. Springs of pure water, crystal streams alive with speckled trout, and quiet lakelets, abound. On the banks of the largest, *Highland Lake*, the North Mountain House has been erected. This delightful sheet of water, three miles in circumference, is fed by springs at its bottom. It abounds in fine varieties of fish, and furnishes withal a pleasing source of amusement. Its outlet forms the head-waters of *Kitchen Creek*, which, with seeming reluctance at first, leaves its parent head; then, as it moves along, gathering strength by fresh accessions, it soon assumes a bolder course, until the mountain's brink is reached, down which it plunges impetuously, forming numerous wild cascades, then, with a fearful leap sheer down the deep abyss, is dashed into snowy spray among the rocks. Rich, luxuriant foliage depends from the overhanging cliffs, through which peers the sparkling sheen of a midsummer's day, clothing all in bright rainbow hues.

"**Ganoga Falls**" are 127 feet high. They are of recent discovery, and must prove a great acquisition to the attractions of North Mountain. They are three miles from the house, by a picturesque woodland road.

Engraved expressly for "Bachelder's Popular Resorts, and How to Reach Them."
GANOGA FALLS, NORTH MOUNTAIN, PA.

The immediate approach is wild in the extreme, affording an opportunity at comparatively small effort to witness Nature clothed in her native dress.

MOUNTAIN STREAM.

The stream continues down the mountain side, forming a series of wild cataracts and charming cascades, and is also a delightful resort for the angler.

The better plan to see this ravine to advantage is, to be taken by carriage a few miles below, turning by a cross-road to the left to a point of intersection with Kitchen Creek at the foot of the mountain, from which place the stream can be followed up to the Ganoga Falls, already described. Although the excursion is full of adventure, there is more romance than pleasure in it, particularly for ladies, unless they are well assured of their physical endurance. At all events, it should not be attempted without a guide, and refreshments should be sent to Ganoga Falls to regale the party on its arrival. Never venture from the main travelled road without a pocket-compass. The importance of this the author has good reason to remember. Late in the autumn of 1873, while visiting Ganoga Falls with the proprietor of this vast domain, we became entangled in dense laurel thickets, confused by a blinding snow-storm, and *absolutely lost* in this unbroken forest of twenty-five thousand acres for more than four hours; and by mere chance avoided a night in the woods with the thermometer at zero. There are undoubtedly still many interesting localities to be discovered in this wild region by future explorers.

Point Look-off is an interesting place, and is much frequented by equestrian and driving parties. The summit of North Mountain is so densely wooded that it is only when, from some opening near the brink, you catch glimpses of the deep, distant view, that you realize your altitude. Such a place is *Point Look-off*.

It is some three miles from the house, and the way is overhung by enormous trees, furnishing a deep, grateful shade. Go, by all means, with the declining sun, when the shadows fall from you in the view you wish to see.

It is surprising how readily visitors to a rural resort like North Mountain drift into the natural freedom of the prevailing customs of the place. Sailing, boating, fishing, hunting, croquet, music, picnic dinners, and promenading, stout shoes and short dresses, exert their equalizing influence, until the whole household seem to forget the conventionalities of society, — forget the outside world, and live the while in an atmosphere purely their own. It is this state of feeling, probably, that gives the view from *Point Look-off* much of its charm. Emerging from a primeval forest, you suddenly catch an unexpected glimpse of the world you left behind. Standing on the eastern escarpment of the Alleghany Ridge, you see ranges of rounded hills and green meadows, thrifty farm-houses and distant towns, and, beyond all, you see rising in the blue distance lines of rugged mountains.

The view extends miles beyond miles, east, south-east, and south, embracing the counties of Luzerne, Columbia, and Montour, which fills the mind with awe and admiration.

The enduring character of the conglomerate and hard sandstone formation protects the surface to the mountain's brink; but the deep gorges through which the maddened streams have cut their way, exposing the formation from surface to base, furnish a field for the geologist, of unusual interest, and to the student and lover of wild and romantic scenery scenes worthy of the artist's pencil.

NORTH MOUNTAIN VIEW.

The "North Mountain View" is one of the most remarkable features of this character. It is on the south-west side of the mountain, less than a half-mile from the house, and is of easy access. To the spectator from the head of this wild and broken cañon, the scene is indescribable by the artist's pencil or writer's pen. It is a singular combination of scenery, possessing the grandest features, clothed with the picturesque and beautiful, over and beyond which — stretching far, far away — is that immeasurable distance which always "lends enchantment to the view."

A visitor writing of it says, "Here, standing on a perpendicular ledge of rocks, you gaze with a mixture of wonder and admiration down upon this magnificent view. Seven distinct mountain ranges dovetail one into the other, forming a long, deep gorge, through which you look for miles beyond miles. . . . Far into the unseen depths of the ravine below is heard the roaring of a creek, of which occasional glimpses are seen sparkling in the noonday sun." The accompanying cut is from an original sketch by the artist, Thomas Hill; and yet the almost magic touch of that celebrated artist fell far short of the sublime grandeur of nature.

There are those who delight to sail on the lake, to fish in its waters, and walk on its banks; there are those who find pleasure in roaming through the groves and penetrating to the unbroken depths of the forest; there are those who, following the sportive streams from crest to base, watch their falling waters, and amid wreaths of snow-white spray linger for hours to tempt the speckled trout; and those who enter joyously into all the sportive games in which the place abounds. But to none is the delight so sparkling, the pleasure so pure, the joy so lasting, as to the devotee who worships at the shrine of the North Mountain "View." It is *the* great feature of the place, and the walk to it is delightful.

The professional hunter and amateur sportsman will be alike interested in this field for the exercise of their favorite amusements. The North Mountain House consists of two buildings, surrounded and connected by long, broad verandas. The older or stone house was built many years ago, before the days of railroads, on the old turnpike leading from Buffalo, N.Y., to Sunbury, Penn., which is still used by drovers in transporting their stock; and the house is the comfortable home where the weary traveller is sure of a cheerful welcome. Its complete isolation, surrounded by many miles of unbounded forest filled with deer and other large game, soon made it the *rendezvous*, in the autumn and winter months, of hunters and sportsmen. During a day spent there by the writer, in November, 1873, there were four deer and a bear killed in the immediate vicinity, — not an unusual circumstance in the hunting season. The laws of the State protect deer during the summer months; but the autumn finds this a rich field for sport.

Wild fowl, in their migratory flight, frequently make "Highland Lake" their resting-place. Others rear their young in its quiet coves.

FOREST LIFE

The North Mountain House embraces among its patrons many of the best families of the State, who select for a summer home this healthful locality, convenient of access, yet far away from the false life of *fashionable* resorts; a spot abounding in the pristine beauties of nature. Here they spend the heated term, and return to their homes in autumn recuperated in strength, with fresh vigor to enter again the battle of life.

PLEASURE ROUTE No. 15.

New-York City to West Point, Saratoga, and the Popular Resorts of Northern and Western New York, Niagara Falls, and the West.

NEW YORK CENTRAL AND HUDSON RIVER RAILROAD.

ONE of the most charming of summer tours is that from **New-York City** to **Niagara Falls**, *via New York Central and Hudson River Railroad*. The general route of travel passes through a portion of the State rich in scenes of beauty and sublimity, and will well repay the time and expense given in exploring the many points of interest. This railroad has four tracks, two for passenger traffic and two for freight, between Albany and Rochester.

Starting from the magnificent **Grand Central Depot** on Forty-second Street, where cars the most palatial of any constructed are to be obtained, the traveller passes, a few miles out, **Irvington**, and **Sunnyside**, the cottage of the great "essayist," and will pause first at *Sing Sing* for a visit to **Croton** and **Rockland Lakes**. The last named, across the river and opposite Sing Sing, is one mile from the Hudson, and one hundred and fifty feet above it. This lake furnishes one of the main sources of supply of ice to the metropolis, more than two hundred thousand tons of which are annually stored in its houses.

Croton Lake is reached by carriage conveyance from *Sing Sing*, and it is the lake from which New-York City is supplied with water. It is about five miles long, and quite narrow, being artificially formed by throwing a dam two hundred and fifty feet long, forty feet high, and seventy feet thick, across the Croton River. Its waters are conducted to the city by an aqueduct forty miles long, with a daily capacity of sixty million gallons.

Eleven miles above Sing Sing, following the railroad, and stopping at *Peekskill*, the traveller leaves the cars for a visit to **Lake Mahopac**, which is fourteen miles east of the Hudson. The lake is eight miles in circumference, and eighteen hundred feet above the sea.

It is the most beautiful of a family of twenty lakes lying within a circle of thirteen miles, all of which are supplies to the *Croton*. Although the surrounding landscape is comparatively quiet, lacking bold, hilly feature, yet its own calm waters, and tiny wooded islands, and the many romantic dells in its neighborhood, together with the broad, macadamized driveway, all serve to give *Lake Mahopac* just claim to special notice. Lake Mahopac is also reached directly by a branch of the Harlem Railroad, which leaves New-York City from the Grand Central depot.

Petrea Island in this lake, nearly round, is a lovely bit of fresh verdure standing just in the centre. For neighbors, it has the *Fairy* and *Grand Islands*.

Returning from Lake Mahopac to the river, at *Peekskill* begin the famous **Highlands of the Hudson**, the most celebrated of American river-scenery. Passing along the east bank, the railroad enters the *Tunnel*, two hundred feet in length, running through the bold promontory of *Anthony's Nose*. On the west side can be seen **Buttermilk Falls**, a series of rapids making a leap of a hundred feet, on a stream joining the Hudson just below West Point. They form a charmingly picturesque feature in the scenery of the river. Farther on to the north is *Garrison Station*, — the point from which the traveller crosses the Hudson to reach the **United States Military School** at **West Point**. The *Academy* here, the most prominent feature, is of stone, while the *Barracks*, the *Library, Observatory, Hospital*, and *Officers' Quarters*, make up a formidable and garrison-like settlement.

Nearly ten miles above West Point, is **Fishkill**, the seat of the fine country home of the Secretary of State. This town is interesting also, as the scene of Cooper's novel "The Spy." Thirteen miles along lies **Poughkeepsie**, of *Vassar College* fame. Passing *Catskill Station*, and the beautiful mountain region, of which mention is made elsewhere (see Catskills), and also **Hudson**, a large town on the west side of the river, we approach **Albany**, the capital of the State. This city lies on the side of a height crowned by the *State Capitol*, rising two hundred and twenty feet above the river.

The *Capitol*, unfinished, stands on the public square, the *State House* and *City Hall* occupying two of the remaining sides.

The *Dudley Observatory* in the northern part of the city, the *State Arsenal*, the *Medical College*, the *State Normal School*, and the *Law School*, are among the most important of public institutions. There are many objects of interest in Albany, connected with the original Dutch settlement and the early history of the Republic, which will well repay an extended examination.

Five miles north of Albany, lies **Troy**, a city of forty-six thousand people. It has many fine churches, public buildings, and residences, and is the largest manufacturing city of the Hudson Valley. Here is the *Watervliet Arsenal*, one of the national depots for the storage, manufacture, and repair of war implements and material; and also large establishments for the working of iron into all known uses.

There are four main lines of railroad centring here, by which the tourist can reach either the famous *Saratoga Springs*, the Lakes *Champlain* and *George*, described elsewhere, the *Adirondack Country*, and on northward, and various **points of interest on the line of the Troy** and

Boston road to *Hoosac Tunnel*, &c. Continuing westward, however, on the New York Central Road, we come by a short detour to **Ballston** on the Saratoga and Schenectady Branch, at which place is the **Artesian Letitia Spring.**

It was first known and made accessible in 1868. These springs are said to cure rheumatism, gout, and several minor diseases.

The **Sans Souci Spring** near by is composed of a fair proportion of lime and magnesia, with a preponderance of chloride of sodium. There are several comparatively unexplored springs in and around this town, promising a busy future for the already thriving village of Ballston.

Returning again to the main line at *Schenectady*, one of the oldest towns in the State, and the seat of *Union College*, and thence to *Palatine Bridge*, we come to the station from which the stage takes the traveller for health to Sharon Springs; for a description of which, see index.

Little Falls, seventy-four miles from Albany, is a large manufacturing town on the north bank of the Mohawk, which here forces a passage through a range of high hills, giving a wild and exceedingly interesting view of the winding river and its surroundings.

Utica is a flourishing city of thirty thousand inhabitants, standing on the south bank of the Mohawk, and was at one time the site of Fort Schuyler, a frontier defence against the Indians. It is the seat of the *State Lunatic Asylum*, and is the depot and outlet for the surrounding rich agricultural region, and its annually returning summer visitors.

BLACK RIVER RAILROAD.

At this station the Black River Railroad diverges to the north, along the line of which are many natural objects of interest and beauty. The most prominent are **Trenton Falls**, seventeen miles north of Utica. "on West Canada Creek, a tributary of the Mohawk River, which consist of a series of cascades of unexcelled picturesqueness and beauty. The principal falls are five in number, and, passing up the stream, are named, successively, Sherman Fall, High Fall, Mill-dam Fall, Alhambra Fall, and Rocky Heart. To appreciate them fully, the tourist should descend the bank, by stairway, to the rocky level at the bottom, and pass up along the left bank on an irregular line of shelf-path, presenting little difficulty, and no danger to the careful. The rock strata of the gorge cannot fail to excite admiration; and the unique collection of fossils and crystals found in the neighborhood, and kept on view at a hotel near the falls, is an interesting subject for examination and study. From a point called the Rural Retreat, a splendid view of the High Fall, from above, can be had."

Still farther north are passed in succession, **Sugar Falls**, a hundred and fifty feet high, falling into a lonely valley filled with bright evergreens and charming cascades. The river sinks at this point, and is hidden under the limestone strata, appearing again about eight hundred feet distant.

Black River Rapids, a mile or two out of the village of *Port Leyden*, are of considerable interest. They flow through a narrow, deep ravine, at one time known as Hellgate.

Brantingham Lake, eight miles from *Lyons Falls*, is a lovely sheet of water, picturesque in islands and extensively wooded shores. It abounds in pickerel, bass, and trout.

Having exhausted the Black River lakes and falls, and returning again to *Utica*, we face the setting sun, and approach *Rome*, at the junction of Black River and the Mohawk. **Rome** will interest the tourist principally as a railway centre where several roads intersect.

ROME, WATERTOWN, AND OGDENSBURG RAILROAD.

The main lines of travel through the Empire State are from east to west; connecting these, and running at an angle from them, are many roads which frequently lie through picturesque regions, and unfold scenes of grandeur and beauty.

The Rome, Watertown, and Ogdensburg Railroad is of this class, although we must wait for another edition for a detailed description of it. Yet a few of the more prominent localities will be noticed.

This road connects with the Central New York at Rome, and extends in a general northerly direction through Watertown to Ogdensburg on the St. Lawrence River, and to Cape Vincent at the eastern terminus of Lake Ontario, with a branch to Oswego. At Watertown the Sackett's Harbor branch of the Utica and Black River Railroad crosses it.

This is the main route of approach to Alexandria Bay on the St. Lawrence, a place of rapid growth and popularity, located in close proximity to the "Thousand Islands." Palace and drawing-room cars run from Albany direct, *via* Rome, to Cape Vincent, connecting with elegant steamers, down the St. Lawrence, through thirty miles of charming river scenery, to *Alexandria Bay*. If Ogdensburg is first made, either from the south by the Rome, Watertown, and Ogdensburg Railroad, or from the east by the Central Vermont, fine steamers are also in readiness to convey the tourist from Ogdensburg up the St. Lawrence, thirty-six miles, to Alexandria Bay.

The **St. Lawrence River** is, without question, one of the grandest and most interesting streams on the American Continent. It receives the waters from half a dozen inland seas, each the receptacle of hundreds of tributary streams. The same turbulent river that madly rushes to the brink of Niagara, and plunges in sublime grandeur into that fearful abyss, now, under another name, playfully meanders through the picturesque channels of the Thousand Islands, sweeps beneath the walls of Quebec, and moves majestically on to the sea. It is beautiful, it is grand, and well deserves its acknowledged popularity.

From Lake Ontario to the Gulf of St. Lawrence, is one continuous chain of attractions for the visitor, which each deserve special description in this book of "Popular Resorts." In the present volume, however, but few will be mentioned, prominent among which is **Alexandria Bay.**

This American port is built upon a massive pile of rocks, and occupies a romantic and highly picturesque situation. Some two or three miles below the village is a position from which a hundred islands can be in view at one time. It is in Jefferson County, and in immediate contiguity to the **Thousand Islands,** which stretch themselves along the centre of the St. Lawrence for a distance of forty miles below the termination of Lake Ontario.

The steamboat-ride from Cape Vincent to Alexandria Bay affords an excellent view of these islands, which are said really to number nearly eighteen hundred. The river is about twelve miles wide, but so closely studded with islands of all shapes and sizes, from the fractional part of an acre to ten miles in length, that there really seems at times a difficulty in trending a channel through them. The water of the St. Lawrence is here of a bright green color, and beautifully clear. The islands are generally rocky, and thickly wooded; and the water in places so deep that steamers can easily run within a few feet of them. The "Rapids" of the St. Lawrence are a short distance below.

Each year adds to the popularity of this beautiful summer resting-place. Twenty years ago it was the quiet rendezvous of a few statesmen and literary gentlemen, — a respite from the excitement of debate and the cares of life; now it is a fashionable watering-place, where thousands come for rest and pleasure. Elegant private villas have been erected, grounds cultivated, and every thing that wealth and taste can accomplish has been and is being done to enhance its attractions.

Steam-yachts for pleasure, and sail and row boats for fishing parties, can always be had, fitted up in the best manner, and manned by experienced boatmen. Fishing here means something, and is worthy the name. And, in the season, the shooting in the neighborhood is unusually fine. The suburban drives are delightful, with ample facilities to enjoy them.

A line of steamers from Toronto to Montreal touch at Alexandria Bay; and it is also reached by the Grand Trunk Railway, which passes on the opposite bank of the river.

Commodious summer hotels have been constructed, assuring the visitor of the very best accommodations. *The Thousand Island House* alone, recently completed, has a capacity for seven hundred guests. Perched upon a rock whose base is laved by the pure waters of the St. Lawrence, this house commands a glorious view of this magnificent scenery. Every variety of boat navigates the waters; the majestic river steamer and large white-winged craft, the graceful sail and row boat, constantly flit across the scene.

Returning again to the line of the Central New York, the route from *Rome* passes *Verona*.

Verona Springs have won some fame as a water-cure establishment. At *Oneida* we cross the line of the NEW YORK and OSWEGO MIDLAND RAILROAD, and here our eyes are feasted by the first glimpse of **Oneida Lake**, a hundred and twenty-five miles west of Albany. It is twenty miles long and six wide, and a hundred and forty-one feet above Lake Ontario. Its name, Oneida, signifies "The people of the stone." The route continues along its northern borders nearly its entire length.

It lies low, and is marshy on its shores; but there is a fine open view to the south of the rich Onondaga County highlands. It is situated in a fertile dairy and stock territory, where the "Inspiration Community," founded by John H. Noyes, is located.

Cazenovia Lake, reached by rail from Syracuse, would scarcely repay for a special trip to its waters, were it not interesting as the headquarters around which are medicinal springs, and also waterfalls of rare beauty. At *Chittenango*, on the Cazenovia and Canastota Railroad, a village situated in the valley of the outlet of the lake, are the *White Sulphur Springs*, strongly favoring the White Sulphur of Virginia. The waters flow from a bold ledge of rocks a mile out of the village; and, although Chittenango is not at all a fashionable resort, it is visited by many invalids, using the waters with much good result.

Yates Spring, at one time the most popular, is now almost entirely neglected. A few pretty summer residences, and one or two quiet hotels, are found here. Three miles from the springs the Chittenango River falls a hundred feet over a limestone cliff; and, two miles east, *Canaserrago Creek* falls a hundred and forty feet.

The Green Lakes are a series of pools in the limestone formation on Limestone Creek, near Manlius Station.

The largest of the group, *Lake Sodom*, is only one-quarter mile across, and one hundred and fifty feet deep, with perfectly clear, limpid water. These lake waters hold in solution large quantities of the sulphate of lime, and are strongly tinctured with sulphuretted hydrogen gas.

Visitors should not consent to leave Syracuse, not having seen *Green Lakes*.

Prof. Silliman says, "The bottom is a grass-green slate; the sides are white shell marl; and the brim, black vegetable mould; the waters per-

THOUSAND ISLAND HOUSE,
Alexandria Bay, N. Y.

fectly limpid. The whole appears to the eye like a rich porcelain bowl filled with limpid nectar."

But the waters are disagreeable to the taste, and the principal lake has received the name of "Sodom."

Syracuse, one hundred and forty-eight miles west of Albany, is a busy city of fifty-four thousand inhabitants. It is the seat of *The Syracuse University*, under the direction of the Methodist denomination, and of several public and private institutions of learning and charity. It is located on **Onondaga Lake**, in the western part of Onondaga County. This lake is six miles long, one mile wide, and three hundred and sixty-one feet above tide-water. On its banks are the most extensive salt-manufactories in the country; in fact, the entire vicinity is among the most fruitful and valuable in the State.

Near Liverpool on the east bank was, formerly the chief fortress of the Onondaga Indians, an important tribe of the Six Nations; and in this locality were fought many of the bloody battles between the French and Indians. A remnant of the Onondaga tribe live on a reservation at the head of the lake.

At Syracuse the railroad divides, forming two separate lines, uniting again at Buffalo. Taking the lower route, the upper being comparatively uninteresting, the tourist enters one of the most fascinating lake regions of the continent, which will well repay his examination.

Skaneateles Lake, the most picturesque of all the lakes of Central New York, is reached from a point midway between Syracuse and Auburn. It is eighteen miles long, one and a half wide, and eight hundred and sixty-five feet above the sea. It is set among imposing hills rising twelve hundred feet above its surface, giving its waters a profound effect of "deepening, darkening, beautiful blue" such as is rarely found in American lake-scenery. This lake does not boast of busy steamboat enterprise or traffic in any shape, although a small passenger-boat steams irregularly on its placid waters, for the accommodation of tourists. About ten miles south-east of Skaneateles, is the little **Otisco Lake**, lying away off quietly by itself, almost hidden among deep valleys and sombre woodland hills.

In this immediate neighborhood is **Owasco Lake**, the smallest of this chain of lake waters. It is eleven miles long, and one wide, is seven hundred and fifty-eight feet above the sea. It is framed by stern, rocky bluffs, alternately diversified by the fertile Owasco Flats.

Seneca Falls is a station on the Seneca River, where it joins **Cayuga Lake**: the latter receives its name from the powerful Cayuga tribe of the Six Nations. It lies between high rolling bluffs and uplands, and rich farming country. It is thirty-eight miles long, and from one to three wide. Its waters flow from Ithaca, north to Cayuga Bridge on the New York Central Railroad.

As the steamer passes along down the lake, a few miles from Ithaca, *Cornell University* is seen crowning one of the adjacent hills. Quiet little villages and peaceful farming lands are passed all the way up toward the end. *Wells College* for ladies is at Aurora on its east bank; the steamer stopping at that village, and also at *Levana* and *Union Springs*, a thriving town with several mills working on the water-power of two exhaustless springs.

Taughannock Falls are located on the west shore of Cayuga Lake, six miles north of Ithaca, and are easily reached from Taughannock Station on the Geneva, Ithaca, and Athens Railway. This fall is two hundred and fifteen feet in perpendicular plunge (fifty feet higher than Niagara), is set in the rarest of romantic surroundings, and is so fascinating in effect as to call from Dr. Cheever the following rhapsody: "The fall is, in truth, the *Staubbach* of Switzerland, most absolutely reproduced, and of concentrated beauty and grandeur."

One mile from the city of Ithaca is a remarkable ravine, containing a large number of cascades and waterfalls supplied by *Fall Creek*, which descends in gradual fall, four hundred feet in one mile. First in this series, the traveller meets **Ithaca Falls**, falling one hundred and sixty feet, the waters at its base forming a wide, deep pool. Following the rocky path up the gorge, the tourist passes successively **The Plateau**, a broad, shelving rock; **The Promontory**, which affords extensive views over the wide valley, taking in the village. *The Tunnel*, cut two hundred feet in length through solid rock, and formed to carry water to the mills below, leads to the beautiful **Forrest Fall**.

Next comes **The Foaming Fall**, a clear leap of thirty feet; then **The Rocky Fall**, fifty-five feet high; and near by **The Trip-hammer Falls**, down which the dark water plunges into a broad basin. Above and overlooking them are **Flume Falls** and the **Beebe Dam**.

The path winding through this romantic dell has been formed by blasting off the rocks, and is frequently protected on the way along at the most dangerous places by strong railings. The whole effect produced by this natural wonder is exceedingly fine, and this romantic dell may be pronounced the charming crown of New York lake-scenery.

The *Gorge* is a favorite resort for all classes, and is especially patronized by picnic parties.

Pulpit Fall and *Buttermilk Fall* on Buttermilk Creek, two miles south of Ithaca, are among the interesting suburban objects of the city; and *Lick Brook*, and *Enfield Glen Falls* one hundred and sixty feet high, are much visited.

Seneca Lake is touched on its northern extremity by this road, and south, at its head at Watkins, by the Northern Central Railroad, to which turn for fuller description. A line of steamers connects Seneca and Watkins by daily trips. The lake is thirty-five miles long, and about four wide. The depth of these waters is remarkable, reaching in some places one thousand feet. They are wonderfully clear, rarely freezing over, even in the coldest weather.

Its hilly shores, like Keuka and Canandaigua, are richly productive of grapes in valuable quality, giving to the general tone of the scenery a rich and pleasing effect. This lake is valuable in commerce, affording extensive transportation for coal. At the southern extremity of the lake is **Watkins Glen**, perhaps the most romantic spot in the country, which is described in detail elsewhere.

Once more proceeding westward from Geneva, **Lake Canandaigua** claims our attention, on the shore of which lies the village of its name. This is one of a cluster of six like waters running from about the same line from north to south; and they are of variable size and importance.

Canandaigua, the most western, is fifteen miles long, and six hundred and sixty-eight feet above the sea. About three thousand vineyards enclose its shores. This region is a quiet, restful spot, and all strangers are fascinated by its charms.

Sodus Point is a recently developed summer resort, noted for its salubrious atmosphere, located on Lake Ontario, at the terminus of the Sodus Point and Southern Railway.

Crooked Lake, or Keuka, farther to the east, is seventeen miles long, and one and a half wide. It is seven hundred feet above the sea, and is protected by high hills cultivated for vineyard use.

The train, here turning northward, reaches **Rochester**, a beautifully shaded, modern city, situated on the Genesee River, seven miles from Lake Ontario. It is the principal manufacturing town of Western New York; its magnificent water-power giving it an immense advantage over other competitive cities. In the vicinity are the largest nurseries of fruit-trees and plants in the country. In Rochester there are many costly buildings, the *Athenæum*, the *University*, and *Theological Seminary*, being among the most imposing. In the lower part of the city is the **Genesee Falls**, one of those wild and romantic assemblages of water battles with rocky obstructions, which are ever novel and of exciting interest, — wonderfully enhanced, in the Genesee, from the skilful mastery of its turbulence, harnessed by the hand of art and scientific devices.

Continuing westward, from Rochester, the **Oak Orchard Acid Springs** at *Medina* claim notice. They are especially medicinal in the cure of cutaneous diseases. They number nine distinct springs within an area of eight hundred feet. Their waters are remarkable as containing in each gallon, by the most exact measurement, eighty-three grains of sulphuric acid. It is said that a similar spring exists in Persia, and that two of the same sort are found in South America. These springs are also reached by the way of the New York Central Railroad at a point six miles north-west of Batavia by the upper route.

Passing **Lockport,** a city of fifteen thousand people, noted for its immense and massive locks, lowering the bed of the Erie Canal sixty-six feet from the lake ridge to the level of the Genesee Valley, we approach the grand culmination of the journey, —

Niagara Falls, and the traveller or wonder-hunter finds himself at last in the presence of that world-renowned glory of the waters. Description of those stupendous falls has been exhausted long ago. No attempt at details will be made here. There are other sublimities of nature, no less equal in their impressiveness and startling grandeur. There are water leaps of loftier heights, and amid scenes that fill the soul with delight and awe. But no such vast volume, no like rush and turmoil and thunder, has yet been discovered in the habitable parts of the globe. We find inexpressible delight and profound emotion in every variety of earth's peculiar, more exceptional, and fanciful aspects, — in the tender and sweet calm of woodland green, and shady solitudes of rills and wild flowers and birds, and the wavy mists of remoter hills ; in the great stabs and scars which mighty convulsions have inflicted ; in the stern and sterile summits where ice and snow forever reign; in the cascades, the sinuous streams, the wave-washed crags of the Atlantic shores. But from none of these come such profound impressions as from the awful plunges and the reverberating thunders of Niagara Falls. With a thousand pretentious rivals, it is the crowned monarch of them all.

PLEASURE ROUTE No. 16.

From New York City to Niagara Falls and the West, via Erie Railway.

ERIE RAILWAY.

THE Erie Railway is one of the main trunk lines connecting the seaboard with the great inland lakes. Reaching from New York to Lake Erie at Dunkirk, with a branch to Lake Ontario at Rochester, and another to Buffalo, connecting with the lines from the other inland seas, it has become one of the great arteries of the country, second to none in carrying capacity. Its passenger-trains are equipped with the finest coaches made; its broad-gauge road-bed challenges comparison for smoothness and safety; and its railway hotels and restaurants are placed at convenient distances, and kept in unexceptionable condition.

The whole extent of the line lies through a country rich in rare scenes of romantic beauty, and wooded, picturesque loveliness, and has long been, and will long be, a favorite route for the summer tourist.

The Erie Depot at Jersey City is reached by ferry from the foot of Chambers Street, or from Twenty-third Street.

In order to observe the many points of interest, the trip should be made by daylight, and the comfort of the traveller will be much increased by taking the beautiful palace-car attached to the morning express-trains. These cars deserve a separate chapter by themselves, especially those of this road.

At 9.15, A.M., you pass from the depot through Bergen Tunnel across the Hackensack Meadows by the smaller suburban towns, until Paterson, N.J., is reached, where a stop is made for a visit to **Passaic Falls**, which are waters of value, and much mercantile and mechanical interest, affording large revenue power.

The falls themselves are in the midst of a small park, where they throw themselves abruptly down about sixty feet into a deep ravine, between boulders of basalt, and flow rapidly off between beds of limestone. Leaving Paterson, after passing several smaller towns, the road enters the State of New York at *Suffern*, from which place a branch road runs direct to the Hudson River at Piermont. Farther on is Ramapo, the point from which the beautiful valley of the *Ramapo River* and *Falls* are reached. The river breaks through the mountains at this place, and passes by an interesting series of successive falls and cascades to join the *Pequannock* at Pompton, which afterwards falls into the *Passaic*. There are many beautiful bits of landscape and water scenery in this charming valley.

Greenwood Lake, or Long Pond, lies among the mountains of Orange County, nine miles south of Chester. It is a wonderfully clear sheet of water, six miles long.

Surrounding Greenwood are several lakelets of pleasing resources, furnishing the best of boating and fishing, and always rewarding the pleasure-seeker with frequent surprises.

It is a special delight to explore the lovely *Macopin*, a perfect gem, and also the more extensive waters of the *Wawayandah*, lying in the valley of the Wawayandah mountains near the boundary line of New York and New Jersey. Its name, of Indian origin, is translated, "winding stream," in characteristic picture of the zigzag nature of the lake, which is at one place almost divided by an island. Its waters are very deep, clear, and abound with trout. The entire region is attractive; and, as it is near New York, it is always, during the heated season, a popular retreat for those whose business cares forbid a far-away resting-place.

Near this station is also the *Seven Springs Mountain House*, a spacious hotel for summer resort.

At *Goshen*, the Wallkill Valley Branch leaves the main line, running through Montgomery and other towns to *Kingston* on the *Hudson;* where connection is made with the New York, Kingston, and Syracuse Railroad for the "Catskills." (See index.) Just above *Montgomery* is *New Paltz*, six miles from which lies **Lake Mohonk**, a favorite goal for the denizens of New York City. This lake is three-fourths of a mile long, but very deep, and with high, precipitous shores. Its waters abound in fish. The neighborhood of the lake supplies many objects of interest, among which are *Sky Top* peak (three hundred feet high), upon the side of which Lake Mohonk is embedded, the *Labyrinth*, *Eagle Rock*, and *Sunset Rock*, each an object of interest.

About five miles off, on Rondout Creek, are *High Falls*, elsewhere described.

Goshen is one of the oldest towns in Orange County, and is noted for its dairy products, "Goshen or Orange-County butter" being the popular and standard brand of New York City. *Middletown*, a flourishing village, slips by, giving place to the view of the **Shawangunk Mountains**, which the train rapidly ascends by a series of "planes" to *Otisville*, gaining the summit of the grade a little farther on. The view from here, as the train rushes along the western slope, is grand indeed. **The Neversink Valley**, the rugged overhanging mountains, and the hamlets of *Millford* and *Port Jervis* in the distance, form a panorama of grandeur impressive to the beholder, and never to be forgotten.

Following down the Neversink Valley, *Port Jervis* is reached, a town of seven thousand people, lying at the confluence of the Neversink and Delaware Rivers. This is the terminus of the eastern division of the

Erie Road. The *Falls of the Sawkill* are six miles away, with stage communication. The gorge through which the stream passes is one of great wildness and grandeur. Three miles beyond Port Jervis the train crosses the river, passing into the State of Pennsylvania, through which it runs for twenty-six miles. A portion of this distance, the road was constructed at an expense of a hundred thousand dollars per mile, being built on broad galleries hewn out of the mountain side. We turn from the rugged overhanging rock, and look down upon the verdant meadows below, Eden-like and charming in their quiet loveliness.

At **Lackawaxen,** situated at the junction of the Lackawaxen and Delaware Rivers, a branch road runs to *Honesdale*, penetrating the rich coal regions of North-western Pennsylvania. This line passes **Hawley** *en route*, the eastern terminus of the Pennsylvania Coal Company's gravity road, or "switch-back" (see index). The wire suspension-bridge which at Lackawaxen spans the Delaware, sustaining the canal-aqueduct and railway trains, is a great feat of engineering skill, and forms a marked feature in the landscape.

Seven miles beyond, the train again crosses the Delaware into New York State. This section, although lacking the boldness of that recently passed, is filled to overflowing with scenes of unassuming but fascinating beauty, with no particularized feature until the road leaves the Delaware Valley near *Deposit*, and, rising by heavy grades, reaches *Summit Station*, thirteen hundred and sixty-six feet above the sea. We soon cross the deep embankment where once rested the famous *Cascade Bridge*, which, by single span of two hundred and fifty-six feet, cleared a gorge a hundred and eighty-four feet in depth. Here the scene was wild indeed. The improvement, while detracting from the picturesqueness of the route, adds materially to its security.

The view now opens to one of great expanse of landscape beauty. The Delaware, whose company we kept so long with pleasant recollections, we have left behind; and we now catch our first view of the Susquehanna, which, throughout a length of over four hundred miles, carries with it a succession of scenic loveliness. We now descend to the famous **Starucca Viaduct,** near Susquehanna. This stupendous structure is the most remarkable engineering achievement of the road. It crosses the *Starucca Valley* by a series of eighteen arches, with a length of twelve hundred feet, and height of a hundred and ten feet. The charming Starucca Valley and the viaduct have been vividly portrayed by Cropsey, the celebrated landscape painter, in his picture, "An American Autumn."

At **Susquehanna** are the repair-shops of the road, comprising large buildings covering eight acres, and fully furnished with the best and latest improved machinery in the country. The train stops here for dinner. There are many landscape views of interest in this vicinity. *Mount Oquago* and *Painted Rock* are among the attractions. From Susquehanna to Binghamton the country is more tame. At *Great Bend*, intersection is made with the Delaware, Lackawanna, and Western, and the Delaware and Hudson Canal Companies' railroads, extending from Scranton northward.

Binghamton is a pleasant city of thirteen thousand inhabitants, situated at the confluence of the *Susquehanna* and *Chenango* Rivers. Its capital railroad facilities, and proximity to the immense coal and iron centres, have developed to itself a large manufacturing interest, which is rapidly placing it in the front rank of the inland cities of the Empire State. It contains many public and private buildings of cost and beauty, among them the *Court House* and *State Inebriate Asylum*, &c.

Binghamton is itself a delightful place of rendezvous, and is also the point of departure for many summer resorts. Leaving Binghamton, if we proceed north-west, on the Albany and Susquehanna Railroad, passing **Vallonia Springs** (whose waters are strongly recommended), and other smaller stations, we reach at last the junction, and, in due time, **Cooperstown**, lying at the foot of **Otsego Lake**, where the Susquehanna River takes its rise. This town is a favorite summer home, and has become, in a certain way, historical by the visits of many eminent writers.

Mohegan Glen, three miles away, contains several waterfalls ; and within an area of twenty miles by pleasant driveway will be found *Hartwick* and *Seminary* Lakes, *Bear Cliff Falls*, *Westford* and *Pierstown Hills*, *Beaver Meadow*, *Rum Hill*, *Cherry Valley*, and *Sharon* and *Richfield Springs*, all objects of much interest. The Cooper House is the popular stopping-place. (See index for fuller description.)

Returning again to Binghamton, we "take up our line of march" westward, entering a rich farming country, through which flows the placid Susquehanna.

At **Owego** the tourist can leave the main line, and, proceeding by the Delaware and Lackawanna and Western Railroad, reach Ithaca and the natural wonders in its vicinity.

Elmira, a city of twenty thousand people, is delightfully situated on the Chemung River. It is a manufacturing and mercantile town, handling large quantities of coal and grain, and containing extensive construction shops of the Erie Road. It has many fine buildings and several good hotels. The Northern Central Railroad here comes in from the south, crossing at right angles, continuing past *Havana* and Watkins Glens and Seneca Lake to Canandaigua.

At **Corning**, seven miles from Elmira, and two hundred and ninety miles from New York City, the road divides. forming the Rochester Division, running north-west by *Painted Post, Bath*, and other smaller stations to *Avon*, where are **Avon Springs,** celebrated as cures for many chronic, and especially cutaneous diseases. Here are three springs, whose waters are taken internally, and also used in bath-form. Avon village is a pretty valley town, a bright and sprightly summer resort, busy with hotels and comfortable boarding-houses.

The "Attica Branch" runs from *Avon* to the Buffalo Division at *Attica*, passing through a beautiful fertile rolling country, and by *Batavia* and other flourishing towns.

Rochester, the terminus of this branch, is a fine city, seven miles from Lake Ontario, elsewhere described. (See index.)

The Buffalo Division branches from the main line at Hornellsville, and passes *Canaseraga* and the Chautauqua valley to *Portage*, where occurs the principal object of interest on this division, the remarkable bridge over the Genesee valley and river, recently burned. This was the longest wooden bridge in the world. It was eight hundred feet long, two hundred and thirty-four feet high, and contained more than sixteen million feet of timber.

The famed **Portage Falls** are just below the bridge, and are reached by crossing on a plank walk, through the timbers, far below the railroad track. A series of surging, boiling rapids leads to the brink of the **Horse Shoe Fall,** down which the river rushes in an unbroken torrent, falling sixty-eight feet. A short distance below are the **Middle Falls,** one hundred and ten feet high, the waters falling into a deep, dark pool, surrounded by perpendicular rocky walls. The action of the water has worn an enormous cavern in the softer rock of the west cliff, called **The Devil's Oven,** which in times of low water is capable of containing one hundred people.

Leaving the imposing *Middle Falls*, the river rushes on its tumultuous way, through a steep, narrow gorge, winding and almost doubling for a distance of two miles, making its final plunge one hundred and fifty feet down a steep series of steps and perpendicular falls to the last final conflict at the bottom, where, compressed to a width of fifteen feet, it hurls itself against the *Sugar Loaf*, a rugged point of rock one hundred feet high, by which it is deflected, and turned off at right angles to its former course.

The Genesee Canal here approaches the river by long galleries cut in the steep bank, and crosses by a high *aqueduct* above the bridge. The whole course of the Genesee from *Portage* to *Mount Morris*, for several miles, is continuous between high banks (three hundred to three hundred and fifty feet), and is a remarkable exhibition of the power of water when confined in narrow limits.

Passing Warsaw, where the beautiful *Otka Valley* presents a scene of charming country life, and Attica, the junction of the "Rochester Branch," the traveller approaches the foot of *Lake Erie* at **Buffalo**, one of the chief cities of Western New York. This city stands on the eastern shore of the lake at the debouchure of the Niagara River. It has the best harbor on the lake, protected by extensive breakwaters; and its fine water front of five miles has been improved by the erection of basins, ship canals, and enormous elevators. The public parks are large and carefully kept, the streets broad and conveniently arranged.

Black Rock, a suburb of Buffalo, is opposite the Canadian town of *Fort Erie*. At this point is the great **International Bridge**, built in the years 1869 to 1873, at a cost of $1,500,000, with a total length of more than thirty-five hundred feet, divided by an island in the river. It is used by the Erie, the New York Central, and three Canadian railroads. Buffalo has communication five times daily, *viâ* Black Rock and Towanda, with *Niagara Falls*.

The main line of the Erie Railroad continues from Hornellsville, up the valley of the Canisteo River to Almond, and then begins to climb the heavy grades of this portion of the road, arriving at *Tip Top Summit*, at an altitude of seventeen hundred and sixty feet above the sea.

Descending the slope into the Genesee Valley to *Genesee*, the road turns sharply to the north-west, running down the river to *Belvidere*; then, bending at right angles, it leaves the valley, and again climbs the watershed of the *Alleghany Mountains*.

Cuba Summit, 1,677 feet above tide-water, is the dividing line between the waters of the lakes and the St. Lawrence River, and those running south to the Gulf of Mexico by the Ohio and Mississippi Rivers.

Salamanca is an important railway junction, and is the eastern terminus of the Atlantic and Great Western Railroad. At *Salamanca* occurs a singular formation, called **Rock City**, occupying one hundred acres on the summit of a high hill. Huge blocks of a white conglomerate stone, with sharply defined outlines, convey the impression of ruins of a regularly built and partially destroyed city.

Salamanca is the point of departure by the Atlantic and Great Western Railroad for the beautiful Chautauqua Lake.

The road from Salamanca descends by long, easy grades to the western slopes of the mountains, passing through extensive forests and a few unimportant stations, gradually approaching the level of Lake Erie, which it strikes at **Dunkirk**, the lake terminus, four hundred and sixty miles from New York City. The harbor of Dunkirk is artificially formed by piers and breakwaters. The town has lost much of its importance as a shipping point, being distanced in its race for prosperity by its more successful neighbor, Buffalo. It has replaced its commercial interests, however, by establishing manufactories of various kinds. The **Gas Spring**, from which the city is lighted, is well worth a visit.

At Dunkirk, connection is made for points west by the Lake Shore and Michigan Southern Railroad, and by the "Buffalo and Erie" for Buffalo, Niagara Falls, and points north.

ATLANTIC AND GREAT WESTERN RAILROAD.

We return to *Salamanca* to start from the eastern terminus of this important "trunk-line." Our direction is now changed towards the south-west, and we follow the course of the Alleghany River a short distance, and through the Conewango Valley to **Jamestown**, located on the rapid outlet of the Chautauqua Lake. Jamestown is a pleasant, enterprising village, which is much visited in summer; its proximity to "Chautauqua" adds greatly to its popularity as a resort.

Chautauqua Lake is thirty-five miles from Salamanca, in the extreme western corner of New York. It is twenty-four miles long, and three wide, and is 723 feet above Lake Erie, and 1,362 feet above the sea. This is said to be the highest navigable water on this continent, and is pronounced one of the most beautiful and valuable of New York lakes.

Hemmed in by high hills, it presents, on a clear day, a charming view of water scenery; and even the mists, frequently arising from its surface, are no serious drawback to its artistic effect. The thrift and importance gained by the busy ply of merchandise steamers, from *Mayville* at its northern extremity, connecting with the Erie Railroad and its outlet at *Jamestown*, descending to the Alleghany River, add to the interest of the scene. There are several summer resorts at different points, which are annually well filled, and, when better known, must become very popular with pleasure-seekers. The steamer "Jamestown" runs in connection with trains on the Atlantic and Great Western Railroad. The line of the road continues through *Meadville*, Penn., on the Venango River.

A "Branch" runs south-east from Meadville, down the valley of French Creek, to *Franklin* and *Oil City*. On the main line the pretty little **Conneaut Lake** is seen by the way. Several unimportant stations are passed; the Jamestown and Franklin division of the Lake Shore Railroad is crossed. At *Greenville* intersection is made with the Erie and Pittsburg, and Shenango and Alleghany Railroads; and beyond the station of **Orangeville** the road enters the State of Ohio for the West.

PHILADELPHIA.

THE city of Philadelphia, in some respects, is the most renowned on the continent; and is about to receive a new crown of honor, in the centennial anniversary of the Declaration of American Independence, "which was read from a stand in the State House yard, July 8, 1776," having been adopted by the Congress of the Colonies on the 4th preceding. This grand event, wherein forty millions of people will hold jubilee on their first national birthday, invests Philadelphia with distinctive interest, and makes her history and characteristics the cynosure of all eyes. In view of the mighty excitement, and the vast numbers who will make pilgrimage to the shrine of liberty at the coming anniversary, such information as may be condensed in our limited space will be important and interesting. Especially will those who may desire to visit the great carnival, from remote points, need instruction as to the immense network facilities of travel to Philadelphia and "how to reach it." There will be more people assembled together, more exciting and remarkable attractions, more wonderful things seen and heard for a lifetime talk, than the world has ever before known; and the reader will do well to study and carefully preserve the details herein given of facilities for reaching Philadelphia from all sections of the country. It may as well be added, that the issue of this work for 1876 will treat this topic in the most thorough manner, furnishing details where this must treat in generalities. In the present volume, however, descriptions of a general character will point the way from distant towns by pleasant routes towards the great centres of travel, connecting each with the main lines leading to Philadelphia; and within the radius of Pennsylvania will give illustrated descriptions of every important road by which the city can be approached, and furnish as well the most desirable short-trip excursions from Philadelphia for strangers desirous of seeing its vicinity.

The settlement of Philadelphia has not only the curious character which generally attaches to such events, but is solitary and notable in one exalted fact. William Penn, its founder, was a Quaker, and his companions were Quakers. Their doctrine was eminently "peace and good-will on earth." In 1682 these men of simple and primitive faith purchased of another simple and primitive race the site on which Philadelphia (brotherly love) was planted. The two met beneath the trees, in the open air, with the sun and the dome of blue above. The treaty made was never signed, and *it was never broken,* — and the only treaty with Indians that never was violated. At the outset, then, the record of Philadelphia is romantic, bright, unsullied, and most teaching.

Philadelphia had a little over twenty thousand inhabitants when the Declaration of Independence was made. The number is now estimated at about six hundred thousand, and its area comprises nearly one hundred and twenty square miles. One of its unique features is the arrangement

FALLS VILLAGE BRIDGE,
Philadelphia and Reading Railroad.

of the streets in the city proper, and the massive and grand bridges which span Philadelphia's noble rivers, — the just pride of the inhabitants. The ambition to be a vast city has led the Quakerites to the anaconda practice, first set by New York, of swallowing the smaller contiguous places from time to time. The original city, as is well known, is on the checker-board order, — streets intersecting at right angles, arranged and numbered upon a plan which leaves it no competitor. But the additions to the place have altered matters, as a whole; so that the suburban streets of Philadelphia now have their full share of the hoop-pole style.

In sight-seeing attractions, no city has a more diversified nor a more delightful variety than Penn's home. Though Nature is constrained to don the garbs and obey the fashions of human art, yet she has been greatly aided by man's ingenuity and ambitious skill. Twenty years ago the Quakerite from Philadelphia was never weary of describing the beauties and the marvels of Fairmount and its colossal water-works. New York, however, — which is the chronic sore of Philadelphians, — having built its Croton wonder, and made its great Central Park, the glory of exclusive Fairmount Water-works was materially dimmed. But Fairmount now blooms in fresh loveliness, in size eclipsing all other parks, and is only one of many gems set in a vast and magnificent new park which well contests the palm with the great "Central."

First of all "the sights" is "Independence Hall," with its rare and precious relics and its hallowed associations. Next we would place Girard College, famed for its architectural beauty; then Fairmount Park, which in itself is a country of romance and beauty not surpassed; next, the cemeteries, of which there are several; and now trips into the shady nooks that lie along the banks of the Schuylkill and Wissahickon; and, finally, the immense number of public buildings, finishing off by promenades on the thronged streets, where a distinctive richness of fashion, good taste, and loveliness abounds when the day and weather are propitious.

It does not come within the scope of this work to deal minutely with the inner attractions of large cities. They are rarely the points of special interest for the summer tourist; but an abundant variety of "sights" can always be found tabled to hand, in detail, in every city. But the mighty multitude who will throng to Philadelphia in the midsummer of next year will all have to reach the place, and get away afterwards. The copious and finely illustrated descriptive routes to be travelled and selected from all quarters, and especially in the admirable and most abundant lines of Pennsylvania, presented in these pages, will set every one at ease in selecting his line of march, — whether for this year or the next. As has been hinted, however, this part of our subject will have very special care in the issue for 1876.

If a day-trip in the country is desired, either to mountains or seashore, these pages will furnish the tourist the fullest directions, with ample illustrations of prominent scenes along all the most desirable routes, "and how to reach them."

ILLUSTRATED PLEASURE ROUTE No. 17.

Philadelphia to Reading, Pottsville, Williamsport, and the Coal Regions of Pennsylvania.

PHILADELPHIA AND READING RAILROAD.

How little the above heading serves to inform the reader of the extent of track controlled by this road, its branches and connections! From Philadelphia to Reading, the original terminus, is only fifty-eight miles; the present line owned, leased, and operated by this company aggregates fifteen hundred miles. From Philadelphia to Reading is still the main line, which has been extended to Pottsville, Tamaqua, and Williamsport. From this, branches, large and small, lead in every direction; some to beautify cities and villages, others into the fastnesses of the mountains, developing a diversity of enterprise and a wealth of scenery of unrivalled interest and beauty. Philadelphia is the starting-point, Reading and Pottsville the centres from which the principal branches diverge.

COLUMBIA BRIDGE.

Leaving Philadelphia from the commodious depot at Broad and Callowhill Streets, we run through the built-up portions of the city for a half-mile to *Fairmount Park*. Skirting this, our course is taken along the *Schuylkill River*, which we cross at Columbia Bridge, and pass Belmont Station, near the entrance of **Belmont Glen**, which is justly known as one of the finest features of Philadelphia's great park; hence

the popularity of this entrance. To accommodate
the thousands who annually visit Fairmount, the
Reading Railroad Company has organized what is
known as "the park accommodation train," which
affords superior facilities to visitors.

Just beyond the Glen lie the grounds on which
are being erected the buildings for the Exposi-
tion, incident to the centennial
anniversary of our nation's birth.
These are pierced by lines of this
road from three directions, affording remarkable facilities and conveniences for visitors; and the
territory which the road drains

THE SCHUYLKILL FROM COLUMBIA BRIDGE.

and its numerous connections would
seem to warrant for it an immense travel
during the continuance of the exhibition. We
are now fairly on the banks of the Schuylkill,
the river of which the poet Moore sang and wrote: on
whose banks he found that rest, though slight, which he
had elsewhere sought in vain; but while we fain would quote
from the writings of one who it might be said immortalized
this beautiful stream, yet, as we have not reached a distance of five miles

from our starting-place, with a passing look at Tom Moore's cottage, an admiring glance at Fairmount Park with Laurel Hill Cemetery infolded in its embrace, we glide along through other scenes of picturesque loveliness. Our route follows the windings of the river through *Falls Village*, Manayunk, and Conshohocken, until we reach *Bridgeport*, opposite *Morristown*. The latter is a pleasant suburb of Philadelphia, with which it is connected by a branch-road, which has its depot at the corner of Ninth and Green Streets. This "branch" follows parallel with our course on the opposite side of the river, and the charming scenery through which it lies renders it an extremely popular route for Philadelphians seeking suburban residences. Bridgeport is the terminus of the "Chester Valley Branch," which leads through a region of great loveliness. From Bridgeport the route continues along the river bank, each turn opening up new attractions and fresh charms of scenery, and each hamlet and village, as we speed along, displaying evidences of the gigantic industry for which this valley is famous. We still follow closely the banks of the Schuylkill, which glides quietly and gracefully along the green fields that come down to meet it, and all flies like a dream of contentment.

VALLEY FORGE.

We now reach **Valley Forge**, memorable in the annals of the nation for the sufferings of the patriot army under Washington during the winter of 1777-8. The place of encampment, the old earthworks, and lines of intrenchment, still exist. This region forms a delightful ramble: the surrounding scenery is picturesque and beautiful, which, in connection with the historical associations, makes Valley Forge a place of great interest to all classes. Many an interesting story, handed down from generation to generation, can be told to a ready listener.

Just above Valley Forge Perkiomen Creek empties into the Schuylkill. The valley through which this stream flows is noted for the quiet beauty of its scenery, its mineral and agricultural resources. It was for many years the abode of Audubon, the great naturalist, in whose works reference is made to many rare birds that seek shelter there.

The Perkiomen "Branch" Railroad, following the line of this creek, extends at present twenty-three miles in a north-easterly direction towards Allentown, its prospective terminus. When completed, this road will form a new and attractive pleasure route between Philadelphia and all points in the Lehigh Valley, one of the most picturesque regions in the State, to Northern Pennsylvania and Central and Western New York.

SCHUYLKILL RIVER ABOVE POTTSTOWN, PENN.

Phœnixville is situated at the mouth of French Creek, which flows through a remarkably fertile valley. This is the terminus of the Pickering Valley "Branch" Railway, extending back eleven miles. At Phœnixville is located one of the largest rolling-mills in the country, which will well repay a visit.

The railway station is of brown stone, artistically designed and conveniently arranged, and is quite attractive in appearance. On approaching the depot the train leaves the main track, and runs under a covered way, affording shelter, and avoiding danger from passing trains.

Shortly after leaving this place we enter a tunnel, from which we emerge to cross to the east bank of the river. **Pottstown** is the next point of interest reached. This is a thriving borough of 4,125 inhabitants. It is pleasantly situated, with pretty surroundings and a fine fertile country beyond. Very extensive repair-shops of the company are located at this place. The train here crosses *Manatawny Creek* over a bridge 1,071 feet long.

At Douglassville we cross in quick succession, near their mouths, the Manatawny and Monocacy Creeks, each adding scenic beauties to the route.

We next pass *Birdsboro'*, a smaller but quite pretty village. Our route is still on the banks of the Schuylkill. Its current grows more rapid as we near the mountains, which rise threateningly in our front: the scenery changes with each mile of advance. Still following the valley which Nature has formed for the river, we suddenly glide around the curves, and find ourselves at the **City of Reading**, surrounded and overlooked by hills, — *Mount Penn*, *Mount Gibraltar*, and *Neversink*, — which converge to shelter this beautiful town, lying within their embrace. The river, after twisting and turning into curves and loops, escapes from the hills at last, and, leaving the city behind, starts on its race for the sea.

We are now at the terminus of the original Philadelphia and Reading Railroad; yet of the present organization we have but reached the centre — *heart* — of the roads owned and operated by that noted corporation. Reading is not only one of the most important railroad centres in the State, but it is one of the most charming towns to visit. Its hills and suburbs are famed for their pleasant walks and delightful drives. Its citizens are courteous and hospitable. A four-years' sojourn enables the author to speak from experience, and to know whereof he speaks. Reading is a flourishing city of forty thousand people, and in size stands third in the State. Here we find rolling-mills, foundries, car and machine shops of gigantic proportions and unrivalled enterprise. And here, too, we find one of the finest union depots in the country, affording superior accommodations to travellers, to which the trains on all converging roads centre. The management is most complete and exemplary. Without system it would be impossible to meet and transact safely this enormous business. The clock in the depot tower is connected by electric wires with the clock of the main office in Philadelphia, and again with the clocks of the several departments within; hence every vibration in Philadelphia has a corresponding response from every other timepiece at this depot. Consequently not only the passengers, but the depot-master, ticket-sellers, train-master, conductors, and engineers, are simultaneously guided by the same standard of time. In a depot like this, where trains are constantly arriving and departing in all directions, the value of such an arrangement is naturally suggestive to the intelligent mind.

Nor does the watchful care for the comfort and interests of guests end here: indeed, this is but an index of the entire organization. The passenger cars are supplied with the Westinghouse brake and Miller platform. Steel rails are giving place to iron, and stone ballast to gravel, and all with the happiest results.

The Lebanon Valley Branch leads westward fifty-four miles to Harrisburg, where connection is made with the Pennsylvania Railroad for Pittsburg and the West; with the "Northern Central" north and south; and also with the Cumberland Valley Railroad. Six miles south of Lebanon are located the famous Cornwall iron-ore deposits, in three elevations, called Glassy Hill, Middle Hill, and Big Hill. The ore lies on the surface to the maximum depth of three hundred and twenty feet in the centre, and covering a surface of one hundred acres; and, though it has been worked more than a century, there are no signs of exhaustion, while the quality constantly improves. It is found in the form of corroded earth, and is shovelled up as readily as garden mould, and transported by rail to the furnaces. At *Sinking Springs*, six miles from Reading, on the road to Lebanon, the Reading and Columbia Railroad diverges to the left to Columbia, on the Susquehanna River, forty-five miles distant. Two very popular summer resorts are located on this route. — **Ephrata** and **Litz** Springs. — affording good accommodations for summer boarders. The line also passes through Lancaster, forty-two miles distant, an old and pleasant town, where connection is made with the Pennsylvania Railroad. The East Penn. Branch leads east from Reading through a delightful agricultural region, thirty-six miles to Allentown, on the Lehigh River. The valleys in which these two branches are located are fertile, and are aptly known as the garden counties of the State. — Lebanon, Lancaster, Berks, and Lehigh, the aggregate value of whose agricultural products was in 1874 upwards of seventy-eight million dollars, while the manufacturing interests are estimated at seventy-five million dollars. Exhibitions of varied interest to this value are rare indeed.

We have left the Schuylkill, and our description has taken us to the Susquehanna and Lehigh. Let us now return to our old companion. Passing north from Reading, the low land begins gradually to struggle with the mountains, the latter getting the final victory, giving an advantage in favor of the picturesque. Penetrating the recesses of the highlands, the road emerges at Port Clinton (seventy-eight miles from Philadelphia), the junction of the Schuylkill and Little Schuylkill Rivers.

The streams rise not far apart in the coal hills to the north, and, describing two great curves, meet again at this point. The station here is of brown stone with turreted roof, and has an exceedingly tasty and antique appearance. Fifteen miles to the northward the main line of this company ends at Pottsville, to reach which, however, we follow the twistings and windings of the river in its tortuous course among the hills, passing

MOUNT CARBON.

the station at Auburn. Schuylkill Haven and Mount Carbon are also centres of considerable importance, from which lead "branches" right and left to the coal regions beyond.

Our next point of interest is Pottsville (ninety-three miles from Philadelphia), with a population of about sixteen thousand. This is the second terminus in the history of this road, and is the centre around which revolves the immense trade of Schuylkill County. Passenger trains leave several times daily in all directions.

Pottsville, where the hotel accommodations are good, we would advise as a headquarters for those who wish to spend several days in the coal regions. The *Mansion House* at Mount Carbon, one mile short of Pottsville, can be heartily recommended as a first-class establishment in every particular. Its situation at the foot of Sharp Mountain, its pic-

LITTLE SCHUYLKILL RIVER
Above Port Clinton.

turesque surroundings, the fine views of the valleys and river, linger long in the memory of the visitor.

We now return to Port Clinton, and in a northerly direction continue by the Little Schuylkill Branch to Tamaqua. The scenery here is wild in the extreme, and especially fine and romantic. The tourist should be thankful that the pursuit of anthracite coal and the love of gain has caused railroads to be built through ravines and valleys which under other circumstances would hardly have been attempted. After travelling twenty miles, Tamaqua, a thriving town of five thousand inhabitants, is reached. Here we find a new and very tasty depot, in which is

MAHANOY PLANE

a well-appointed restaurant, with twenty minutes allowed for dinner; this is quite refreshing after a four hours' ride, the time it takes the through Williamsport Express to run from Philadelphia to this point.

We are now in the centre of the coal interests of Schuylkill County. To reach the principal mines, and to see the manner of mining and transporting coal, we now leave the direct route from Philadelphia to Williamsport, and strike to the west by the East Mahanoy "Branch," passing Mahanoy City, Girardville, Mahanoy Plane (from which a branch to Shenandoah City diverges), Ashland, and Gordon, and, thence out through Shamokin to Herndon, a waterside town on the Susquehanna.

At **Mahanoy Plane**, as its name implies, is an inclined plane for raising coal-cars from the valley to the top of the mountain, from which they run by continuous down-grade to Mount Carbon. This plane is 2,410 feet long, rising in perpendicular height 354 feet. At the head of the plane we are

BROOKSIDE.

but twelve miles from Pottsville by the descending grade, by which coal is transported. The object of the plane is thus apparent when we consider the distance to that point around through the valleys by which we came.

The towns enumerated have a mixed population, estimated at forty thousand people. We soon reach *Gordon*, at the foot of the Gordon Planes. The lower plane has a length of 4,755 feet, and a rise of 404, placing you 1,206 feet above tide; the upper is somewhat shorter, though steeper. From the top, or head, of this plane, coal-cars are run down nineteen miles to Schuylkill Haven.

The picturesque station of Brookside is reached by leaving the main line at Auburn, between Port Clinton and Schuylkill Haven, and continuing *viâ* " Susquehanna Branch," through Pine Grove; the tourist will not only pass through the exceedingly interesting geological regions marked by the Pine Grove and Lorberry coal-fields, but, continuing through Tower City, will meet continually scenes of great scenic interest, culminating at the terminus of this "branch" at *Brookside*, — a spot which

SUSQUEHANNA RIVER,
At Herndon.

cannot fail to please the lover of Nature in her wildest moods. The view which here awaits the gaze of the visitor is one of singular combination. Artificial hillocks, the dust and *débris* of mines, rise thick and high about you; coal-breakers, like enormous black spectres, rear their dizzy heights, to the very top of which the dull mule clambers with his freight of coal. The oddity of the scene is attractive to the stranger; and this is in the midst of, and surrounded by, the most striking landscapes. As you gaze straight down the perspective of the valley, and mile beyond mile fades in blue distance, you feel that the vision must reach the distant Susquehanna at Harrisburg. Here is the artist's opportunity for toil and pleasure: the practical and the ideal are most completely and artistically blended.

There is still another section lying farther to the north, rivalling this, both in the magnitude of its coal operations and the grandeur of its scenery, through which lies the Mahanoy and Shamokin Branch Railroad, which passes through and thoroughly permeates the Mahanoy coalfield. This route terminates on the Susquehanna River at Herndon, amid scenery of the most lovely character.

In connection with the movements of the coal and general carrying trade of this company, it would be well to give the tourist some idea of the machinery and material required for this enormous traffic. There are in use 405 locomotives, 15,073 coal-cars, 3,819 freight and 279 passenger cars; and during the fiscal year ending November, 1874, the tonnage of the road was, in coal, over six million tons, while that of merchandise was 3,088,000 tons. The number of passengers carried amounted to 6,965,000; and when we state that since this company has been in operation, it has carried over forty-one millions of passengers, *and never killed one*, we say what we feel quite able to maintain, that this is more than can be truly said of any other railroad.

Let us now return to the main, and from Tamaqua start again northward towards Williamsport, feeling that our time has been well spent in our peregrinations in the wilds of Pennsylvania, in viewing its wonderful scenery, and studying its mighty coal interests. In this frame of mind, we are in good condition to appreciate the startling grandeur of the scenes through which runs the Catawissa Branch Road.

In the autumn of 1873, and again in the summer of 1874, it was the privilege of the author to run leisurely over this portion of the line, accompanied by D. C. Reinhart, Esq., local superintendent, under circumstances which enabled him to study carefully the artistic merits of the route. As a "special," with locomotive, parlor car, conductor, and pilot, our party had the road to itself for the day. We had sketching, berrying, and picnicking to our hearts' content.

Seated in the last car, watching the changing diorama of the landscape, we almost fail to notice that a gradual ascent is being made. We pass East Mahanoy Station, and reach **Tamanend,** where the "through car" from New York is attached, having come over the Central Railroad of New Jersey and its Lehigh and Susquehanna and Nesquehoning Branches, *viâ* Easton, Allentown, and Mauch Chunk. (See index.)

Leaving Tamanend we pass Quakake; and now, in place of running along the valleys, we are getting up among the mountain tops, passing through tunnels, winding around curves, on some of which it appears as if the rear of the train was chasing the engine, and in danger of making a collision. In the original survey of this road, it was located on the highlands, while in later years it would have taken the valleys; thus it required deep cuts, heavy piles, and tunnels. Here may be seen an American forest preserved in all its wildness ; and as we wind around the hills, climbing higher and still higher, the landscape widens, and objects in the valleys below grow small in the distance. Still upward and onward goes the train, twisting around the curves, and darting through the tunnels, until the summit is reached, and with which comes a feeling of relief; for we have unconsciously been laboring and straining to help the engine up the mountain side. Every puff seemed to find a corresponding echo within us, an inclination to push or help in some way; but now we are at rest, and drink in the wide-spread view before us.

Some of the lowering hills, completely cleared to their summits, have been cultivated; others in all their natural loveliness are covered with forest verdure. The valleys look deep, dark, and lonesome, with here and there a cultivated spot, with a snug little farmhouse nestled under the hillside, sheltered from the cold blasts of winter. The blue smoke that so gracefully curls from the chimney shows, that, although in the midst of coal, wood still holds sway as fuel, and promises to do so for years to come.

In the valley, close to Girard Station, is a fine old homestead, which at one time was the residence of Stephen Girard, to whom Philadelphia is so greatly indebted for his munificent charities. The ground on which it stands is, we believe, part of that which was placed in trust to the city at the time of his death.

Passing Mahanoy and Krebs Stations, we reach Ringtown, and near by is the first trestle-bridge over which the road crosses. It is a narrow valley, nearly 150 feet deep. From this point up to Catawissa there is a succession of these valleys which are spanned in a similar manner, excepting that at Mainville, where the trestle has been replaced by a first-class truss-bridge.

We here get the first good view of Catawissa Creek, as it meanders through the valley far below us at our right. The scene is wild and picturesque. The creek seems but a narrow thread winding along the base of the mountains, — here and there lost to sight, and again widening to a pleasant vista. It is a subject for the artist's pencil; indeed, for many miles we watched with pleasure the changing views, unfolded like a vast panorama of selected scenes.

We are now approaching **Mainville Water Gap**; although less grand than those of the Delaware and Lehigh, it still forms a bold and enjoyable landscape. The valley is highly cultivated, but is shut in by bold hills, among which the creek winds until lost in their shadows.

MAINVILLE WATER GAP.

We seem completely hemmed in; and, while wondering which way the train will find egress, we suddenly glide around the mountain, and emerge to the open country beyond, while the scene fades from view. We cross another trestle, obtaining a fine view of the McAuley and Nescopec Mountains in the distance. A few miles beyond, Catawissa is reached. This quaint old town is situated on the North Branch of the Susquehanna, at the mouth of Catawissa Creek. Nature has done much for this place, — all, in fact, that an artist could ask in combining the beautiful with the grand for a painting: bold mountain bluffs, deep wooded valleys, a brawling stream, a noble river spanned by bridges,

CATAWISSA.
Junction of the Catawissa Creek with the Susquehanna

every thing, indeed. But the town exhibits a want of thrift and energy that is painful to the stranger, who looks with pity upon a community upon which such fine natural advantages and artificial improvements are thrown away; for this is a railway centre of no mean advantages. Thomas Moran, one of Philadelphia's best artists, once sketched and painted this enchanting scene.

From Catawissa we cross the Susquehanna to Rupert, a junction station, where connection is made with trains on the Lackawanna and Bloomsburg Railroad, westward to Northumberland, and north-eastward to Wilkes Barre and Scranton, and by stage to Bloomsburg, a large manufacturing town, three miles distant. Visitors to *North Mountain* (see index) connect here by train for Shickshinny. From Rupert we follow the North Branch of the Susquehanna to *Danville*, an important iron manufacturing town, having a population of ten thousand. The value of the industrial products of Montour County for 1874 is estimated at seven million dollars, most of which was produced at Danville.

We are now fairly down into the valley; and moving north-westerly through Montour and Northumberland Counties, both good agricultural districts, we reach the West Branch of the Susquehanna at Milton, an attractive place, pleasantly situated for those who desire a few weeks' rest and quiet. The hotels are fair in size, and very well kept, with charges moderate.

The towns of New Columbia and several river stations are now passed in rapid succession. We have gradually left the coal regions, and without apparent knowledge have been creeping into the very heart of the largest lumber region in the State.

Muncy is a fast-growing waterside town, the outlet for the lumber brought by rail and water from the Muncy Creek region. From Milton we pass Montoursville, another lumber manufacturing town, and, gliding along the river bank, surrounded by landscapes the most charming, we rapidly approach the city of Williamsport, where this road has its northern terminus, and where it connects with the Philadelphia and Erie Railway for Lock Haven, the oil regions, Erie, and the West; also with the "Northern Central" for Elmira, Watkins Glen, and Canandaigua, connecting at the former with the "Erie," and at the latter place with the New York Central Railroads, for Buffalo, Rochester, Niagara Falls, and all principal points in Western New York, the Dominion of Canada, and the West.

Williamsport has a population of sixteen thousand; is beautifully located on the West Branch of the Susquehanna, has a number of large hotels and fine private residences, and excels in these particulars any city of its size in Pennsylvania. The lumber business is the leading feature: a walk through what is called the "basin" among the mills will cause you to wonder at the immensity of that interest.

The stock of lumber, lath, and pickets on hand in this region Jan. 1, 1874, amounted to 363,947,165 feet.

This great corporation, over whose track we have passed through so many scenes of industry, wealth, and landscape beauty, has many other interests than those described in this article. It owns or controls 153 miles of canal, has an immense coal shipping depot in the northern or Richmond District of Philadelphia; it owns fourteen steam colliers, having an aggregate carrying capacity of 15,500 tons, in which it transports, together with canal barges, large quantities of coal to the Eastern markets. It has its own shipyard for building and repairing the colliers, and within itself manufactures nearly all of the principal material used in the operation of a railroad.

COAL TRANSPORT.

In tracing the combined lines of the Philadelphia and Reading Railroad our course has led through many and varied scenes of rural and mountain scenery; but Philadelphians are not confined to mountain scenery alone: the whole New Jersey coast is at their doors, than which there are no more delightful seaside resorts in the land. The excellent railway facilities furnished by the Camden and Atlantic Railroad have contributed largely to the popularity of *Atlantic City*, specially noted for its dry, salubrious climate.

Long Branch is also easily reached, though not as popular with Philadelphians as Atlantic City and Cape May.

Cape May is also readily accessible, and is the fashionable resort of Philadelphians.

Long Branch, "one of the most popular seaside resorts in America, is in Monmouth County, N.J. It was visited for health and recreation previous to 1812; and, soon after the termination of the war with Great Britain, hotels were opened for the accommodation of visitors. Still its magnitude is of recent growth; and the last fifteen years have done more for its development and improvement than the preceding fifty had accomplished. Its hotel accommodations are sufficient for fifteen thousand persons; yet each recurring season crowds them to their full capacity. Elegant and spacious cottages, owned and occupied by persons of distinction, line the principal avenues for long distances, some of them being surrounded with extensive grounds, highly ornamented and carefully kept.

"The beach at Long Branch is famous for its natural grandeur, as well as for its artificial attractions. It is an open bluff, rising some twenty feet or more above the tide-line, and extending a distance of five miles. Along this, the grand drive is constructed, and the principal hotels are erected. Here, during the season, showy and elegant equipages dash, in passing and repassing lines, while the verandas and porticoes are thronged with spectators. No view could well be more animated or attractive than this, with its life, gayety, and beauty, relieved by the wide and restless ocean, swelling and rolling in boundless perspective. Some of the inland drives are equally pleasant; and, in fact, the facilities everywhere offered for this exhilarating enjoyment may be ranked as one of the greatest attractions of the place.

"The Monmouth Park race-course is a few miles from Long Branch; and its annual meetings rank among the most popular in America. These take place during the 'season;' and the list of entries generally embraces all the famous horses on the turf. Nothing on this side of the Atlantic so nearly approaches an English 'Derby Day' as a sweepstake at Monmouth Park, when the multitudes from New York, Philadelphia, and all other adjacent cities and towns, pour out to witness the famed steeds contend for the championship."

The State of New Jersey fails to offer any natural wonders to attract its share of the sight-hunting and money-spending thousands, who afford a summer harvest for more favored States. But her beaches are among the finest, as elsewhere described. Doubtless spots of landscape exist worthy of some delay to those flitting birds of passage; but the Jerseyites make their summer jaunts to more attractive regions than they can find at home. Their broad, extended beaches, which prove such sources of pleasure to the citizens of other States, fail to interest them. Fortunately, a few hours' ride will take them to the picturesque region of Pennsylvania, where they will find mountains and rocks to their heart's content.

CAPE MAY.

This old, established, and most justly celebrated watering-place is situated at the extreme southern point of New Jersey, on a narrow peninsula extending a distance of ten miles, bounded by the waters of the Atlantic Ocean on one side, and the Delaware Bay on the other.

For more than half a century it has been the resort of persons seeking health and pleasure during the heated term of the summer months.

The perfect safety of the surf-bathing, and the firmness of its broad, even, and unbroken beach, are unequalled at any other seaside resort.

The difficulty in reaching *Cape May* prevented for many years the rapid improvements its admirable location seemed to warrant, until the **West-Jersey Railroad Company**, appreciating the immense advantages to be gained, have from time to time extended their lines, and in 1863 opened an all-rail route from Philadelphia.

Pleasure tourists, at once becoming acquainted with the facilities thus afforded, flocked to the Cape. Property increased in value; handsome and costly cottages were erected, large and commodious hotels built, novelties introduced; and great improvements were manifest.

The Railroad Company did not confine its efforts merely to building the road, but aided generously with its capital, not only individual enterprises, but those to develop the natural advantages of the place.

It now became the resort for the *élite* and fashionables of Baltimore, Washington, Pittsburg, and Philadelphia; and, among its many regular sojourners, Chicago, St. Louis, New Orleans, and San Francisco have their representatives.

The hotels at this most popular watering-place are conducted in every manner equal to the principal hotels of our largest cities: the leading houses, the "*Stockton*," "*Congress Hall*," and "*Columbia*," each accommodating comfortably from one thousand to twelve hundred guests.

The Stockton Hotel, under the management of Charles Duffy, Esq., of the Continental Hotel, Philadelphia, is without doubt the most attractive and commodious house to be found at any of our seaside resorts; and, as a combination of mechanical and architectural beauty, it cannot be surpassed.

Congress Hall is the successful rival of the Stockton, and is under the proprietorship of Col. Cake, the popular landlord of Willard's Hotel, Washington, D.C. These houses, with the "Columbia" and fifteen or twenty others, furnish excellent accommodations to visitors.

Cape May can be reached direct from New York and Philadelphia by rail, *viâ* the Pennsylvania and West Jersey Railroads. Palace-cars are run through from New York without changes. Three express-trains are run daily from Philadelphia, the time being about two and one-half hours.

ILLUSTRATED PLEASURE ROUTE No. 19.

Philadelphia to the "Garden of the Atlantic Coast," Wilmington and Baltimore.

PHILADELPHIA, WILMINGTON, AND BALTIMORE RAILROAD.

THE GARDEN OF THE ATLANTIC COAST.

In a country like the United States of America, occupying twenty-five degrees of latitude and nearly sixty degrees of longitude, divided from north to south by several ranges of lofty mountains, and drained by many mighty rivers, it is natural to suppose that every variety of climate, every quality of soil, and every class of production, will be found; affording ample scope for the most diversified tastes, whether for permanent homes or temporary stopping places. Hence it is that mountain and sea-shore, waterfalls and springs, each have their admirers.

There is a section, however, which, for the many points of excellence it embraces, is comparatively unknown, — yet a section, which for

STRAWBERRY CULTURE.

MIDDLETOWN, DEL. PEACH GATHERING

the fertility of its soil, the salubrity of its climate, and the abundance of its fruits, has won for it the appellation of the *Garden of the Atlantic Coast.*

It is here that the early luxuries are grown that grace the tables of every city north of Mason and Dixon's Line. It is here that the luscious peach is plucked and strawberry gathered almost before the blossom is shed in a more northern climate.

If the reader will turn to the map of the United States, this garden can be easily pointed out, for its bounds were set by the Great Architect of the universe when the world began. The waters of Delaware Bay and the broad Atlantic lave its eastern shores, and the Chesapeake separates it from the main land.

It is here that our choicest luxuries are grown.

The production of this peninsula, in garden fruits, is

simply enormous. A "strawberry patch" of fifty or one hundred acres, or a "fruit yard" of thirty or forty thousand peach-trees, is no uncommon sight, upon which hundreds of busy hands are engaged in gathering and shipping the fruit. Nor do the attractions of this peninsula end with the fruits it produces; for it is the paradise of sportsmen. Both fish and game abound. Canvas-back ducks are specially plentiful in the autumn; and the immense yield of oysters in its bays is a notable feature.

At Cristfield, the great oyster mart, fleets of vessels are constantly arriving and departing during the shipping season. Nearly surrounded by water, with the restless current of the Gulf-Stream flowing along its coast, the temperature is warmer in winter and cooler in summer than in other sections of the same latitude.

CRISTFIELD, MD. OYSTER SHIPMENT.

The "Branch" to the Peninsula diverges from the main line at Delaware Junction, thirty miles from the City of Philadelphia.

This road, commenced in 1832 from Baltimore to Port Deposit, was consolidated as the Philadelphia, Wilmington, and Baltimore, and finished to Philadelphia in 1838, since which time it has steadily improved in its character and equipments, and advanced in popular favor.

The cars leave Philadelphia from the Company's depot, corner of Broad Street and Washington Avenue, and also at its New York connec-

RIDLEY STATION

tion, corner of Thirty-first and Market Streets. Within the last decade the suburbs along this line have been remarkably improved.

The stations along the road are a noticeable feature. These are not only placed at convenient intervals, but great artistic taste has been displayed in their design and decoration. The result is obvious. Persons of like tastes — wealthy citizens of Philadelphia, seeking suburban homes — choose localities of which they can speak with pleasure; and the emphasis with which they allude to "our depot" is always interesting.

As a result of this liberal movement on the part of the Company, many elegant villas and country residences have been erected along the line of the road. That of Mr. F. O. C. Darley, the celebrated artist delineator of American life and scenery, whose illustrations have gladdened so many hearts, is subjoined as a sample of the picturesque character of these structures.

The scenery along the line of the Philadelphia, Wilmington, and Baltimore Railroad is a fine, rolling country, divided by woodland and field. Many stately mansions, with their numerous outlying buildings, crown the ridges, about which lingers an air of independence and comfort. The through train glides along, flashing past farms and stations, affording many broad and pleasing views of the Delaware River at the left.

RESIDENCE OF F. O. C. DARLEY, ESQ.

Wilmington is the chief city of the State of Delaware, has 30,840 inhabitants, and is pleasantly located on land sloping down to the river. The town is regularly built, and has many churches and public buildings, and is the seat of considerable manufacturing interest.

There are many pleasing localities in the neighborhood. The *Brandywine Springs* are near, and as a summer resort have a fine local fame. The waters are recommended by the medical faculty; and good hotel accommodations will be found.

THE WILMINGTON AND READING RAILROAD diverges from Wilmington, and, continuing up the valley of the Brandywine, intersecting the Philadelphia and Reading Railroad at Birdsboro', opens up a route of great historic interest and scenic beauty. Eighteen miles beyond Wilmington, at **Elkton**, we cross the famous *Mason and Dixon's Line*. "Little did Charles Mason and Jeremiah Dixon dream, as they set that tangent

MOUNT ARARAT.

point for the determination of boundary-lines of three States, how famous they would become." We are now rapidly approaching the Susquehanna at Havre de Grace, a region filled with game, and the shooting-ground for sportsmen. The Chesapeake Bay here meets the Susquehanna River; and through hundreds of estuaries the bay intwines with the outlets of the river: the coast line is often lost in winding bays, or among the projecting headlands and foliage-clad shores. A good locality for camping-out parties.

Havre de Grace is also noted for its magnificent bridge over the Susquehanna, 3,273 feet long, which in itself is worthy a chapter; and its scenery possesses many points of interest, particularly if the tourist has the time for a short detour up the *Port Deposit Branch Road* to Perryville. The route leads under an overhanging cliff called *Mount Ararat*, which strengthens and gives fresh variety to the scenery.

Continuing from Havre de Grace to Baltimore, the rolling character of the scenery increases, with many fine water views, and much of interest to the tourist. This is the direct line from the North to the South.

ILLUSTRATED PLEASURE ROUTE No. 20.

Baltimore and Washington to Fortress Monroe, Hampton Roads, and Vicinity.

FORTRESS MONROE.

FORTRESS MONROE and its immediate surroundings to Americans certainly, and to a large number of Europeans, has now become classic. It was here the Army of the Potomac first landed in Virginia, on its memorable march up the Peninsula, undergoing all the terrible trials and sufferings of a protracted and bloody war; it was here that four years later this same army embarked for the homes they so longed to reach, with peace again ruling o'er the nation; here, too, under the eyes of thousands of anxious watchers, the great battle of the "Merrimack" and "Monitor" was fought; many yet resident at Old Point give vivid descriptions of every event in that mighty conflict, and mark the steps of its progress to the listener with vivid scenes of it before him.

The fortress itself — the largest in the United States — is a grand feature in the attractiveness of the locality, and contains within it many objects of the greatest interest to the visitor. Its extensive and beautiful parade, shaded with live-oaks; its slopes, coated with green from March until November; and its garrison, the famed Artillery School, with the music of an excellent band at the morning guard mount and the evening dress parade, give to the visitor pleasures to be found at very few resorts in our country.

There is within the fortress, also, a museum containing objects of great interest to the civilian as well as to the soldier; and many hours may be pleasantly and profitably passed by the visitor in looking over the collection.

The drives in the vicinity to the Hampton Normal School, the **National Military Home**, the National Cemetery, and to and through the town of Hampton, are over good roads, and also command many exceedingly interesting landscape views. In the town of Hampton is one of the oldest churches in our country, the inscriptions on some of the tombs in its cemetery bearing date as early as 1658.

For the invalid, as well as the robust pleasure-seeker, the climate at Old Point Comfort is unsurpassed for salubrity, and is exceptionally free from great and sudden changes in temperature, in support of which is the range of the thermometer here, as taken from the notes of the Meteorological Observatory. These show an average for the past ten years of 48°, 52°, and 63° for the spring; 60°, 74°, and 76°, for the summer; 70°, 59, and 46°, for the autumn; and 45, 44°, and 42° for the winter months.

The invalid *en route* for the warmer climate of Florida to recuperate, or returning therefrom, and fearing to face the rigors of an uncertain month in spring at the North, may find a resting place at Fortress Monroe, free from all danger of sudden and violent changes in temperature. Boating and fishing may be enjoyed fully on and in the broad waters of Hampton Roads and Chesapeake Bay, and the fish are very plentiful and excellent in character. The Hampton Bar and Lynhaven oysters, deservedly celebrated wherever this luxury is known, are here found in abundance. The bathing also is very fine, the beach being of an easy and continuous slope, and unusually free from large pebbles.

The *Hygeia Hotel*, lately built at Fortress Monroe, or *Old Point Comfort*, as it was generally named by visitors in ante-bellum times, takes the place of the one of that name which was in existence before the war, and was patronized by many of the best of our people from all sections of the Union. This hotel is most thoroughly built and elegantly furnished, and its situation is admirable, far superior to the old Hygeia, and is in all its appointments every way worthy of its beautiful locality. It stands upon the beach, at the head of the broad and substantial landing provided by the National Government for the various steamers which touch here daily to land their passengers and mails. From its balconies and corridors, the view of Hampton Roads and Chesapeake Bay is unsurpassed. Even Cape Henry and Cape Charles light-houses may be seen on a clear day, or their lights by night, without the aid of a glass. Vessels of all classes, steam and sail, American and foreign, are passing at all times or riding at their anchors in sight from every room. The ever-changing scenes from the balconies are a source of never-ending interest and pleasure. Fortress Monroe may be reached daily from Baltimore by the splendid steamers of the *Baltimore Steam Packet Company* (Old Bay Line), which connect at Baltimore with through trains to and from Philadelphia, New York, and all northern points, and from New York by the elegant steamships of the *Old Dominion Steamship Company*. The steamers for Washington, Richmond, Norfolk, Cherrystone, Yorktown, and Cobb's Island, also touch here, both going and returning, affording unsurpassed mail facilities from every section of the country. In conducting the hotel, every effort is made by the proprietor to insure the comfort and pleasure of his guests, and to make the "Hygeia" in every respect worthy of patronage. The table is supplied in abundance with every delicacy of a locality rich in edibles rare in more northern latitudes; and facilities are provided for *bathing, boating,* and *fishing,* all of which can be reached literally at the doorstep of the house.

ILLUSTRATED PLEASURE ROUTE No. 21.

Philadelphia, New York, and the East, to Harrisburg, Pittsburg, the West, and South-west.

THE PENNSYLVANIA RAILROAD.

THE Pennsylvania Railroad may well be said to stand at the head of the railway system of America. Like the aorta of the human body, which connects the heart with other important organs, thence ramifying into the extremities, this road, with its branches and connections, forms the *great central line of the country.* It not only extends a greater number of

COATESVILLE BRIDGE.

miles, uses more running stock, and employs more men, than any other, but in point of construction it is considered the *model* railroad of the United States. The bridges alone, in the scientific knowledge and artistic beauty displayed in their construction, would form an exhaustive subject for description. The Coatesville Bridge, at the village of Coatesville, is a beautiful and imposing structure. It stretches eight hundred and fifty feet across a chasm, and is seventy-five feet high. Like many of the bridges along this road, it is built of iron supported by stone pillars, thus securing strength and durability with architectural beauty combined. Indeed, this is a distinctive feature of the Pennsylvania Railroad; and throughout the entire line no expense is spared in the construction of bridges and culverts.

CONNECTING RAILROAD BRIDGE,
Fairmount Park, Philadelphia.

"So far as scenery goes, no lines of railroad on the continent can surpass those running through Pennsylvania. Magnificent agricultural panoramas, beautiful river views, splendid mountain pictures, picturesque hills and valleys, lovely villages, and flourishing towns and cities, are seen in quick succession. A ride of twelve hours between Philadelphia and Pittsburg shows more interesting variety than can be seen in the same time and distance anywhere else in the United States.

"On, on, on, goes this tireless train, over a clear track, carrying the traveller by a panorama, the like of which can be found nowhere else on this continent, and probably not in the world. After having breakfasted in the Mississippi Valley, and dined at the capital of Pennsylvania, the passenger finds himself seated at supper in the metropolis of New York, where the Atlantic throbs and swells in its ceaseless activity.

"It is no new thing to say that the scenery on the line of the Pennsylvania road is beautiful, and in many places grand. Every American

BRYN MAWR.
Pennsylvania Railroad.

who travels or reads has seen or heard of it; and the pencils of many artists have labored lovingly to portray, for popular gratification, the attractions of the Alleghany mountains; the Juniata, Susquehanna, and Conemaugh Rivers, and the wonderful agricultural vales of Lancaster and Chester Counties, through which this road runs. Long sweeps of wooded hills; lofty mountains and dark ravines; picturesque valleys opening into each other; sparkling and placid waters; wide, rolling, pastoral landscapes,—follow in rapid succession. The magnificent rivers are crossed by corresponding bridges. The bold mountain ranges and wild ravines, which would have disheartened a less enterprising company, are overcome by feats of engineering skill, which, combined with the natural artistic features of the country, make this the most interesting route in America.

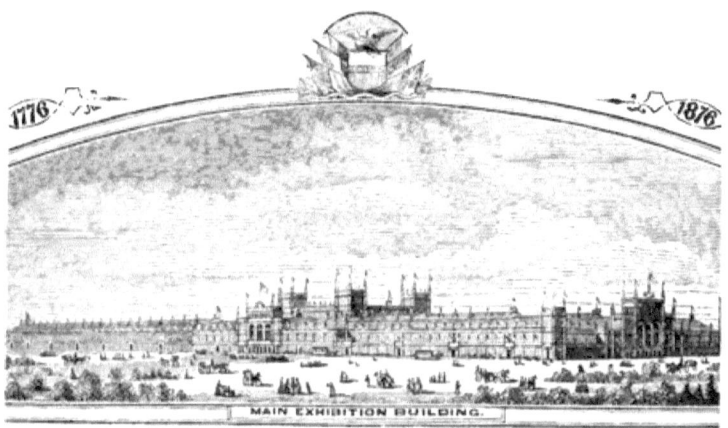

INTERNATIONAL EXHIBITION.
Philadelphia, April 19, 1876.

In the approaching anniversary of our nation's birth, when Philadelphia shall become the central point of attraction, and tourists by thousands leave their distant homes to view the spot where a few honored men dared to brave the anger of a British Crown and proclaim freedom to a nation, this line, more than any other, will bear them on their way, affording opportunity to thousands to test its accommodations and attractions. As each mile passed brings them nearer the city they will remark with surprise the remarkable rapidity of the train and the complete arrangements with which every detail of the work is accomplished. A few years ago a man who would have predicted such improvements would have been pronounced hopelessly insane, not even the most sanguine enthusiast on railroads, when their construction was commenced, dreamed of overcoming distance so rapidly and at such a rate; and it is only because of the perfection of machinery and the inventions of science that it can be done now.

FAIRMOUNT PARK, PHILADELPHIA.

Tourists, especially for pleasure, can scarcely find a more desirable spot for a day's enjoyment than Fairmount Park. Central Park of New York is older, and has received more artificial embellishments; but in size, or in the character of natural attractions, Fairmount has no equal in America. It contains nearly three thousand acres, — more than three times as large as Central Park. The beautiful Schuylkill flows through it, affording a great variety of fine water views, with admirable facilities for boating. In addition to this, the Wissahickon — famous for its charming, picturesque scenery — contributes six miles of leafy banks to its adornment. The Fairmount Water-works, which have been in operation since 1822; the celebrated Wire Bridge; the bronze statue of Lincoln; the mansion of Robert Morris of Revolutionary fame, once the resort of illustrious men, now fallen to decay, — are among the attractions. But nothing short of a personal visit can convey an idea of its beauties.

The charming Wissahickon of itself indelibly fixes for Fairmount Park a pre-eminence over any of its competitors. It is rare, indeed, that a city of the magnitude of Philadelphia can boast within its borders a "retreat" of such rich variety and exquisite loveliness. A more pleasing combination of the wild and picturesque, the grand and beautiful, cannot be found in America, than are presented on the banks of the Wissahickon.

HESTONVILLE, NEAR PHILADELPHIA

Fairmount can be reached by several lines of street railway, or *via* Pennsylvania Railroad to Hestonville, West Philadelphia, which is three miles from the depot at Thirty-second and Market streets.

STATISTICAL DATA.

The Pennsylvania Railroad (main line) formerly extended from Philadelphia to Pittsburg, a distance of three hundred and fifty-four miles. Now it has its eastern termini at New York, Philadelphia, Baltimore, and Washington; and unites them, by its own direct lines, with Pittsburg, Erie, Cleveland, Toledo, Chicago, Cincinnati, Indianapolis, Louisville, and St. Louis. Connections are also made with St. Paul, Duluth, Omaha, Denver, the cities of California, and with Memphis, Mobile, and New Orleans.

To transact its extended and diversified business, the Company now owns, and runs upon its own lines, eleven hundred locomotives, one thousand passenger cars, and twenty-six thousand freight cars. It owns two thousand miles of completed road, besides the other thousands which it controls. Its workshops cover an area of more than five hundred acres. It employs a vast army of men, many of whom are mechanics and experts of the highest skill. It has two hundred and twenty-two foreign ticket-offices (and agents, independent of those at its own stations), established in thirteen different States. Its chief officers have been civil engineers; and they employ in their service thoroughly practical men.

It is from the proceeds of a business of such mammoth proportions that this Company are able to overcome difficulties along portions of their line, which it would be folly for a less wealthy corporation to undertake. A knowledge of these advantages induced the formation of a new organization, known as the "Pennsylvania Company," having for its object the consolidation and harmonious management of all roads under its control. This company, with a capital of $12,000,000, was organized by the election of Thomas A. Scott, Esq., as president. In 1873 it had nearly five thousand miles of railroad under its control.

CONSTRUCTION.

Ordinarily companies are satisfied if their roads are graded with sand or gravel. The Pennsylvania uses, in addition, eighteen inches of broken stone, in which the ties are embedded. This insures a dry, elastic, permanent bed, free from dust, allowing the car doors and windows to be kept open in summer. Steel rails of maximum weight are used, connected at the ends by plates, bolted to the sides, but so arranged that expansion or contraction will not cramp the rail. The joints are made *between* ties; thus insuring an agreeable elasticity, which rails secured in a "chair," *on* the ties, never have.

The Pennsylvania Railroad Company, in the construction of its road, employs the highest grade of engineering talent, and the best skilled labor: hence, although its course along the streams and through the mountains follows a tortuous route, a rate of speed can be maintained with impunity which would be actually dangerous on most roads in the country. Although the original cost of construction is much greater, the Pennsylvania Company finds it more satisfactory, and believes it to be cheaper in the end. The "stone ballast" allows the water from the heavy rains of autumn to percolate through it, leaving a dry bed for winter, and therefore free from the annoying frost upheavals, and consequent displacement of rails, as well as from the dust of summer, to which clay and gravel ballasted roads are subjected. "Jumping the track" is never known on the Pennsylvania Road. This is believed to be due to the thorough construction of its bed.

The accommodations provided for summer tourists on the line of the Pennsylvania Railroad are unsurpassed. Good hotels in all the towns reached by it are the rule, not the exception; and many of them are elegant in all their appointments. It would be difficult to select any highway of travel anywhere that can compare, in the essentials of comfort, safety, expedition, and interest, with the magnificent system of railroads managed by this Company.

ARDMORE STATION,
Pennsylvania Railroad

The day express from Pittsburg to New York is a wonderful result of engineering skill.

This magnificent run of four hundred and forty-four miles is made with but three stoppages, — the first, of only five minutes, at Altoona, after a stretch of one hundred and seventeen miles; the second, of twenty minutes for dinner, at Harrisburg, after an unbroken dash of one hundred and thirty-two miles; and the third and last, of only five minutes, at Philadelphia, after a run of one hundred and five miles, leaving a single stretch of ninety miles across New Jersey to destination. No time being lost in stopping, the wonderful locomotive-engines work away with the regularity of fixed machinery, — *taking their supply of water from the track-tanks as they go, and carrying their fuel with them;* and the time is made by uniformity of progress more than by an increased rate of speed. The train is made up of Pullman parlor cars and the best of the Company's day coaches, all splendidly upholstered, mounted on combination springs, and furnished with plate-glass windows, through which the landscape can be distinctly seen.

But it is, perhaps, more natural to take the tourist, in imagination, from the eastern to the western termini. A description of the route from Washington and Baltimore will be found elsewhere, in an article on the Northern Central Railroad. That train joins us.

The tourist will early remark the tasteful yet substantial character of the stations on the Pennsylvania Railroad. These are usually built of stone; and for artistic effect several kinds are frequently used.

The thorough construction of the road-bed will also be noticed; the "stone ballast" and other features tending to give strength and durability to the structure. But the first peculiar *sensation* will be experienced as the train, with unabated speed, dashes around a curve in the road; and the oscillating movement of the car instinctively causes you to attempt to overcome it. But confidence soon assumes her sway; and what was at first a cause of fear becomes a source of pleasure. Chester Valley furnishes the first grand view of landscape scenery. The cars pass along an elevated ridge on the outer rim of this magnificent amphitheatre, affording a landscape of peculiar grandeur and loveliness. From Philadelphia to Lancaster the road leads through an undulating country, interspersed with fine cultivated farms, fields, and forests, while thriving villages and flourishing manufactories enliven the scene.

The Columbia Branch, which intersects the Pennsylvania Railroad at Lancaster, connects at Columbia with the York Branch, and at York with the Northern Central Railroad, affording the most direct route from Philadelphia to Gettysburg, a very popular resort.

A few minutes spent at Lancaster for refreshments, and the train is hurrying on to the banks of the Susquehanna, which we strike a short distance below Harrisburg; and until we reach that place the river is constantly in view. On the opposite banks the trains on the Northern Central Railroad, with passengers from Baltimore and Washington, can be seen.

A few miles above Harrisburg, the Pennsylvania Railroad crosses the Susquehanna over a bridge 3,845 feet long, affording an imposing view of the river and surrounding scenery. At this point the mountain barrier has been forced asunder by some mighty convulsion; and the grand old stream, having united its forces above, moves calmly on to the sea. This mountain gorge is the gate which opens to fields beyond of untold wealth and beauty. From this point, by the Northern Central and Philadelphia and Erie Railroads, on the east bank of the river, we can continue a hundred miles along this beautiful stream, with scenes ever changing ever new, and yet so beautiful that every turn presents a charming picture. Here bold, precipitous mountains, with overhanging rocks, crowd down to the river's bank, around which we quickly glide: again the hills recede to the blue distance, giving place to rich, cultivated fields and cosey farm-houses. But our course leads up the *west* bank of the Susquehanna, the most magnificent river in Pennsylvania, to the mouth of

THE BEAUTIFUL JUNIATA.

Turning up this lovely stream, whose praises have been sung by the poet's muse, we flash along its banks, around the hills, and through the valleys, catching, as we go, glimpses of picturesque villages, quiet vistas, and charming landscapes, stopping at last at Altoona.

"Altoona is situated at the head of Logan Valley, immediately at the base of the main Alleghany Mountains, and is the location of the principal construction and repair shops of the Pennsylvania Railroad Company. This Company has, in fact, created the town; and the business it concentrates here not only sustains it, but stimulates it into prosperity and rapid growth. The shops of the Company are among the most extensive and complete on the continent, and in themselves are objects of more than ordinary interest, illustrating, as they do, the perfection of American railroad management."

LOGAN HOUSE, ALTOONA, PENN.
Pennsylvania Railroad.

Southward from Altoona runs a system of branch railroads, penetrating the extensive iron deposits and rich agricultural valleys existing in Blair County. The manufacture of iron is extensively carried on all through this region, some of the establishments being large and complete. The limestone valleys are highly improved, and among the most productive in Pennsylvania. Wedged in between the eastern spurs of the Alleghanies, they are surrounded by picturesque scenery, and enjoy an atmosphere of more than ordinary purity. The streams flowing through them are fed by mountain springs, and are deliciously cool and clear, affording favorite homes for trout; and the angler finds the locality one of his paradises.

At Altoona, our sharpened appetite appeased at the **Logan House**, one of the many excellent hotels built by the Company for the accommodation of travellers, and re-enforced by an additional locomotive, with undiminished speed we dash up a grade of ninety-five feet to the mile. Bold precipices and deep chasms threaten our further course; yet up, up the mountain-side we climb,— this load of living freight, along a route which the frightened deer would have shunned a half century ago. But now our course is surely barred: high precipices tower above our heads; and the roar of a wild torrent can be heard through the mists of the deepening valley below. As we enter a gorge, the mountain flings itself in our path; but, turning to the left, on we fly. We pass the famous "Horseshoe Bend," and the race is won. By engineering skill, that charm of science, this seemingly impassable gulf is safely passed. The train soon enters a tunnel of 3,872 feet, and emerges on the western face of the mountains. We stop at "**Cresson**" the popular summer resort, near the summit of the Alleghanies, and leave the visitors to this delightful "retreat."

MOUNTAIN HOUSE, CRESSON, PENN.
Pennsylvania Railroad.

"**Cresson**— situated almost on the summit of the Alleghany Mountains, where they are crossed by the Pennsylvania Railroad, at an altitude of two thousand feet above the level of the sea— is a very popular resort during the hot months of summer.

The accommodations provided are of the best kind; the surroundings are attractive; the atmosphere is deliciously cool and pure." The primeval forests, with which the place is surrounded, are permeated by a labyrinth of paths, cosey nooks, and rustic seats. Berries of the most luxuriant growth abound; "and, in brief, it would be difficult to find a more delightful retreat from the stifling heat of cities in midsummer, than is here provided by nature and art combined. Several springs of medicinal waters flow from the mountain in the vicinity; and pleasant drives lead away through the almost unbroken forests, where the laurel spreads its wreath of blossoms in spring, and resinous hemlocks and pines give forth their aroma, and sigh their ceaseless music. The old Portage Road, with its ten inclined planes — once an American wonder, but now abandoned — crossed the mountain very near to Cresson, and in its ruins possesses great interest for all who note the advance of improvement. A short distance from the place, and accessible by stages, is **Loretto**, a centre of Catholic faith and education, founded by Father Gallitzen, a prince of the noble house of that name in Russia, who retired into this wilderness, and devoted his fortune and his life to the cause of religion."

On resuming our westward course with rapid speed, the downward grade is passed: yet the Westinghouse Air Brake controls the train, giving security to its movements, and ease and confidence to the tourist. The scenery from the Alleghanies to Pittsburg, though fine, does not compare with the remarkable combination of the beautiful, the grand, and the sublime to be found east of the mountains.

Pittsburg is located at the junction of the Monongahela and Alleghany Rivers, the head waters of the Ohio. It commands an immense inland navigation, and possesses remarkable geographical advantages. It is situated in the heart of the bituminous coal formation; and the location of extensive beds of iron ore is equally favorable. Pittsburg is also an important railway centre, besides being one of the most thriving manufacturing cities in the Union. The scenery here is bold and striking.

The **Branch Roads** of this Company reach some of the most delightful summer resorts in the United States, and carry the traveller through scenery as beautiful as can be found on the continent. At Harrisburg connection is made with the Cumberland Valley Railroad, which passes through the lovely Cumberland and into the great Shenandoah Valley of Virginia. At Huntingdon trains are in waiting to convey visitors to the famed Bedford Springs; and at Tyrone connection is made with the Bald Eagle Valley and Clearfield Roads, which run through regions unsurpassed in picturesqueness. Indeed, as before stated, to give any thing like a connected sketch of the scenery reached by the Pennsylvania Railroad and its branches, would require a book of itself.

ILLUSTRATED PLEASURE ROUTE No. 22.

From Washington and the South, through Baltimore, Harrisburg, Williamsport, and Elmira, to Watkins Glen and Niagara Falls.

WASHINGTON, D.C., FROM ARLINGTON HEIGHTS.

A BIRD'S-EYE view of the metropolis of the nation, from a spot associated with so many historic memories, can scarcely fail to interest the tourist. Beneath the central dome of the picture are supposed to congregate the assembled wisdom of the nation, drawn to Washington to deliberate upon the important questions of the day. Here, too, assemble the beauty and gayety of the country, whose encouraging presence gives zest to the debate, and whose cheering hospitality gives relief from the labors of the hour.

THE CAPITOL stands on an eminence ninety feet above tide-water. The site was selected by George Washington; and the corner-stone was laid by him Sept. 18, 1793. It was first occupied by Congress on the third Monday of November, 1800. On the 24th of August, 1814, the interior was destroyed by fire by British troops. In 1850 Congress voted an extension, the corner-stone of which was laid by President Fillmore, July 4, 1851. The statue of America, which surmounts the present dome, is two hundred and ninety-six feet and six inches above the ground. The plan of the city embodies two sets of streets, one set running with the four *cardinal points,* across which the avenues run diagonally.

The stranger should also visit the Executive Mansion; State, Treasury, War, Navy, Post-Office, and Interior Departments; Patent Office; Smithsonian Institute, &c.

In times past a visit to Washington from remote sections of the country was an event of no ordinary character. Weeks were consumed in performing what a few hours will now accomplish. Perhaps the most important improvement of recent date is the completion of the **Baltimore and Potomac Railway**, by the **Northern Central**, and its great ally the **Pennsylvania Railroad**, from *Baltimore* through *Washington* to *Quantico,* Va., a point of junction with the "Richmond, Fredericksburg, and Potomac" Railroad, securing an unbroken railway connection with the South Atlantic and Gulf States, the North, and the Great West.

ENTRANCE TO BALTIMORE TUNNEL.

This line connects with the Northern Central, by tunnel through Baltimore, at the northern limit of the city. Here also connection is made with the Union Railroad, designed as a connecting link with the Philadelphia, Wilmington, and Baltimore Railroad, and through it with the railroad system of the Middle and New-England States, by which trains now pass through Baltimore without the aid of horses.

The **Northern Central Railroad**, which commences at Baltimore, and runs almost due north through Maryland and Pennsylvania, penetrating into the State of New York as far as Canandaigua, offers to the tourist and traveller a variety of beautiful scenery unsurpassed on the American Continent.

The Excursion.

Leaving Baltimore at 7.30 in the morning, provided with elegant coaches, Westinghouse air-brakes, and all the modern appliances which add to the comfort or safety of the passenger, we move through the northern suburbs of the city, and pass along "Jones Falls," a small stream, seemingly insignificant as it creeps lazily over its rocky bed, but which has caused the city fathers of Baltimore much anxiety as to "what they will do with it." Seven miles from Baltimore we break suddenly upon "Lake Roland," a small but beautiful sheet of water, the source from which a populous city draws its supply. For twenty miles we pass through a rich limestone valley, where the abundant crops give token of the richness of the soil. The road follows the tortuous course of the stream, affording at every turn new and pleasing views. Thirty-five miles from Baltimore we cross the Maryland line, and enter Pennsylvania. Passing through a rich agricultural district, filled with substantial farm-houses and small villages we arrive at Hanover Junction. At this point connection is made with the railroad to Gettysburg, thirty miles distant, where that memorable battle was fought that obtained for it the name of the "Waterloo of America," and which will make it a place of interest for all time.

FROM THE TIDE WATERS OF THE SOUTH

TO THE PICTURESQUE REGIONS OF THE NORTH.

Soon after leaving Hanover Junction the spires of the borough of York are seen in the distance. We are whirled rapidly along, now and again catching glimpses of the substantially built portion of the town. York is a thriving borough of some twelve thousand inhabitants, the county-seat of York County, and the centre of a rich farming district.

During the invasion of Pennsylvania in June, 1863, it was laid under contribution by Gen. Early. Twelve bridges on the line of the Northern Central Road were burned at the same time, and nineteen on the "Wrightsville Branch," which extends from York to Wrightsville, and thence by the "Columbia Branch" to Lancaster, where connection is made with the Pennsylvania Railroad.

After a stop at York of five minutes we are away again; and ere long we reach the banks of the beautiful Susquehanna. On we speed, the noble river on our right flowing calmly onward to the sea. We are now approaching Harrisburg, the capital of Pennsylvania. The river, as we glide along its banks, with its attractive scenery, its islands and rocks, with the town beyond, affords a view of unusual beauty.

THE SUSQUEHANNA.

The numerous islands and huge masses of rock with which the broad bosom of the river is studded lend the charm of variety to the scenery. Now and again long bridges span the noble stream; villages are seen on the opposite shore; and for fifty miles we have a changing panorama of river scenery.

Harrisburg is reached by the Northern Central, over a bridge a mile in length; the trains running on the top. Nothing obstructs the prospect up or down the river; and the slow rate of speed allows a good five minutes' view of rare loveliness from a car window. Harrisburg is the point of connection with the "Pennsylvania," the "Cumberland Valley," and the "Philadelphia and Reading" Railroads. Here a magnificent Pullman Parlor Car is added to our train, and a coach from Philadelphia. These are run daily between Philadelphia, and Watkins and Rochester, N.Y.

HARRISBURG, PA
Northern Central Railroad.

The city of *Harrisburg* is pleasantly located, overlooking the Susquehanna, which is here spanned by two fine bridges a mile in length, connecting Harrisburg with *Bridgeport*, from which point the above view was sketched.

Having made connection with the train from Philadelphia, we again continue our course up the banks of the Susquehanna. The scenery is one unbroken panorama of loveliness, — a combination of views, either of which would make the reputation of any popular resort. Approaching *Sunbury*, the conductor calls out, "Passengers for Shamokin, Mt. Carmel, &c., change cars." This is the terminus of the main line of the Northern Central Railway, one hundred and thirty-eight miles from Baltimore; and here connection is made with the "Philadelphia and the Erie" Railroad, extending in a northwesterly direction through Pennsylvania to *Erie*, a distance of two hundred and eighty-eight miles. This road, forty miles of which unite the main line of the Northern Central with its leased roads north of Williamsport, is leased and operated by the Pennsylvania Railroad. From Sunbury we pass up the valley of the West Branch of the Susquehanna to Williamsport. A sumptuous dinner awaits us at the "Herdic House," which is one of the best hotels in Pennsylvania. Fifty-two miles above we reach *Renova Springs*, a resort of great beauty and growing popularity.

RENOVA HOUSE.

At *Williamsport* the "Northern Central" road leaves the Susquehanna, continuing northward up *Lycoming Creek*, which it crosses nineteen times in twenty-six miles. High hills broken into a thousand forms hem it in, many rising into mountain-peaks which cut sharply against the sky. Waterfalls spring from their rugged sides, and are lost in the tangled growth below. The valley is narrow, in places a mere cañon, yet rich in the washings of ages. *Ralston* is the first place of note on this division.

The cool and invigorating atmosphere, the grand scenery, excellent trout-fishing, and good fare have already secured for this locality an unusually large number of guests during the summer months.

This is a good illustration of the scenery in the vicinity. This stream is famous for the number, variety, and beauty of its waterfalls. The whole region possesses great attractions for the artist and lover of nature. It is also noted for the number of its trout brooks and abundance of fish, affording an excellent opportunity for the follower of Izaak Walton to beguile the lonely hours far away from the haunts of men; also an abundance of deer and other game in their season. Is it strange, then, that this picturesque region is rapidly growing in favor with tourists and pleasure-seekers?

Four miles above Ralston we reach *Roaring Creek*, a name significant of the wild scenery of the neighborhood. A good hotel furnishes accommodations for visitors. A more productive soil covers the hills; many are cultivated far up their slopes. Occasionally wild torrents are seen hurrying down the mountain sides, at places forming beautiful cascades. We are nearing the summit; the country opens; and broader fields meet the eye. The water with the descending grade now turns northward; and the train flies rapidly on. The village of *Canton* is next reached. We have now fairly left the mountains; broad cultivated fields stretch far up the neighboring hills, which with gentle undulations surround the town. A picturesque stream finds its way through the valley opening to the east.

Minnequa Springs, a popular summer resort, is only two miles beyond.

DUTCHMAN'S RUN, RALSTON, PA.

Minnequa is chiefly noted for the medicinal character of its waters. *Alba*, the next station, is a quiet hamlet nestled among the hills. From this place to *Troy* the scenery possesses little to interest the traveller; but *Troy* is a delightful village, handsomely located, and contains many beautiful private residences and several churches. *Sugar Creek* flows through the village, which adds to the picturesque beauty of the place.

From Troy to *Elmira* the railroad runs through an agricultural country with valleys flanked by high hills. These become less abrupt as they meet those bounding the valley of the *Chemung*, which, crossing at right angles, extends nearly fifty miles east and west, between the *Susquehanna River* and *Painted Post*; *Elmira* being about midway, the largest and most thrifty city in Southern New York. It has a population of twenty thousand. The city is built on an extended plain, bounded by chains of lofty hills, some of which admit of cultivation to their very summits, while others are crowned by heavy woodland. The streets are broad, crossing each other at right angles, and are lined with shade-trees. The *Chemung River* flows through the midst of the city, and is spanned by three iron bridges, one of which is completed. Elmira, the seat of several very prosperous manufacturing interests, is also surrounded by a rich agricultural region, and is the great railway centre of Southern New York. The "Erie" running east and west, and the "Lehigh Valley" and "Northern Central" from the south, unite here, and continue north and westward, connecting with the "Great Western." The "Utica, Ithaca, and Elmira" Railway will also have its terminus here. Many features of mechanical industry will interest the tourist. The Elmira Rolling Mills are the most important in the State, and turn out daily large quantities of railroad rails and merchantable iron. The La France Rotary Pump and Steam Engine Manufactory is attracting great attention among practical men, and will repay an examination from those interested. The Pullman Car, which contributes to the comfort and enjoyment of so many, is built here; also the Erie Car Shops, and many wholesale boot and shoe manufactories, all testifying to the growing importance of this manufacturing centre.

Elmira is also an immense coal-distributing point, for both anthracite and bituminous from the "Lehigh Valley" and the "Northern Central" Railroads. The Pittston and Elmira Coal Company, and Langdon & Co., with principal offices in this city, handle over half a million tons of anthracite annually; while the MacIntyre Coal Company mine and ship through this city three hundred thousand tons annually of bituminous coal from Ralston, Penn., fifty miles south.

Elmira's educational institutions are celebrated. The Female College has a national reputation; and its buildings and grounds are an ornament to the city.

We are rapidly approaching a section of country noted for its deep gorges, or glens, cut far down the solid rock by the action of running streams. These form a great variety of water views, of which the accompanying cut of **Empire Fall**, *Glen Excelsior*, is a fine illustration. These falls are on the east side of the valley, near the head of Seneca Lake, not in sight from the road, but are visited from Watkins in a small steamer which plies on the lake for the accommodation of excursion parties.

EMPIRE FALL — GLEN EXCELSIOR.

Empire Fall consists of a series of cascades, falling 106 feet to the valley below. *Havana* is the most important village passed between Elmira and Watkins, and is but four miles from the latter place. It is a quiet, pleasant village, and is remarkable for its glens, waterfalls, and cascades. The falls leap from a great height almost into the streets of the village. There is also a mineral spring at Havana. The intricacies of its glens contain many interesting features; and, whenever the hotels are improved, this must become a place of great resort.

We are moving directly north. The narrow valley from which we have debouched has opened to a mile in width, and is so level that one can but think that the lake, near at hand, at some time covered it. The ground rises on either side precipitously, beyond which evidently lies a table-land drained by streams, which, finding their way to the brink.

Engraved expressly for "Bachelder's Popular Resorts, and How to Reach Them."

RAINBOW FALLS, WATKINS GLEN, N.Y.

Rainbow Falls is, perhaps, one of the most interesting features of the Glen. With the bright sheen of a summer's day playing in the rising mists, the scene is frequently clothed in rainbow tints, but nowhere with such brilliant hues or perfect arch as at *Rainbow Falls;* and the hour of four on every afternoon finds a crowd of guests worshipping at its shrine far in the depths of *Watkins Glen.*

Above the house, the Glen extends for miles, embodying many remarkable features. The "Cathedral" is the most imposing. This is an immense amphitheatre, with walls of solid rock rising to the perpendicular height of three hundred feet, while the forest trees with which the top is fringed stretch their arms far over the yawning gulf. Into this mighty chasm the waters spring with a frightful leap, bathing its sides with feathery spray, then quietly spreading over the rocky floor. The atmosphere, even in the hottest day, is cool and moist. Trees of primeval growth, hardy shrubs, and luxuriant vines cling with wild forms of beauty from the interstices of the rock, reflecting their rich foliage in the emerald pools beneath, while far above is seen the bright blue sky; and at times the rich sunlight, reflecting from cliff to cliff, clothes all with a soft, mellow glow.

HECTOR FALLS, SENECA LAKE, N.Y.

The interest in this region is by no means exhausted with a visit to Watkins Glen. *Havana Glen*, already alluded to, is reached by coaches from the Glen Mountain House. It possesses many curious and interesting features, and will well repay a visit. *Hector Falls* is also a point of interest, and should be included in the visit to Empire Falls.

These are situated on the east side of Seneca Lake, but a few miles

GLEN MOUNTAIN HOUSE,
Watkins Glen, N.Y.

distant, and are reached by a small steamer. Neither should the sojourner at Watkins Glen miss a sail on Seneca Lake, one of the most beautiful bodies of water in the world, varying from one-half to six miles in width, and forty miles long. It is of remarkable depth and purity, and in the coldest weather never freezes over.

SENECA LAKE, N.Y.

From Watkins, trains on the "Northern Central" continue along the western shore of Seneca Lake forty-seven miles to Canandaigua, N.Y., where connection is made with the New-York Central Railroad for Albany or Niagara Falls. The route is pleasant, and possesses many points of interest, especially along the shore of the lake, where several waterfalls will be pointed out by the intelligent conductors, if requested. By an admirable arrangement, the cars run through from Baltimore to Rochester. From Rochester to Niagara Falls the country is level, and has few attractions for the traveller. Soon we hear the "roar of Niagara;" and, if our journey is uninterrupted, in sixteen hours after leaving Baltimore, we may be domiciled at some of the mammoth hotels for which the place is celebrated.

THE SOUTHERN STATES.

The proper and just representation of the pleasure routes and resorts of the "Great South," in this volume, is attended with no little difficulty, from the fact that heretofore comparatively little has been done by the "powers that be" to develop by illustrated descriptions the scenic merits of that section of the country; hence it is that many a Southern tourist who leaves his native State to study the beauties of popular Northern resorts, discovers with chagrin that the place he has come hundreds of miles to see possesses no more natural interest than another in his own neighborhood, whose attractions no one had fully observed or thought to develop. This fact is also applicable to many other sections of the country. The most wonderful as well as beautiful features often remain a long time unnoticed. This will be perfectly illustrated by the following incident of the author's experience. In the autumn of 1873, while making a sketch of Hooksett Falls, N.H. (see index), a farmer, engaged near by, left his work to watch the progress of the drawing; and as line after line developed the scene, — the river, the falls, the mill, and passing train, — he became exceedingly interested; and, when finally a few bold dashes of the brush clothed it with effect, he exclaimed excitedly. "By 'gosh!' if that ain't a picter! I was born here, and have lived here all my days, and have looked up at that river a thousand times, but never knew it was a *picter* before." And so it chances that many of the most delightful localities are yet hidden from the admiring gaze of pleasure-seekers, because those most interested in their development have not yet discovered a "picter" in them.*

VIRGINIA.

Beyond the fact, that there is "a great Natural Bridge," and "the White Sulphur Springs," and that "Virginia is (or was) the Mother of Presidents," the world knows little of the inviting features and boundless charms of that great and justly famed State. Indeed, heretofore the people of Virginia, from their familiarity with the scenery, realized but little of the innumerable treasures of romantic beauty and pictures of sublime grandeur their State in fact contains. But one of the results of the late war has been, to change very much of all this

* When the author commenced to write and re-arrange the present volume, it was with bright anticipations and hopes that the Southern States would be well and fully illustrated, as he had assistance promised from the officials of ten different Southern railroads. But, alas for human expectations! nine of them, for just and good reasons undoubtedly, "went back on him;" but fortunately one (the Chesapeake and Ohio) remained, enough to prove his assertion that the pleasure resorts and pleasure routes of the South only need development to compare favorably with any in the land.

ignorance and indifference. Virginia is fast rising, as a star of the first magnitude, in the horizon of "summer resorts." Few States, in fact, equal her in the diversity, surprising character, and interest she affords to the lovers of the marvellous and the picturesque ; and fewer still can excel her claims in this respect.

UNIVERSITY OF VIRGINIA

Sharp eyes watched the wonderful resources developed by the South in the emergencies of the late struggle. The realities of their lines of communication, and their undeveloped possibilities, were then seen and pondered. The result is, that abundant capital has made Virginia a new battle-ground since peace came among us ; and the struggle for victory is now upon the fields of keen competition for the carrying trade of our mighty granaries in the West, and even of the Orient. Necessarily, the attractions to be found along the routes, and contiguous to them, now springing into a more vigorous life, become a part of the incentives to travel, — the sources of business and profit. Hence, among other new things, the old neglected delights of mountain and forest, river and glen, are becoming fresh to the knowledge of our people. And this, as yet, is but an entering wedge to greater things for proud "Old Virginia."

Richmond will be the starting point for the great body of visitors who may decide to "do" the natural wonders of Virginia. That historic city is a place of rare beauty of location and natural surroundings. It is really built — as was Boston — upon three hills: Union, Shockoe, and Gamble's Hills. Between Union and the other two hills there is a deep valley. James River runs between Richmond and Manchester, with a moaning wail that is never at rest. It is here that fourteen miles of "falls," as they are termed, are ended. In summer the river is sometimes swollen to a great height, by vast volumes of water poured into it from tributaries above during heavy rains or freshets. From Union and Shockoe Hills, comparatively little opportunity to view the better scenic effects can be had. But from the farther verge of Gamble's Hill, — and, better still, from some of the verdure-clad margins of the canal above the city, — views can be found that baffle all description. The waters of the river dashing and foaming over its rugged and rocky bed, the inter-

EARTHWORKS ON THE CHICKAHOMINY (NEAR RICHMOND)

spersed islands loaded with luxuriant vegetation, the winding canal, the great smoking Tredegar Works, the city in graceful outline and beauty upon the left, with the soft bland zephyrs peculiar to mornings and evenings in that climate, make up as sweet a scene and picture as eye could desire. Those who can obtain access to the top of the Capitol will find a grander view, but not a lovelier one, than that from Gamble's Hill.

Richmond and vicinity is now stocked with wonderful things, such as only become famous by the tragedies of great wars. Visitors will wish to see the stupendous cordon of fortifications with which the city was encircled, the Capitol and the delightful grounds around it, and its statuary. They will stand amazed at the immense Tredegar Iron Works; look at "the Armory" where formerly the "State Guard" were held in readiness against negro insurrections, and at the churches, and other public buildings; but the chief and deep interest will be in Libby Prison, in Jefferson Davis's residence, and in all the novelties which are stamped with the terrible impress of the great war. The trip from New York City, by the Old Dominion Line of steamers, *via* Norfolk, by the James River, is a sinuous and interesting route, interspersed with pleasing views. From Washington the course is along the Potomac by boat, and cars through battle-scarred Fredericksburg. From the south the entrance is upon the great line passing through Petersburg, also the scene of terrible and bloody struggles.

VIRGINIA SPRINGS.

Elsewhere we give a synopsis of the most generally known, yet known too little, mineral springs of Virginia. Of the deservedly high claims many of these healing waters have upon public confidence and patronage, there is not the least question. The "White Sulphur" have had a long and wide fame; and it is *the* place of fashionable favor in the State, and is not likely soon to lose its prestige. Of the other mineral springs the general public know little. When the great capitalists find time to cipher over the promise of profit in that region, or when the proprietors decide to develop them by illustrated descriptions, the million will come to know all about them, and take them into favor. As yet, however, many of these springs are not prepared to entertain sumptuously. Besides, several of them are to be reached, in the main, by stages over Virginia roads; and, while those roads may promote health among the dyspeptic, they are not favorable to temper-al sweetness, as all people of experience thereon will declare. Still there is a sociability and good cheer in stage-coach riding never found in a railway car.

Virginia also has her full share of mountains, rivers, falls, caves, and glens. The *Natural Bridge* in Rockbridge County has long been a source of wonder and admiration, yet has failed hitherto to command the attention from tourists that its grand and impressive character would seem to inspire. This is undoubtedly due largely to the want of proper and popular hotel accommodations. The *Peaks of Otter*, Bedford County, having an altitude of 5,307 feet above tide, each year increase in popularity. The *Hawk's Nest*, Fayette County, with its precipitous face, falling sheer down a thousand feet to the bed of the stream that flows at its base; *Weyer's Cave*, Augusta County, with its wonderful sights; the *Natural Tunnel*, Scott County, passing hundreds of feet through solid rock, with its high vaulted roof, stupendous precipices, and weird caverns; and thousands of other natural features, — promise for Virginia a popular future with the sight-seeing public. Many of these have been brought to notice by the opening of the Chesapeake and Ohio Railroad.

PLEASURE ROUTE No. 23.

Washington, Norfolk, and Richmond, to Staunton, the Mineral Springs of Virginia, Charleston, Huntington, Cincinnati, and the West.

CHESAPEAKE AND OHIO RAILROAD.

It should be remembered that Virginia is pierced by two ranges of mountains, the Alleghanies and the Blue Ridge, and between these lies

COMMISSARY DEPARTMENT.

a great valley, three hundred and sixty miles long. Such a region, so vast, could only be a great storehouse of Nature's curiosities.

Commercially studied in connection with the future of Virginia as a famous resort for travellers and summer tourists, it strikes the eye at once, that the leading artery of communication across the State will make some port on the Chesapeake its grand depot. In the waters of that magnificent bay, Virginia possesses the only complete harbor facilities on the entire South Atlantic seaboard. Heretofore, no determined purpose to seek commercial power can be said to have animated

VIRGINIA SCHOOLHOUSE OF THE OLDEN TIME.

Virginia councils. Jefferson discountenanced commerce, and favored agriculture. He sneered at great cities as "great sores." And Jefferson's authority was undisputed. Hence no great commercial port, or marine, ever grew there. But the **Chesapeake and Ohio Railroad** has taken the field, with its headquarters at Richmond; and a new life is rapidly dawning upon the State. This line will cut through the very heart of those attractions for summer money-spenders, which so profusely abound in the mountainous regions of Virginia and West Virginia. Tourists from Washington can reach Richmond by rail, and from New York and the North by rail or the Old Dominion Line of steamers; or they can strike the line of the "Chesapeake and Ohio" over the Orange and Alexandria Railroad at Gordonsville, west of which lie most of the great natural resorts of Virginia. From Gordonsville to Charlottesville (twenty-one miles), the cars of the Chesapeake and Ohio and the Orange and Alexandria Railroads, running upon a double track, follow the same route. From Charlottesville the Chesapeake and Ohio strikes due west thirty-nine miles to Staunton, the central point first to be reached by visitors to the principal resorts in this section of Virginia.

THE PASSAGE OF THE BLUE RIDGE.

This range stretches from north-east to south-west, across Virginia, cutting the line of the Chesapeake and Ohio Railroad, its crest generally twenty-five hundred feet above tide; and it offers no "water gap" between the Potomac and the James. To the west lies the valley of Virginia, its streams on this line eight hundred feet above the affluents of the James which flow at its eastern base. It was a formidable obstacle to the railroad. In 1849 the State of Virginia undertook to build the Blue Ridge Railroad from its eastern base at Mechum's River to Waynesboro' in the valley, a distance of seventeen miles. The main feature was a tunnel 4,260 feet long at the summit of the grade, which required seven years to complete, and which until recently was the largest finished tunnel in the country. The engineer was Claudius Crozet, a Frenchman of great intellect and attainments, and this work is a monument of his skill. The road cost about a hundred thousand dollars per mile, and is now a part of the Chesapeake and Ohio Railroad.

Leaving Mechum's River, the road ascends along a spur of the main ridge, which it reaches in about five miles; thence, encountering deep ravines and projecting spurs, it climbs along the steep mountain slope, making three lesser tunnels, and still ascending through another, reaches an elevation of fifteen hundred feet above tide at its western portal. The grade of the road is seventy feet per mile, less than this on the curves and greater on direct lines; the tunnel is straight, and passes seven hundred and fifty feet below the crest, and is so well ventilated that one can often see entirely through it from the rear of passenger trains. As you approach the tunnel, winding up the mountain's side, the view is surpassingly grand. You look down upon cultivated fields eight hundred feet below, stretching away for twenty miles, dotted here and there with plantation-houses and villages, and bounded in the distance by rugged mountains, a panorama the eye never tires in gazing upon, and which is by common consent one of the grandest views on the line, and surpassed nowhere. The road descends from the tunnel for three miles to the western base, across the South Branch of the Shenandoah, and crosses with undulating grades the Valley of Virginia to the North Mountain, its western boundary.

At Staunton, a hundred and thirty-six miles from Richmond, a connection is made with the Valley Railroad extending to Harper's Ferry. This neighborhood is noted for the picturesqueness of its scenery, and

TUNNEL.

ample hotel accommodations are furnished visitors at the Virginia House.

Augusta or **Stribling** Springs are only twelve miles away by a romantic road reached by stage. This is also the point from which to visit **Weyer's Cave**, seventeen miles distant, one of Virginia's greatest curiosities. This cave is among the fantastic eccentricities of nature,

and as such stands unrivalled. It is one of the discovered wonders of
our country, and affords a prolific field for thought and conjecture. Possibly, in the time whereof no man knoweth, this, and like subterranean
glories, may have been palaces for nymphs and their families. Indeed,
as one looks over the torch-lit scene, and sees a million eyes glittering
and flashing seemingly upon him, it is easy to fancy the wondering
spirits of the place still there and marvelling at the strange intrusion.

ROCKBRIDGE ALUM SPRINGS, VA.

The **Passage of North Mountain** was less costly than that of the
Blue Ridge; it is crossed without a tunnel. At the summit the grade
reaches an elevation of 2,074 feet above tide; there are pleasant landscape views, some of great beauty, some very rugged; and on the right
towers *Elliot's Knob*, 4,450 feet above the level of the sea, probably the
highest point in the State. At Goshen, thirty-two miles west of Staunton, and one hundred sixty-eight from Richmond, the road, after running
nearly parallel to the North Mountain for many miles, turns abruptly
westward, and passes through *Panther Gap* in the Mill Mountain, known
farther north as the Shenandoah. The scenery here is wild and interesting. At *Goshen* passengers leave the railroad for *Lexington*, twenty miles
distant, and **Natural Bridge**, thirty-five miles; and also for **Rockbridge Alum Springs**, nine miles distant; *Rockbridge Baths*, nine

miles. These resorts are all reached by stage over picturesque roads, and each worthy a visit from the tourist, and a fuller description. Seven miles west of Goshen, and one hundred and seventy-five from Richmond, is Millboro'. Here passengers for the **Bath Alum** Springs, ten miles; **Warm**, fifteen miles; **Hot**, eighteen miles; and **Healing Springs**, fifteen miles,—may take coaches; although the two latter places may be more easily reached from Covington, thirty miles farther west. Leaving Millboro', at an elevation 1,680 feet above tide water, the road encounters very heavy work. Within a few hundred feet of the station it passes through a tunnel 1,300 feet long, and with a short interval through a second; then crosses a great ravine by an embankment one hundred and sixty feet in depth; then through cuts of eighty feet or more, and over another embankment about one hundred feet in depth all of rock; and through a third tunnel,—all within three miles, one of the most costly sections of the line, and containing a series of magnificent views. It then descends a long and wild valley to the *Cow Pasture River*. Here at a sharp bend the river has been turned through a new channel, which cost sixty thousand

GRIFFITH'S KNOB AND COW PASTURE RIVER.

dollars, saving two bridges for the road, which cuts across another sharp bend by a tunnel, and immediately crosses the river. After some more heavy grading Jackson's River is reached at Clifton Forge. Jackson's

FALLING SPRING FALLS.

River with the Cow Pasture forms the James, a few miles below. It has here cut its way through the mountains forming, one of the wildest of gorges, always admired, and which is well worth a special visit. Indeed, it is difficult for the reader, guided by imagination alone, to realize the scenic grandeur of this route. Continuing up Jackson's River, and crossing it twice, it reaches *Covington* at the base of the Alleghany, two hundred and five miles from Richmond. Passengers for the **Healing Springs**, fifteen miles; **Hot**, eighteen miles; and **Warm Springs**, — may here take coaches, and pass on the way the **Falling Spring**, a lovely cascade. These stage-routes are through grand mountain scenery, with occasional views of Jackson River and in full view of the cascade of Falling Springs.

The stream which forms this cascade rises in the *Warm Springs Mountain*. About three-fourths of a mile from its source (says Pollard), it falls over a rock two hundred feet into the valley below. The scene is broken and exceedingly picturesque.

PASSAGE OF THE ALLEGHANY.

Covington was the eastern terminus of the Covington and Ohio Railroad, a work undertaken by the State as a connection between the East and the West. Its construction was suspended during the civil war; and in 1868 the amount expended on it (over three million dollars) was given to the Chesapeake and Ohio Railroad Company, on condition that the latter should complete it. It now forms an important part of the line. The passage of the Alleghany is a stupendous work, and always excites admiration. After crossing Jackson's River a third time, the road follows the valley of Dunlap's Creek for five miles, crossing the creek four times with bridges of a hundred and thirty feet opening, and makes two tunnels and several very heavy cuttings. It then ascends by a grade of sixty feet per mile a rugged mountain slope, with excavations and embankments following each other in rapid succession, unequalled, it is believed, in this or any other country. There are a great number of cuts of sixty feet in depth, many over a hundred, and the slopes of some reach even to a hundred and fifty feet above grade. Each of these interesting result of engineering skill adds new and fresh artistic features for the tourist's gratification. The embankments are equal in magnitude. That over "Moss Run" is a hundred and forty feet in depth; over Jerry's Run, a hundred and eighty-five feet. The lower slope stakes of the latter were over two hundred feet below grade. These streams cross the road by tunnels cut in the rocky sides of the ravines, tunnels ample in size to pass a railroad train. The embankment at Jerry's Run contains over a million cubic yards of material. It is on a gentle curve, which is produced on either side for some distance, and views of it can be had from several points on the road; but even this huge mass is dwarfed by the mighty hills which surround it, so insignificant are the works of man when brought face to face with those of Nature. After crossing the great embankment, the road follows a ravine, and in a short time enters Lewis Tunnel. This is about four thousand feet long, and is cut through hard rock, through a spur called the Little Alleghany, and, including its approaches, is seven thousand feet from grade to grade. It was worked from two shafts as well as from the ends, and is a continuous grade of sixty feet per mile. After passing this tunnel the road encounters almost immediately the main Alleghany, which is pierced with a tunnel nearly forty-eight hundred feet in length, passing four hundred and fifty feet belows its crest. The summit level is on the eastern side 2,000 feet above tide, and from its ditches the water flows at will, either to the Atlantic or the Gulf of Mexico.

GREENBRIER WHITE SULPHUR SPRINGS, WEST VIRGINIA

The grade descends through the tunnel at the rate of thirty feet per mile; and the road, now on the affluents of the *Kanawha*, descends with the streams, losing nothing by undulations, nowhere exceeding thirty feet per mile; and after clearing the Alleghany twelve miles west of the tunnel, with nothing over twenty feet per mile until steamboat navigation is reached (until sixteen miles below Charleston, in fact), a distance of a hundred and sixty miles, with gentle curvature, forming one of the grandest and best constructed roads between the East and the West.

At *Alleghany Station* passengers leave the railroad by stage for the **Sweet Chalybeate Springs**, nine miles; and **Old Sweet Springs**, nine miles; and **Red Sweet Springs**, eight miles,—each popular with its patrons. Four miles west of the tunnel are the **White Sulphur Springs**, twenty-two miles from Covington and two hundred and twenty-seven from Richmond. The White Sulphur Springs, now opened to easy access from every part of the country, will long maintain their royal prerogative, as the high domain of Virginia aristocracy and fashion. Compared with Saratoga in landscape beauties, the latter sinks out of all pretence of just comparison. Saratoga is as barren of natural graces and woodland beauty, as the White Sulphur region is profusely endowed with rural delights and a sweet atmosphere.

The hotel accommodations here are simply enormous, but with increased patronage, induced by improved railway facilities, even these must fall short of the demand. In the twenty-two miles before reaching White Sulphur Springs, the road has passed through eight tunnels, aggregating two miles and a quarter in length; besides through and over other work encountered at every step, of the heaviest character, which has cost over four million dollars, a considerable portion, however, including most of the tunnel, being constructed for a second track. This work was planned and mostly executed by Charles B. Fisk, State Engineer, and was completed by the Chesapeake and Ohio Railroad Company. From the Alleghany tunnel, the road follows the waters of *Howard's Creek* to the *Greenbrier River*, with two tunnels and some heavy embankments and bridges, showing conclusively that the grand scenic character of the route is still maintained. It then follows the Greenbrier for thirty-seven miles, cutting off bends here and there, three of them by tunnels. The first, near Second Creek, is sixteen hundred feet long, and on a curve of sixteen hundred and thirty feet radius through a bold limestone cliff; the second, eleven hundred feet long, was in rock and earth; the third, one and a quarter miles in length, the longest on the road, saving over eight miles in distance at the "Great Bend." This tunnel was worked from two shafts and from the portals; it passes four hundred and seventy feet under the crest; one of its shafts was three hundred and seventy feet in depth; the tunnel was completed in two years and four months. The lines of the engineer, carried from the shafts and portals, varied from the true line *only a fraction of an inch*. There are three fine iron bridges over the Greenbrier, of from four hundred to five hundred feet opening. There are many striking views on this part of the line, which, alternating with the rugged mountain scenes, give additional interest to the route; the road skirts frowning limestone cliffs of great height, then dashes through quiet meadows giving charming views of the beautiful river buried among the mountains. For most of the distance the road is but a few feet above extreme high water, but on emerging from the Great Bend tunnel the traveller finds that the river by its greater length has fallen to nearly one hundred feet below him. This elevation is reduced as he approaches the junction with New River about six miles below.

THE START DOWN THE GREENBRIER.

The New River (more properly the Kanawha) rises in North Carolina, and where we join it is a large stream, broader than the James at Richmond. The road follows it for sixty miles, until it loses itself in the Kanawha, never losing sight of it save where it passes through a tunnel or a deep cut. Its valley is from six hundred to one thousand feet below the general level of the country; and it cuts below the bases of mountain ranges, whose axes cross its course. The minor streams, which, running parallel with these ranges, drain the country, plunge down precipices to reach its level. It crosses thus these mountain ranges, a mighty river often fifteen hundred feet wide, and sometimes not a hundred, falling an average of ten feet per mile, and for twenty consecutive miles of its course seventeen feet to the mile, in places a roaring torrent, its bed obstructed with gigantic bowlders, which have rolled from the cliffs above; bowlders often measuring thousands of cubic yards. At first it is a wide noble river, with here and there a strip of meadow on its bank, and here and there a cliff, but always buried in the mountains. Mountains clothed with forest, with here and there a gray crag jutting out: this is its character for ten miles. Suddenly it plunges twenty-four feet at the Great Falls, a grand waterfall. Its valley is here contracted: it is bounded by overhanging sandstone cliffs more than a hundred feet in height, and falling back from these are still the rugged slopes of great mountains. Here the engineers had some trouble. The traveller will see it, and he will see rocky glens opening to his view as he passes,

RICHMOND FALLS, NEW RIVER, VA.

lighted by falling cascades, and unfolding scenes rich in landscape beauty; but it would require too much space to describe such details. The valley becomes narrower as we descend, the mountains more elevated; but the bends of the river are gentle, and for more than fifty miles but one tunnel is passed, this at Stretcher's Neck, where a tunnel of nineteen hundred feet saves four miles in distance. This tunnel is on a curve. The western end gave great trouble in its construction. After the arch was completed, a slide from the mountain crushed portions of it and the stone portal, and an interior arch became necessary. It is now perfectly safe, but reduced to single track dimensions for eighty feet. At the western portal of this tunnel may be observed the greater height above the river due to its longer course and the gentle grade of the railroad. One mile below on the right is a beautiful cascade where Dowdy Creek ("What's in a name?") joins the river. At Suttle's Cliff, still lower down, there is a hundred and thirty feet cutting, but it is embankment on the river side. At *Sewell*, formerly known as Boyer's Ferry, forty miles below the mouth of Greenbrier, the valley becomes a true cañon. For twenty miles below there is not a strip of arable land in the valley, and at points the cliffs are perpendicular from the river edge. Here the scenery is wild indeed. Such slopes as these are generally covered with bowlders, some of immense size; and along these slopes and under

WHITCOMB'S BOWLDER

frowning cliffs the railroad gropes its way under their shadow; one of which, Whitcomb's Bowlder, it was proposed to tunnel as the easiest solution of a location of the road, and it was actually under-cut on one side for trains to pass. The passenger looks upwards a thousand feet or more as the train sweeps a graceful curve around some concave bend, and sees the beetling cliffs of many colored sandstone, looking with their great angles like gigantic castles and fortresses erected by Nature to guard these her penetralia.

The running of New River Rapids, once a common occurrence, will soon become one of the things of the past, and its recital placed among the events of "a good while ago." Before the completion of the railroad, this means of transportation was much in vogue. The accompanying illustration will convey a fair idea of the passage. "The boats are admirably adapted to the work they have to do.

RUNNING NEW RIVER RAPIDS.

More than sixty feet long by less than six feet wide, they accommodate themselves very well to the impetuous torrent, and are for these waters excellent seaboats. The steersman, clear-eyed, skilful, cool-headed, as he needs to be, winds a parting blast on his horn as your boat is let go. Three negroes make up the whole crew: the headsman, who is in fact the captain, stands in the bow to direct the steersman by waving his arms; the steersman, who guides the boat with a long and powerful oar; and a third man who accelerates its speed with oars, in the eddies and smooth parts of the river. The headsman or captain knows the river, and he not only directs the course of the boat, but has besides under his own control an oar projecting from the bow with which in sudden to slue the boat more quickly around." Tourists will, "for the novelty of the thing," occasionally try the passage of New River Rapids; for although seldom attended with serious results, the occasion is one of excitement and adventure.

MILLER'S FERRY, FROM THE HAWK'S NEST.

Among the beautiful cascades, that of Fern Spring. about six miles below Sewell, is much admired. The stream, often very large, pours down the mountain side for hundreds of feet in volumes of spray and foam. The river is very rapid: at low water, viewed from the cliffs. as from the Hawk's Nest, a white thread in the cañon; in floods, a raging torrent of astonishing power, rising at points, thirty, fifty, and even seventy feet above its low-water level. At Hawk's Nest the river is crossed by a fine iron bridge, with six hundred and fifty feet opening and fifty-three feet above low water. It cost, with its masonry, one hundred and forty thousand dollars. From this point for eight miles the rugged character of the cañon is intensified: the Hawk's Nest towers six hundred feet above the road. The roadway is formed by blasting down precipices, the grade being from fifty to seventy feet above the river. It passes through tunnels. Streams of water falling from the mountain sides form beautiful cascades. The whole forms the grandest sight to be seen on this line of road, and one never to be forgotten. From this rugged pass the river glides into a lakelike opening. It soon receives the Gauly, and loses its name in the Kanawha.

CHARLESTON, WEST VIRGINIA.

Charleston, the capital of West Virginia, is a thriving and rapidly growing town, which promises to become in time an important manufacturing place, for which it has good natural facilities. After leaving the Kanawha, sixteen miles west of Charleston, the road crosses to the Ohio, and terminates at the Big Sandy River, the boundary of Kentucky and West Virginia.

HUNTINGTON, OHIO RIVER.

In its course the road passes through twenty-six tunnels, five of which are four thousand feet and more in length, the total amount of tunnelling being over seven miles.

The reader who has followed this article cannot fail to notice the wonderful richness in popular resorts of the region through which the Chesapeake and Ohio Railroad runs. Although so fully described, there are still many places having great local popularity, which have been barely alluded to, and others for want of space entirely omitted, which in a future edition should receive more attention, both in illustration and description.

This new avenue of communication will develop untold sources of wealth from the hitherto land-locked seclusions of both Eastern and Western Virginia.

PLEASURE ROUTE No. 24.

Richmond, Washington, and the North, to Wilmington, N. C., Charleston, Savannah, Jacksonville, and New Orleans.

ATLANTIC COAST LINE.

THE Southern States are intersected by three great arteries of travel, leading directly from the north-east, so that the traveller or tourist, with his objective point as far distant as Mobile or New Orleans, and his starting point at New York or Washington, may have his choice out of the three, with the certainty in either case of rapid transit and a passage through an interesting section of country. Each through line is made up by the consolidation of lesser ones, and is connected with the others by branch lines. They are severally known as the **Atlantic Coast Line**, the route being by the way of Richmond, Wilmington, N.C., Charleston, Savannah, and Jacksonville, or New Orleans; the **Piedmont Air Line** by way of Danville, Charlotte, Atlanta, and Mobile to New Orleans; and the **Great Southern Mail Route** by the way of Bristol, Knoxville, and Chattanooga, to New Orleans and the South-west.

There are few places of popular resort on the first-named line before reaching Charleston, although the traveller with plenty of time will find it well employed in stopping at some of the more important stations for a day or longer. Among the watering places of lesser note, which lie scattered through the eastern part of North Carolina, either upon the Coast Line, or attainable by the various intersecting branches, are **Lack's Springs**, about two miles distant from Macon Depot, on the Raleigh and Gaston Railroad; **Panacea Springs**, about the same distance from Littleton Station, upon the same road; **Cleveland Mineral Springs, Causilor's Springs**, and **Sparkling Catawba**, in Lincoln County, and **Sulphur Springs**, in Gaston County, along the line of the Wilmington, Charlotte, and Raleigh Railroad. These last are tolerably well developed. **Lewis' Mineral Springs** lie on the Wilmington and Weldon Railroad, five miles from Faison Station.

Spring Church, locally celebrated for its medicinal waters, lies three miles distant from Pleasant Hill Station, on the Petersburg and Weldon Railroad, just over the North Carolina border. There are here two or three good boarding-houses, but no hotel. The village is famous as having once been a popular duelling ground for the North Carolina chivalry.

The tourist who wishes to visit the battlefields of the State can stop at **Mount Olive**, on the Wilmington and Weldon link of the line.

Within twelve miles of this station were fought the battles of Bentonsville, Whitehall, and Neuse River Bridge, the former quite an important engagement.

Wilmington lies twenty miles from the sea, on Cape Fear River, and attracts a portion of the Southern winter travel. At the mouth of the river are situated *Forts Fisher* and *Caswell*, both of which played an important part in the scenes of the Rebellion. Besides communication by rail, there is a weekly line of steamships to New York.

Camden, on the Wateree River, is a beautiful South Carolina town, possessed of considerable historic interest. Two battles were fought here in the time of the Revolution: one between Gates and Cornwallis, and the other between Greene and Rawdon. The remains of Baron De Kalb, who was killed in the former battle, lie under a handsome monument on the village green.

Columbia, the capital of South Carolina, is a pleasant little city. Nearly half its present residents are from the North. It was burnt by Sherman on his "march toward the sea," but has been rebuilt in a far more tasteful manner. The **Congaree Falls** are a few miles distant, and attract a large number of visitors annually. There are several good hotels in Columbia, the principal being the Nickerson House and the Columbia Hotel. The more prominent stations between Columbia and Charleston are interesting to the tourist, mainly on account of their Revolutionary associations. At **Orangeburg** are to be seen the remains of the works erected by Lord Rawdon, for defence, after the fall of Charleston.

The **Healing Springs**, three miles from Blackhill Court House, after passing Orangeburg, are widely noted for their remedial qualities. They are three in number: an iron, a sulphur, and a seltzer spring. The climate is remarkably equable and healthful, resembling very much that of *Aiken*. The accommodations are fair. At the Court House there are three hotels, churches, telegraph-offices, &c.

Farther on is **Eutaw Springs**, the locality where a well-known Revolutionary battle was fought. At Branchville, sixty-eight miles above Charleston, the Charleston and Augusta Railway takes the tourist or invalid to **Aiken**, a place which within the past few years has become extensively known for the salubrity and remarkable healing qualities of its climate. A dozen years ago it was a mere village; now it is a large and flourishing town. Hundreds of families from the North make it their residence during the winter months, while the permanent population has doubled within the past five years. The climate seems to have an extraordinary effect on those afflicted with pulmonary diseases, and many remarkable cures have been effected by that alone. The visitor will find an excellent hotel and numerous boarding-houses of a select character. There are many delightful views in the vicinity, and no better society can be found anywhere in the South.

Charleston is considered by many the most beautiful city of the South. It lies at the confluence of the Cooper and Ashley Rivers, both broad and beautiful streams, and seen from the harbor presents a charming appearance. In the summer some of the streets are perfect forests of palmettos and magnolias, which cast a grateful shade over the stately old-fashioned mansions of the city, rendering them delightfully cool and pleasant in the hottest of the weather. The public buildings are none of them worthy of especial mention. The visitor will find the older part of the city interesting, its general character being unlike that of any other in the country. The best general view of the old quarter, and, in fact, of the whole city itself, is from the tower of *St. Michael's Church*, on the corner of Broad and Meeting Streets. This, the oldest church in Charleston, was built in 1752, and its tower is regarded as the finest in the South. Its chimes are peculiarly sweet, and can be heard for a long distance. Another fine view is to be had from *St. Phillip's*, on Church Street, a structure nearly as old as St. Michael's. In the ancient graveyard connected with St. Philip's Church are buried some of the most famous men of the State: Calhoun, Christopher Gadsden, William Moultrie, the Pinckneys, Rutledges, Middletons, and others equally well known.

The two most popular hotels of the city are the *Charleston* and the *Mills House*, both first class in every respect. There are others less stylish, but with quarters as comfortable and tables as abundant.

The **Forts** are objects of attraction to tourists, especially since the war. Castle Pinckney stands at the entrance of the harbor. South of it lies Fort Ripley, built during the first years of the war, principally of palmetto logs; beyond lies Moultrie. Inside the harbor, grim, battered, and still defiant, stands Sumter. On Sullivan's Island may still be seen the ruins of works erected before the Revolution, as a defence against the French and Spaniards.

Sullivan's Island, in Charleston Harbor, is the great fashionable watering place of the State, and it is difficult to imagine any thing more delightful than a sojourn here in the heated season. Before the war there was a splendid hotel upon the island, the Moultrie House; but it was destroyed during the struggle, and has not yet been rebuilt. There are scores of beautiful villas and cottages, however, already erected, and every season sees the number largely added to. A steamer runs hourly between the island and Charleston during the "season." The beach is admirably adapted for sea-bathing, and there are many fine drives in the neighborhood. *Mount Pleasant* is another favorite resort of the Charlestonians.

Between Charleston and Savannah there is no place of note which would attract the tourist. **Savannah** is one of the finest and most rapidly-growing of all the Southern cities. Its streets are much broader than those of Charleston, and its squares, or *plazas*,—twenty-four in number,—filled with trees and adorned with beautiful shrubs and flowers, render it a paradise in all seasons. The climate is exceedingly healthful; and many from the North, who are troubled with pulmonary complaints, resort there during the winter months. The suburbs are delightful, and are permeated by scores of beautiful drives. Four miles out is *Bonaventure Cemetery*, on the Warsaw River, renowned for its picturesque beauty; and a mile farther is **Thunderbolt**, a famous summer resort for the residents of the city. *Fort Pulaski*, on the south side of the entrance to Savannah River, is well worth a visit. The principal hotels are the *Pulaski House*, the *Marshall House*, and the *Screven House*.

The tourist over this line will hardly care to stop between Savannah and Jacksonville, if his purpose is to visit Florida. Trains with palace cars attached run through without change, or passage can be taken by steamship which runs weekly between the two ports. The only resort on the railroad worthy of mention is **Suwannee Springs**. These have become within the past ten years very popular, and large numbers of invalids and pleasure-seekers gather there every summer. There are a hotel and numerous boarding-houses. The springs are located directly on the line of the Florida division of the Atlantic and Gulf Railroad, at Suwannee Station.

Jacksonville is situated on the western bank of the St. John's River, twenty-five miles from its mouth, and has a population of about twelve thousand. Since the war it has grown rapidly, large numbers having emigrated from the North. There are a dozen or more hotels of fair character, and many boarding-houses.

The St. John's River is the principal attraction of Florida, and is much frequented by tourists. Large excursion steamers run up from Jacksonville as far as Pilatka, seventy-five miles, daily, and from there light-draft steamers continue to Enterprise, and to the interior lakes and tributaries.

Green Cove Springs, on the west bank of St. John's River, thirty miles from Jacksonville, is the best known watering-place in Florida, and is being rapidly developed. The spring, which gives its name to the place, discharges a volume of water which is estimated at three thousand gallons per minute. The temperature never changes, summer or winter, remaining always at 78° Fahrenheit. The basin of the spring is about forty feet in diameter, the water, which is very strongly impregnated with sulphur, being from twenty to twenty-five feet deep in the centre. A line of steamers run from Jacksonville daily, and there are excellent hotel accommodations for those who wish to prolong their stay.

Silver Spring is one of the most remarkable localities along the river. It is large enough to be navigable, and the water is more transparent than glass.

Lake George, formed by a sudden widening of the river a hundred miles above Jacksonville, is noted for its romantic beauty. It is dotted with picturesque islands, which are covered with foliage of tropical growth.

Green Spring, an extraordinary sulphur spring of a green color, is located at Enterprise, on the east bank of the St. John's River, 205 miles from Jacksonville, and at the head of steamboat navigation of the river. Enterprise is becoming one of the most important points in the South for tourists and sportsmen. Here is the grand gathering-place for fishing and hunting parties who propose to visit the Indian River country. The fishing and hunting of the region is unsurpassed by that of any other locality in the South, or even in the United States. There is one really first-class hotel, and several smaller, with boarding-houses. Enterprise is reached only by steamer from Jacksonville, or by the way of St. Augustine and Tocoi.

Fernandina has the reputation of being one of the most bracing and healthful places in the South, and has for years been resorted to by invalids and consumptives. It has three or four respectable hotels and many boarding-houses.

St. Augustine is one of the most interesting places in the whole South for the tourist to visit. It is the oldest settlement in the United States, having been founded by the Spaniards in 1565. The city and suburbs are rich in historical associations, many of the buildings dating nearly three centuries back. The climate is magnificent. There are several hotels, the best of which is the *St. Augustine*. The approach is from Jacksonville to Tocoi, on the St. John, and from thence by rail.

Florida is one of the most delightful camping-out grounds for pleasure parties in the country, a subject well worthy the attention of those fond of rural sports.

The amount of pleasure travel to Florida the past season is remarkable, with a prospect of steady increase.

PLEASURE ROUTE No. 25.

Richmond, Washington, and the North, to Danville, Charlotte, Atlanta, Mobile, New Orleans, and the South.

THE PIEDMONT AIR LINE.

THE tourist South who, taking the Piedmont Air Line from Richmond, desires to visit the principal places of resort, will find himself constantly obliged to leave the main road for branches and stage-lines leading to them. As a general thing, these deviations are pleasant, as they afford a change of scenery and sometimes of locomotion, and enable one to see more of the real character of the country than would otherwise be the case. The Piedmont Air Line combines several important roads under one general management, and, with its connections, furnishes an uninterrupted route of travel from New York to Florida, passing through Virginia, North and South Carolina, Georgia, and Alabama. At prominent points it connects with lines running east, west, north and south, affording innumerable opportunities for digression.

The first point of divergence after entering North Carolina, to reach special points of interest, is at **Greensboro'**, a thriving manufacturing and mining town. The tourist can here take the North Carolina Railway for Goldsboro', and beyond, to Newbern and Beaufort, points which, though possessing no particular attractions, are yet much visited since the war. **Raleigh**, the capital of the State, lies on this road, and is well worth a brief visit. The State House, which is modelled after the Greek Parthenon, was at one time considered, next to the Capitol in Washington, the finest legislative structure in the United States. Two excellent hotels are to be found here. **Hillsborough** is another interesting stopping place. Here the Provincial Congress at one time held its sessions, and later it was occupied as the State capital. Lord Cornwallis made his headquarters here for some time, and the house occupied by him is still standing. The locality is yet pointed out where several of the early North Carolina patriots were executed by order of Gov. Tryon. With one exception, Hillsborough is the oldest town in the State.

Returning to the trunk line at Greensboro', if the tourist wishes to visit the springs and mountain resorts of Eastern North Carolina, stage can be taken for *Salem*, a romantic little town, from which the best known of them are accessible. The scenery of the region is very fine. **Pilot Mountain** is in the immediate vicinity of Salem, and attracts many summer visitors. It stands alone, and rises like a pyramid three thousand feet above the level plain.

At Salisbury, fifty miles south from Greensboro', the tourist will once more diverge from the direct route for the wild mountain region of this State, *viâ* the Western North Carolina Railroad. This section of the State is famous for its springs, a score or more of which enjoy a more or less extended reputation for their medicinal qualities.

Morganton is one of the most beautiful of the villages of North Carolina, and affords a good base for the visitor who wishes to make excursions to the various gaps, valleys, and secluded places of the region. Magnificent views are obtained from the highlands in the vicinity.

Piedmont Springs lie only about fifteen miles distant. The waters are highly esteemed by the residents of that portion of the State, and are said to really possess remarkable curative properties. A stage runs daily from Morganton to the springs. The hotel accommodations are not extensive, but are of good quality.

Twenty-five miles from Morganton, and attainable by stage from that place, are the **Linnville Falls**, which are the most remarkable in some respects of any in the country. As yet they are little visited, save by the more adventurous of tourists. "They are about 150 feet wide at the first fall, the water making a succession of leaps before reaching a fathomless basin, three thousand feet from the summit. The view is said to be grand beyond description."

The objective point of most pleasure-tourists over this route is *Ashville*, just west of the Blue Ridge. The terminus of the railroad is at *Old Fort*, 114 miles from Salisbury. From this point stages run regularly, passing through *Swananda Gap*, a romantic route. Another means of approach is by leaving the trunk line at Charlotte, forty-three miles below Salisbury, going by way of Lincolnton and Shelby. Near the latter station is **Wilson's Springs**, a very popular watering-place in the State, and largely patronized during the hot season. By taking this route, the traveller passes through the famous **Hickory Nut Gap**, one of the most delightful mountain passages in the whole South, and destined in time to become a grand swarming-place for tourists. It is nine miles in length, and is flanked on each side by immense cliffs, rising, in places, perpendicularly from a thousand to fifteen hundred feet above the roadway.

Ashville is in no sense a watering-place, although several well-known springs are in the immediate vicinity. All the railways west of the Blue Ridge converge here. Some of the finest scenery in America lies within forty miles of the station, and the visitor can make this his base for hunting, fishing, and exploring excursions. There are two or three first-class hotels, and several excellent boarding-houses. Twenty miles from here is the famous **Black Mountain**, with its wild and attractive scenery.

Whiteside Mountain, in the valley of the French Broad River, on the route to Ashville by the way of *Jones' Gap,* is well worth the attention of the tourist. It rises abruptly to the height of fifteen hundred feet, and contains a cave which is accessible only from the top. **Cæsar's Head,** on the same route, is a yet loftier peak. *Mitchell's Peak,* within easy distance of Ashville, is the highest point east of the Rocky Mountains. It rises seven thousand feet above the level of the sea, and affords one of the finest and most extensive views in the South. In Burke County are several remarkable mountain peaks, more or less noted. **Hawk's Bill** is a huge cliff overhanging a rapidly-running stream. Table Rock and Ginger Cake Rock are both places of local resort. The whole course of the French Broad River, from Ashville to the Tennessee State line, is wildly romantic. Thirty miles from the former point are the **Painted Rocks,** a perpendicular cliff three hundred feet in height, on which are traces of figures popularly supposed to have been made by the Indians. The **Chimneys,** a series of precipitous crags, lie in close proximity to the rocks, along the banks of the river.

The celebrated **Warm Springs** of North Carolina are located in this region, on the banks of the French Broad, some thirty miles west of Asheville. The waters are said to be a sovereign cure for rheumatic and neuralgic diseases. They are very clear and of a high temperature. Game and fish abound in the region, and the scenery is especially fine. There are two good hotels and several boarding-houses at the Springs, so that the visitor will find no lack of accommodation. There are two routes of approach from the North, — one by the way of Asheville, the other by taking the East Tennessee and Virginia Railroad, leaving the cars at Greenville. The stage route both ways leads through a beautiful and romantic region, and is not so long as to be tiresome.

Below Charlotte, on the direct line, and just over the South Carolina border, is the old battlefield of *King's Mountain.* This vicinity is a favorite resort for tourists, not only on account of the magnificent mountain scenery, but for its sporting facilities. Among the interesting localities and objects of visit in the neighborhood are the **Great Falls of the Catawba, Hanging Rock,** and **Crowder's Knob,** the latter being the highest peak of King's Mountain.

Greenville, the best known of all the inland resorts of the State, lies at the foot of the Saluda Mountain, on Reedy River. Near it are some of the most famous mountain peaks whose names are familiar to tourists. **Table Mountain,** twenty miles away, rises 4,500 feet above the sea level, and shows on one side a dizzy precipice one thousand feet in height. Magnificent views are obtained from the summit, nearly every bold elevation of the mountain region of the State being visible,

— Cæsar's Head, Bald Mountain, Stool Mountain, Pinnacle Rock, Saluda Mountain, and a dozen others. The visitor will find ample accommodations, and of a fair character.

The **White Water Cataracts**, near Greenville, attract a large number of visitors yearly. They are exceedingly beautiful, and though not grand, are surrounded by the most charming expanse of country. **Slicking Falls,** in the same vicinity, have a deserved popularity, and are much frequented by pleasure-seekers during the season. Both are reached by stage from Greenville. Within easy distance is the famous *Keowee* region, and the *Jocassee Valley*, the latter being regarded by visitors as one of the most romantic and charming little nooks in the South.

Crossing into Georgia, the Air Line passes through the romantic mountains of that State, concerning which so much has been said of late years. The entire mountain region of Georgia abounds with beauties, especially that of the north-eastern portion of the State. The key to this section of the country, and the grand starting-point for tourists' excursions, is **Clarksville,** in Habersham County. Within a circuit of fifty miles lie some of the most noted mountain resorts of Georgia, — peaks, waterfalls, springs, and valleys, — most of them mentioned under their appropriate heads in another portion of this book. In the western part of the broken region is **Rock Mountain,** a summer resort of considerable local repute. It is between two thousand and three thousand feet in height, and on one side is a solid, precipitous mass of rock over nine hundred feet high. Upon the summit are the ruins of an ancient fortification, the builders of which are unknown.

Twelve miles from Clarksville are the famous **Falls of Tallulah,** or cataracts, as they are sometimes called. There are a dozen or more of them, each possessing its peculiar attractions. They are more like furious rapids than falls, the river which furnishes the water flowing through an immense gap, a thousand feet in depth, in some places completely hidden by the immense cliffs which bend over it. **Toccoa Falls,** equally celebrated as a resort for tourists, lies only about five miles distant. Both places furnish fair accommodations for visitors.

The **Falls of the Eastatia,** in Rabun County, twelve miles from Tallulah, are coming into notice, but on account of the difficulty of access are as yet little visited. The **Falls of Amicalolah,** in Lumpkin County, in the same region, consist of a succession of cascades from twenty to eighty feet in height. The **Falls of the Towalaga.** on the line of the Macon and Western Railway, are said to be particularly beautiful, and preparations are being made to furnish accommodation to visitors.

The whole mountain region of the State is thickly permeated with medicinal springs of greater or less popularity. Most of them are accessible by stage from the various stations along this line and its connections. They need to be developed to become popular.

The **Red Sulphur** — comprising twenty different springs within an area half a mile square, consisting of magnesia, chalybeate, red, white, and black sulphur — situated in the north-western part of the State, are becoming justly popular. The scenery is celebrated for its beauty, and the opportunities for hunting and fishing are better than are usually enjoyed in the vicinity of watering-places. Lookout Mountain in Tennessee, elsewhere described, is in the immediate neighborhood, and forms an objective point for summer parties and tourists. The best method of approach is from Trenton, on the Alabama and Chattanooga Railroad, the second station south of Chattanooga.

Warm Springs are located in Merriwether County, thirty-six miles north-west from Columbus. Visitors should leave the Atlanta and West Point at Lagrange Station, taking the stage for the remainder of the journey. Accommodations fair.

The **Chalybeate Springs** lie about seven miles south of the warm springs, and are reached by stage from Geneva, on the Muscogee Railroad.

Madison Springs, in Madison County, have a high local reputation, and bid fair in time to become one of the most popular watering-places in the State. There are several good boarding-houses, and the waters undoubtedly possess medicinal value. The visitor should take the stage which leaves Athens, on the Georgia Railway, for the waterfall region of Habersham County.

The **Sulphur Springs**, in the northern part of the State, are reached by the Athens Branch of the Georgia Railway, or by the Air Line, leaving the train at Gainesville. The distance by stage is seven miles.

Indian Springs, which have been favorably known for many years, and which of late have been rapidly gaining in popularity, are located in Buttes County, thirty miles south of Covington. They are reached by stage from Forsyth, a station on the Macon and Western Railway.

Powder Springs, twenty miles from Atlanta, magnesia and sulphur, enjoy some local reputation. They are reached by stage from Atlanta. **Rowland's Mineral Springs**, in Bartow County, are reached by stage, five miles from Cartersville, on the Western and Atlantic Railway. **Catoosa Springs**, a well-known summer resort, are four miles, and **Cherokee Spring** two miles, from Ringgold, on the same line of railway.

Atlanta, aside from its importance as a railroad centre, and from the war associations with which it is connected, has little to recommend it to the traveller or tourist. It is now the State capital, and contains some fine public buildings. Its principal hotel, the *Kimball House*, is said to be the finest in the Southern States. Atlanta is connected with Augusta by the Georgia Railway, with Macon by the Macon and Western Railway, and with Chattanooga by the Western and Atlantic Railway. The two former, with their extensions, connect with the Atlantic Coast Line.

There is little of interest for the general traveller between Atlanta and Montgomery, the capital of Alabama, nor, in fact, in the latter city. From Montgomery the tourist can either proceed to Mobile and New Orleans by the Mobile and Montgomery and Western Railway, or keep directly on to Jackson and Vicksburg.

Mobile, 186 miles south-west, is a busy and very pleasant city. Excepting the forts, there is little to see. The place is poor in public buildings and hotels, although the *Battle House* has a fair reputation. The **Gulf Shell Road** is a magnificent drive along the bay, between eight and nine miles in length.

Talladega Springs, the best sulphur springs in the State, are located on the line of the Selma, Rome, and Dalton road, eight miles from Wilsonville. **Shelby Springs** are about the same distance from the same station. **Chandler Springs** and **Sulphur Springs**, in the same county and on the same line of road, are distant from Munford Station, respectively, twelve and eight miles. **Bladen Springs**, well known to tourists, lie on the Tombigbee River, in Choctaw County, and can be reached by packet from Demopolis, on the line of the Alabama Central, tri-weekly. All these furnish fair accommodations for visitors, and will doubtless improve in the future.

The journey from Mobile to New Orleans is through a flat and uninteresting country. Biloxi, Pascagoula, Pass Christian, and Bay St. Louis, are resorted to more or less by the people of both cities during the season, though neither of them can be properly classed under the head of popular resorts. Considerable effort is now being made to improve resorts, and add to their attractions.

New Orleans is in many respects one of the most interesting of the cities of the South. Its early history is familiar to every schoolboy, and the prominent part it played during the late war has made it and its surroundings almost as familiar to the residents of the North as Philadelphia or Chicago. Its limits enclose innumerable places of interest. Many of the old relics of Spanish rule still exist in the older part of the city, while the numerous cemeteries, parks, walks, and drives offer a succession of attractions of which the visitor never tires. Since the war it has become popular as a winter for families from the North.

Lake Pontchartrain, five miles north of the city, is a great place of resort. Large parties go there to hunt and fish, and portions of its shores are peculiarly eligible for picnics and pleasure excursions. There is a single hotel at the terminus of the railway; but it is by no means first-class, and affords very limited facilities to permanent boarders. Trains run from the city hourly.

The **Plains of Chalmette**, the scene of the battle between the American troops under Gen. Jackson and the British under Gen. Pakenham, Jan. 8, 1815, is one of the prominent points of interest to be visited. The plains lie south of the city, on the banks of the Mississippi, and are reached by the street-cars. **Gretna** is a pleasant little suburb, lying across the river, and contains many stylish residences. The celebrated **Metaire** track, the great race-course of the South, lies two miles out of the city, on the Lake Pontchartrain Railroad.

The New Orleans hotels are among the best in the country. The most noted are the *St. Charles*, the *St. Louis*, the *St. James*, and the *City Hotel*. Besides these there are several prominent restaurants, two or three of which have a local reputation equal to that of Delmonico's in New York.

PLEASURE NOUTE No. 26.

Richmond, Washington, and the North, to Lynchburg, Bristol, Knoxville, Chattanooga, Memphis, New Orleans, Little Rock, the South and South-west.

THE GREAT SOUTHERN MAIL ROUTE.

THE third great route to the South is by the way of Lynchburg. Knoxville, Chattanooga, and Memphis.

Leaving Richmond, over the Great Southern Mail Route, the tourist passes through a portion of Virginia not only noted for its magnificent scenery, but for its historic associations, both in the past history of the country, and connected with the late war. After crossing the James River, the road traverses the great tobacco section of the State. The **Mid-Lothian** coal-mines, thirteen miles from Richmond, are worthy a brief visit by those who are interested in the natural and mineral resources of the State. *Burksville, Farmville,* and *Appomattox* are rapidly passed; the latter forever famous in American annals as the closing event of the war of the Rebellion. Those who wish to make a personal survey of the country where the grand final scene between the two armies took place can leave the train at either station, and procure a private team and guide for the purpose.

Lynchburg, 125 miles from Richmond, possesses no particular interest for the traveller. It is built upon several small but very steep hills, and from its being a railroad centre, conveniently located to the "springs region" of Virginia, has acquired considerable importance with tourists as a starting-point for the popular resorts in this section of the State. At the time of Gen. Lee's retreat from Petersburg, Lynchburg was the point aimed to be reached; and had the attempt been successful, and the western mountain passes gained, the war might have lasted a year longer. West of Lynchburg, a short ride, are the **Bedford Alum Springs,** which are considered valuable in cases of dyspepsia. One of the principal elements of the waters is sulphurate of iron. The accommodations for visitors are very good.

From Lynchburg the route extends through the south-western portion of the State, intersecting the springs region, and with its stage and rail connections giving access to most of the popular resorts and places of interest south of Richmond. South-western Virginia is famous for its wild and sublime scenery, and for its numerous natural curiosities. It is intersected by the great Alleghany range, whose slopes, precipices, and "gaps" furnish a never-ending series of surprises for the traveller who has an eye for the grand or picturesque.

Twenty-five miles below Lynchburg is the thriving little village of

Liberty, a point of divergence for travellers who wish to visit the **Peaks of Otter**, ten miles distant. These magnificent elevations are cut off from the rest of the range, rising in lofty grandeur five thousand feet above the sea level, and visible more than a hundred miles away. A portion of the ascent is made on horseback. The last half-mile is accomplished by climbing over huge precipitous rocks, bare of verdure, and looking as if ready at any moment to tumble into the depths below. The Southern historian Pollard, in describing a visit to the Peaks, says, "There are many mountains higher; there are others, it may be, with more merit or interest in the surroundings; but none which produce so terribly sublime an emotion of suspension in the sky."

Bonsack's Station, twenty-two miles beyond Liberty, is a familiar locality to tourists who have visited the springs region. From here there are several lines of stages, connecting with the Natural Bridge, thirty miles distant; Lexington, forty-two miles; Sweet Springs (the "Old" and "Red"), forty-seven miles; and the White Sulphur, sixty-four miles. **Coyner's Springs** are within a mile of Bonsack's. The waters — black and white sulphur — are considered specifics for several diseases. Coyner's is a very popular resort during the hot season, and the scenery in the vicinity is remarkably fine. Beside boarding-houses there is an excellent hotel within three minutes' walk of the station, and equally near to the springs.

Alleghany Springs, the waters of which are celebrated throughout the country as a reliable specific for dyspepsia, lie about three miles from Alleghany Station, thirty miles beyond Bonsack's. They are delightfully situated on the Roanoke River, at the eastern foot of the Alleghany Mountains, and are largely patronized. The hotel accommodations are sufficient for a thousand persons. Among the places of interest in the vicinity are the famous **Puncheon Run Falls**, where, through a huge rent high up the precipitous sides of a mountain, a foamy series of cascades leap from rock to rock, plunging two thousand feet into a black abyss below.

Fisher's View, only five miles from the springs, is a locality much favored by visitors, and the frequent scene of picnic and excursion parties. From its summit, which is easily reached, an extended view is obtained of the surrounding mountain region.

Ten miles farther on are the **Montgomery White Sulphur Springs**, one of the most popular summer retreats in the South, and patronized by the most fashionable society. They are only a mile distant from the railroad, horse-cars and coaches making constant connection between trains and hotels. Their situation is highly romantic. A broad lawn, through which flows a rippling brook, is backed by a ragged spur of the Alleghanies, over which in various directions run excellent roads which are scarcely ever dusty. Forests, groves, valleys, and mountain peaks are within easy distance. Fish are plentiful in the streams, and the sportsman occasionally is able to "draw a bead" on a deer. The hotel buildings are spacious and elegant, and within the past two years have been largely added to. One of the favorite pastimes at this resort is the "tournament," which is always an event of interest throughout the country surrounding, and is largely attended.

The Yellow Sulphur Spring is less than five miles distant from the Montgomery, and can be reached, either by stage from that place, or from Christiansburg, a station on the Virginia and Tennessee division of the road. This is noted as a quiet resort for families, the waters being celebrated for the cure of children's diseases. The spring is noted for being the loftiest in the State, — not more than sixty feet below the summit level of the mountain from which it flows. The hotel accommodations are as yet somewhat limited, but very comfortable.

Thirty-two miles from the Montgomery, and about twenty-three miles from Christiansburg, is the famous **Salt Pond**, a small but unfathomable lake among the mountains, 4,500 feet above the level of the sea. No stream leads into or out of it. **Bald Knob** is the highest peak of the ridge on whose shoulder Salt Pond is situated, and nearly a thousand feet higher. From its summit one can look into five States, — Virginia, West Virginia, Kentucky, Tennessee, and North Carolina.

It may well be asked why tourists should seek the mountain regions of the North with such a desirable locality near at hand. Bald Knob needs but to be better known to become the goal of mountain pleasure seekers. Its hotel accommodations, heretofore limited, are improving with the demands of travel, and this has already become a favorite summer home for families. The attractions are enhanced by many and various natural curiosities in the vicinity, — streams, falls, glens, and mountain views. The unwritten history of this region, as a popular resting-place, is pleasing to contemplate.

A short distance from Salt Pond are **Eggleston Springs**, too far from the railroad to be largely patronized, yet possessing peculiar attractions. Tourists who visit the region generally make this their base, deflecting to the right and left to the numerous localities of interest which lie within a radius of twenty miles.

Little Stony Falls, seven miles from Eggleston's, have a descent of sixty feet clear, and then break into a succession of short falls, an equal distance. The speed of the water is terrific, and it seems as if projected over the precipice by some tremendous unseen power high up the mountain side.

Pompey's Pillar is a curious-shaped cliff rising sheer out of New River, near the Springs, three hundred feet in height. The scenery along the banks of this stream for a hundred miles is of the most romantic character.

At Newbern, stages run to **Red Sulphur, White Sulphur,** and **Salt Sulphur Springs,** already described in the Chesapeake and Ohio route, and to the **Pulaski Alum Springs,** only ten miles distant.

At Wytheville, stages run to the **Sharon Alum Springs,** twenty-five miles distant. At Glade Spring Station, connection is made with the **Washington** and **Seven Mine Springs,** two miles distant; and from Saltville, the next station, **Red** and **White Rocks,** on Clinch Mountain, **White Top Mountain,** and **Chilhowee Springs,** are easily reached.

Bristol, which lies partly in Virginia and partly in Tennessee, is the diverging point from which to reach the **Holston Springs,** in Scott County. The distance, over a good stage road, is twenty-eight miles. Under proper management this would in time become one of the best known watering-places in the country. Its development thus far has been very slow. There are four kinds of water, common limestone, chalybeate, a thermal, and white sulphur; and the different springs are so close together that the visitor can in turn drink from each with taking a step. The hotel accommodations are fair.

Ten miles to the west of the Springs is the **Natural Tunnel,** one of the greatest curiosities in Virginia. It is reached by stage from Estillville, fourteen miles distant, on the Holston River. The tunnel is five hundred feet in length, piercing the solid rock, and in shape like the letter S. The height of the arch varies from twenty to ninety feet. The scenery in the vicinity is wildly and magnificently grand. The tourist will find excellent accommodations at Bristol, and a fair hotel at Estillville.

After leaving Virginia, and entering Tennessee, the road passes through a great variety of landscape, varying from the rugged and majestic scenery of the mountain region to the softly rolling meadows of the valleys which intersect it.

Cumberland Gap, the great highway between South-western Virginia and the adjoining States, is a wonderful work of Nature, — a huge gash through the backbone of the towering range of mountains, six miles in length, and of a width varying from twenty to several hundred feet. Precipitous mountains rise on either side to a vast height; and the traveller who journeys along the narrow roadway may, in places, look sheer up the sides of the rocky wall a thousand feet before his gaze rests on the topmost crag. During the war this gap was strongly fortified by the Confederates, who held it for months against the strategy and attacks of a vastly superior force; thus defending the railroad communications between Richmond and the South. The gap lies in **Knox**

County, Kentucky, upon the borders of East Tennessee, and can be reached by leaving the line at Rogersville Junction, taking the Jefferson Railroad to its terminus, and staging it from there.

The Cumberland Mountains in this region are honeycombed with caves, the best known of which is **Big Bone Cave**, called so from the fact that a large number of fossil bones of an immense size were found in its interior when it was first discovered.

At **Knoxville**, in Tennessee, the tourist will find himself well repaid for a day's stop. It is a small but delightful city, and was formerly the capital of the State. Some of the most important strategic movements of the war were made around Knoxville, the best remembered, perhaps, being the siege of the city by the Confederate forces in November, 1863. Between this point and Chattanooga are scattered most of the medicinal springs of the State, although none of them are on the direct line of the road.

They are many in number, yet few of them have attained any particular celebrity. Some of them, if the accounts of visitors are to be relied upon, possess remarkable medicinal qualities, and when properly developed will doubtless prove more attractive. All of those mentioned in the list below are comparatively easy of access, and possess accommodations of a respectable character. Two or three have good hotels, while the rest are supplied with boarding-houses, sufficient in numbers and capacity to fulfil all present demands upon them.

Montvale Springs, nine miles south of Maryville station, on the Knoxville and Charleston Railroad, have an excellent reputation, and are well patronized. **Sulphur Springs** lie three miles from the station of that name on the Cincinnati and Cumberland Gap Railroad. **Bon Aqua Springs**, whose virtues are of recent discovery, are situated near the line of the Nashville, Chattanooga, and St. Louis Railroad, sixteen miles from McEwen's station. **White Cliff Springs** and **Chilhowee** may be reached from Riceville station on the Eastern Tennessee and Virginia road; the stage-route to the former is twenty-four, and to the latter sixteen miles. Along the same road are **Tate Springs**, twelve miles from Morristown station; **Warm Springs**, eight miles from Granville, and **Chalybeate** and **Black** and **White Sulphur Springs**, four miles from Union depot, in Sullivan County. Stages run from all the stations named to the various places.

No part of Tennessee has, since the war, been so much frequented by tourists as that about Chattanooga, the natural gateway of the South, and in the very heart of the mountain region. The city lies on the south bank of the Tennessee, and in regard to climate is unsurpassed by any place in the Union. Hundreds of invalids spend their winters there, invariably with benefit. Within easy distance are **Lookout**

Mountain, the **Tumbling Shoals, Lake Seclusion, Walden's Ridge, City of Rocks, Sulah Falls,** and other objects of interest. Of these, Lookout Mountain is the best known and the most famous. It shoots up three thousand feet feet, pushing the Tennessee from its straight course several miles to the north, and, although six miles distant, seems to overshadow the town. From its lofty summit the tourist can, on a clear day, see into seven States. Along its sides was fought the celebrated "battle above the clouds," in November, 1863, between the Federal forces under Hooker and the Confederates under Bragg, the scene of a magnificent historical painting by the celebrated artist, Walker. The hospital buildings erected by Government on Lookout are yet standing, and are utilized in the summer by the swarms of visitors who make the place nearly as lively as in the days of the "great unpleasantness," though in a different way. Five great railways converge at Chattanooga, — the Nashville, the East Tennessee and Georgia, the Alabama and Chattanooga, the Atlanta, and the new Cincinnati Southern.

Chattanooga is connected with Atlanta by the Western and Atlanta Railway. Along this line were fought some of the bloodiest and most desperate battles of the late war, the entire distance of 138 miles having been contested step by step by the two great armies.

Huntsville, Alabama, is the next station on the main line claiming attention. It is considered by travellers one of the most beautiful towns in the South. Here is the largest spring on the continent. Nine miles from Huntsville are **Johnson's Wells,** a popular resort for invalids and pleasure-seekers. At Stevenson, the junction of the Memphis and Charleston and the Nashville and Chattanooga Railroads, the tourist can diverge from the trunk line to visit **Beersheba Springs** or the romantic **Suwanee.** A short distance below Decatur, on the Alabama Railroad, are **Blount's Springs,** whose waters and hotel accommodations are growing in favor every year. On the same road are the beautifully situated **Valhermosa Springs.** At Tuscumbia, the passenger wishing to reach **Bailey's Springs** takes a branch road to Florence, and stage thence. These springs have had a high reputation for many years, and are annually patronized by large numbers.

The **Mineral Springs** at Iuka, on the main line, are famous throughout the South for their important medicinal qualities. They are of varied character, — chalybeate, sulphur, alum, &c. There is an excellent hotel within five minutes' walk of the principal spring, with croquet ground, music stand, and various outdoor conveniences for visitors and parties. The village itself is very pleasant, and contains good schools, several churches, telegraph office, &c. It will be remembered as the scene of one of the fiercest-fought battles of the war of the Rebellion, between Price and Rosecrans, in the autumn of 1862.

Memphis is the principal city of Tennessee, and, with a population already of fifty thousand, is growing rapidly. From here the tourist may select any one of a dozen different routes. Five railroads have their termini here; while daily lines of steamers connect with Vicksburg, Cairo, Cincinnati, Louisville, and New Orleans. Its railroad lines west extend to Texas and the Indian Territory.

The three great trunk lines already described are crossed nearly at right angles by routes equally important, — the Mobile and Ohio, the New Orleans, St. Louis, and Chicago, and the two great lines from Nashville south, one terminating at Mobile, and the other, by connections, at Charleston, Brunswick, or Savannah. On the Mobile and Ohio Railway are to be found the principal watering-places of Mississippi. The best known of these are the **Lauderdale Springs,** eighteen miles north of Meridian, on the line of the Mobile and Ohio Railroad. They have many visitors in the summer months, and are highly spoken of by medical authorities, and are reached by stage from the above station. The accommodations are very good.

Three miles from Durant station, on the same road, are the **Mineral Springs,** much resorted to by those who are troubled with diseases of the liver and kidneys. Greenwood Springs, whose reputation is rather more extended, are reached by stage from Aberdeen, a distance of eighteen miles, the terminus of Aberdeen branch of the Mobile and Ohio Railroad. **White Sulphur Springs,** twenty-two miles from Tupelo, on the Mobile and Ohio road, are becoming a favorite resort for those troubled with diseases of the skin and stomach. **Godbald's Mineral Wells,** two miles from Summit station, on the New Orleans, Judson, and Great Northern Railroad, are regarded as of great efficacy in curing certain diseases.

Crossing the Mississippi, a line of railway one hundred and thirty-five miles in length connects Memphis with Little Rock, the capital of Arkansas. Sixty miles to the south-west from the latter point, in the very heart of an almost inaccessible mountain range, lie the famous **Hot Springs** of Arkansas, over a hundred in number. Some of them are only moderately warm, while in others an egg can be cooked in a few minutes. A little village has been built, and, in the summer, parties of tourists make the locality tolerably lively. It is being rapidly developed, and in a few years will take its place as the Saratoga of the State. Eight miles distant, on the Ouachita River, is what is known as the **Sulphur Spring,** where patients are sent to recover from the debilitating effects of the hot baths and sweats. The accommodations are fair. Twelve miles to the north-east are the **Mountain Valley Springs,** where a new hotel has been recently erected with accommodations for a hundred patients. The waters here are said to be an infallible cure for dropsy, gravel, and Bright's disease of the kidneys.

PLEASURE ROUTE No. 27.

From Cincinnati and the North-West to Mobile, New Orleans, and the South.

LOUISVILLE AND GREAT SOUTHERN ROUTE.

THE tourist South who has Chicago or Cincinnati as a starting-point, or who takes either of those cities in his way, finds his best route by the way of the Louisville and Great Southern line, with its extensions and connections. Several important roads from the east, the north-east, north, north-west, and west converge at Louisville as a grand central point. From here the direct southern line extends through Nashville, Decatur, Montgomery, and Mobile to New Orleans, while the southwestern line branches below Bowling Green at Memphis Junction, and extends to Memphis, with direct connection by means of the Memphis and Little Rock, Cairo and Fulton, and the International Railroads, to different points in Arkansas and Texas. From Humboldt and Memphis in Tennessee important roads connect with Mobile and New Orleans.

Leaving Cincinnati by the Louisville, Cincinnati, and Lexington Railroad, the tourist crosses the magnificent suspension bridge which connects that city with Covington, on the opposite side of the Ohio River. The span between the two towers is 1,057 feet, the most extensive in the world. The entire length of the bridge is nearly 2,300 feet. Covington is very pleasantly situated, and is one of the busiest and most prosperous cities in Kentucky.

The **Littonian Springs** are only four miles from Covington, and hardly farther than that from Cincinnati. Of late years considerable pains have been taken to add to the natural attractions of the place; and during the season the waters are drank daily by excursion parties from Cincinnati, Covington, and Newport.

At Lagrange, eighty miles below Covington, a branch road leads to Lexington, from which point many beautiful and romantic spots in Eastern Kentucky can be reached by various stage lines. Frankfort, the capital of the State, lies upon this branch. Thirty miles below Frankfort, by stage, is **Harrodsburg**, once the great fashionable watering place of Kentucky, and, indeed, of the South. It is romantically situated upon high ground near Salt River. Formerly it was greatly affected by Southern politicians, and the outline of many a campaign has been laid in the private parlors of the fashionable hotel of the village. The waters of the springs here are famous for their tonic and curative properties. The scenery in the neighborhood is lovely, the people eminently social, and the accommodations for visitors equal to those of any other place in the South.

Lexington can also be reached by the Kentucky Central Railway, from Covington, the route leading through an important section of the State. The famous **Blue Lick Springs** are reached by this road, the tourist taking stage at Paris, eighty miles south of Covington. One of the principal points of interest in the immediate vicinity of Lexington is **Ashland**, the former home of the great Kentucky statesman, Henry Clay. It resembles very little the place of thirty years ago, but still retains numberless charms for admirers of the beauties of nature.

Louisville, one of the finest and most agreeably situated cities of the middle range of States, is located on the Ohio River, at its junction with Bear Grass Creek. The streets are handsomely laid out and beautifully shaded; and some of the public buildings, as well as many of the private residences, are very fine. It has several excellent hotels, the best known of which are the *Louisville Hotel*, the *United States*, and the *National*.

The **Tar** and **White Sulphur Springs** are four miles from Cloverport, on the Ohio, and are easily accessible from that point. The **Drennon Black Sulphur Springs** lie near the Kentucky River, in Henry County. In the summer there is a regular steamer from Louisville.

At **Mumfordsville**, seventy-three miles below Louisville, is a remarkable spring and other natural curiosities which annually attract a large number of tourists. In this vicinity occurred a number of skirmishes and one or two battles between Generals Buell and Bragg, in 1862.

At Elizabethtown, if the tourist wishes to visit the **White Sulphur Springs**, somewhat famous in Western Kentucky, he will diverge from the trunk line, on the Elizabeth and Paducah Railroad, taking stage at Litchfield station, from which place the springs are but four miles distant. The waters possess considerable virtue, and during the season the place is well patronized. The accommodations are not extensive, but of a fair character.

MAMMOTH CAVE.

Returning again to the grand route, the tourist's next stop is at *Cave City*, eighty-five miles south of Louisville, and one hundred north of Nashville, on the Green River. This is the nearest point of approach to the **Mammoth Cave**, the most wonderful curiosity in the South, and, we might perhaps add, in the whole world. At Cave City will be found an excellent hotel, furnishing ample accommodations to the largest parties. Two hours' drive takes the visitor to the *Cave Hotel*, a large and well-kept house, hardly more than two hundred paces from the entrance to the cavern. There are yet other methods of reaching the locality. If the tourist choose, he can land within a mile of the cave by taking the excursion steamer from Louisville; or he may take stage at Bowling Green, passing over twenty miles of excellent road and through a most romantic section of country.

There is an immediate change of temperature noticed on entering the cave, the air feeling pleasantly cool, but not chilly. Summer and winter, the thermometer always stands at 59 Fahrenheit. A person has never been known to take cold from a visit to the cave. It is a singular fact, that no impure atmosphere exists in any part of it, while decomposition of animal and vegetable matter is unknown. It is estimated that there are over three hundred miles of walks, avenues, and galleries already explored in connection with the cave, while there is a possibility of a much larger territory yet untrodden by the foot of man.

The *Great Vestibule*, a short distance from the entrance, is an immense hall, nearly two acres in extent, over a hundred feet high, and with not a single central support. Its dome is lost in inky blackness, which the torches of the guides fail to wholly dispel. A mile farther on the *Church* is reached, a saloon three hundred feet in circumference, with a height of nearly seventy feet. Here is a natural pulpit, and behind it a large gallery, as if especially designed for a choir. Divine worship has many times been held in this subterraneous cathedral, and concerts are often given by visiting parties. A short distance beyond the church the avenue divides into two passages, one leading to *Gothic Avenue*, and the other a continuation of the main cave. The avenue derives its name from its resemblance to the interior of a Gothic cathedral. It is fifty feet wide, fifteen feet high, and two miles long. For a large part of the way the ceiling is as smooth and white as if laid on by the most skilful trowel. The purity of the atmosphere of this portion of the cave renders it a most desirable promenade for invalids Fifty years ago two mummies were found in a niche at the farther extremity of the avenue, in a perfect state of preservation. One was a female in a sitting posture, with her arms folded across her breast. They were clandestinely removed, but by whom, or where they were placed, was never discovered. The *Register Rooms* are next reached, the low ceilings of which are as white as if made so by the hand of art. Here countless visitors have written their names with the smoke of candles. A little farther on the visitor enters the *Gothic Chapel*, a hall of surpassing beauty and grandeur, fifty feet by eighty feet in width, and elliptical in form. At each end are immense stalagmites, and, in addition, double rows of smaller pillars depend from the ceiling on each side, at about equal distances, giving it the appearance of one of the religious structures of mediæval

Europe. It is often lighted up for the gratification of visitors, and presents on such occasions a scene of surpassing magnificence. The *Devil's Armchair*, an immense, solid stalagmite, is a special object of interest; so, also, is the *Lover's Leap*, a huge pointed rock, ninety feet above the floor of the cavern. Nearly every projection, pit and pillar, room and avenue, has its peculiar name; and, as there are nearly three hundred avenues and passages alone, the impossibility of naming each in so brief a description will readily be seen. The lower branch of Gothic Avenue may be reached by a detour just beyond Lover's Leap, though nothing especially remarkable is to be seen.

Returning to the *Grand Avenue*, the *Ball Room* is the first locality where the visitor will feel inclined to linger. Although no regular ball has ever been given here, improvised dances frequently take place, and the sound of laughter and gay voices echo strangely through the vaulted passages and chambers adjoining. Leaving here, the visitor is suddenly surprised by a huge black object to which has been given the name of *Giant's Coffin*. It occupies a very conspicuous position, the path making a bend around it by which the main cave is left to visit the rivers and other wonders beyond. Next comes the *Star Chamber*, the most brilliant of the many apartments of the cave. Splendid formations of transparent gypsum incrust the walls and ceiling. When the light of many torches illumines the sparkling crystallizations which ornament the magnificent dome, it is as if myriads of stars had made their appearance in the arch above. As the visitors move about, waving their lights, the whole cave seems filled with glittering objects, which, with the vast size of the chamber and the snowy whiteness of the sides and floor, makes it seem like one of the gorgeous halls described in the Arabian Nights. The *Chief City* or *Temple*, farther along, just beyond *Rocky Pass*, covers an area of nearly two acres: over it springs an immense dome a hundred feet high. It is larger than the celebrated cave of Staffa, and very nearly as large as the grotto of Antiparos, said to be the most extensive in the world. A huge pile of rocks, called *The Mountain*, rises nearly in the centre, up which adventurous visitors frequently climb, after kindling fires at the foot. The effect is said to be magnificent in the extreme. *Gorin's Dome* used to be regarded as one of the chief wonders of the cave. It is reached by the *Labyrinth*. The dome is apparently two hundred feet high, its sides presenting the appearance of having been fluted and polished, while a huge rock depending from the summit is so disposed as to resemble a curtain. The *Bandit's Hall* forms a sort of antechamber to a number of interesting apartments. An avenue to the left leads to a dozen or more domed apartments, one very much like another; another to the left leads to the *Mammoth Dome*. This is one of the most sublime spectacles to be witnessed in this wonderful cavern. The dome is four hundred feet above the floor, and is only visible under the intensest light. The *River Hall* and the *Dead Sea* are among the most prominent attractions of the cave. The former is a huge slope up which the visitor ascends until he reaches a steep precipice, over which he looks down, by the aid of torches, upon a broad, black sheet of water eighty feet below, named appropriately the Dead Sea. The sight is dark, lurid, and terrific as the throat of the infernal pool.

There are several rivers in the cave, the smallest of which is the *Styx*. Crossing it by a secure bridge, the visitor finds himself on the banks of *Lake Lethe*, where boats are in waiting to convey such as may wish on an aquatic excursion over its waters. On the opposite bank the *Great Walk* leads to *Echo River*, three-quarters of a mile in length, and sufficiently large and deep to float an ocean steamer. Four miles from Echo River is *Cleveland's Avenue*, various places of interest being passed on the way. Close by is the *Holy Sepulchre* and *Mary's Vineyard*. The avenue is three miles in length, and from fifty to seventy feet in height, and may be designated as one of the most wonderful objects in the world. It is replete with formations that are to be found nowhere else in the entire cave, and of surpassing beauty. *Cleveland's Cabinet* is a perfect arch of fifty feet span, extending nearly two miles, and incrusted from end to end with the most beautiful formations, and of every variety of form. The *Snow-ball Room*, a large apartment, looks as if a crowd of sportive schoolboys had been engaged in pelting the sides and walls with snowballs. Beside these named there are hundreds of other places and objects of interest, — the *Rocky* Mountains, *Croghan's Hall*, *Sereno's Arbor*, the *Dining Hall*, *Fat Man's Misery*, the *Relief Room*, &c., which space forbids us to mention.

Language unaided by cuts is inadequate to a satisfactory description of a place so intricate and peculiar, and yet so vast. It is hoped, therefore, the next edition of "Popular Resorts" may present a descriptive article upon this wonderful freak of nature, fully illustrated.

There are several smaller caves in the vicinity, the finest of which, perhaps, is **Indian Cave**, about half way between Cave City and the hotel. The formations are even finer than those to be found in its larger neighbor. **Diamond** and **White's** Caves both lie within a short distance of each other, and near the Mammoth. A few miles away is the **Natural Bridge**, nearly one hundred and fifty feet high, with a span of seventy feet. **Indian Rock** and **Pilot Rock**, both curious formations, are also within easy distance of the cave, and are notable attractions of the region.

The **Ancient Mounds**, which are scattered through the length and breadth of the State, are objects of interest which the tourist cannot afford to neglect. Some of them are of immense extent. No trace has ever been found by which to identify the builders. The most interesting of these mysterious works, perhaps, is that in Allen County, a few miles from Bowling Green, which attracts large numbers of visitors the year round. Others, which in their day must have been fortresses of immense strength, are found in Boone, Warren, Bourbon, Montgomery, and other counties.

Nashville lies at the head of steamboat navigation on the south side of the Cumberland River, the better portion of the city lying on a bluff above the river. During the summer months there is a large influx of visitors from the lower States. The climate is exceedingly mild and healthful. The Capitol building is, with one or two exceptions, the finest in the United States. President Jackson's former residence, the *Hermitage*, lies in the immediate vicinity, and is visited by thousands of tourists every season.

Elkmont Springs, a favorite summer resort, lie upon the same line, just over the Alabama border, ninety-five miles south of Nashville. **Blue Springs** are located one mile, and **Valhermosa** forty miles, from Decatur, the most important town of Northern Alabama. **Blount Springs,** the most fashionable watering-place in the State, are located twenty-one miles from Birmingham station, on the North and South Alabama Railroad portion of the line, and near the banks of the Black Warrior River. For many years it was the best known resort in the Gulf States, and was largely patronized. The place is coming into notice again, and visitors are beginning to make it an objective point. Stages leave *Birmingham Station* daily for the springs. The accommodations, though hardly what they were some years before the war, are yet of a fair order.

There are to be found upon the branches and connections of the Great Southern, as well as upon the other North and South routes mentioned, various places of more or less local note, at present undeveloped, but of future promise. As yet they are hardly of sufficient importance to claim special mention. Illustrated descriptions, as are given in other sections of the country, would add to their interest and popularity.

THE WESTERN STATES.

No section of the Union embodies such a variety of scenery as the Western States. Every grade is represented, from the beautiful level prairie, the deep stately river, to the broad inland sea, and the bold, snow-capped mountain. The scenic features of the West must always be described by superlatives. Here are the highest mountains, the deepest cañons, the largest lakes, and the longest rivers; and here, too, are the broadest prairies, the highest waterfalls, the largest trees, — in a word, the grandest elements of the country are here. Her pleasure resorts, though new, are rapidly improving in character, and increasing in popularity. At present those of the great North-West are the best known; but the startling grandeur of the Far West is attracting the attention of tourists, and each year adds its thousands of visitors. Illustrated descriptions of Western scenery, heretofore less common than in many parts of the country, will do much to increase its national popularity. This will require time; yet it is hoped that the next edition of "Popular Resorts" will represent fully this interesting section of our country. Language, so lavishly used in the description of tamer scenes, utterly fails to convey to the understanding a just conception of the grandeur and sublimity of many features in the Far West. Even the most successful attempts of artists are tame when brought in comparison with these works of the Great Architect.

MICHIGAN. — The Peninsulas of Michigan present many attractions to summer tourists, — attractions which each year increase in popularity, and must soon become favorite spots for summer recreations.

Detroit and **Toledo** are the principal cities, and will command the attention of visitors.

The **Magnetic Spring,** recently opened at *Alpina*, Thunder Bay, on the north-eastern shore of Michigan, is attracting much attention. Its chemical and magnetic properties are very remarkable, and will well repay an investigation from the tourist.

Another **Magnetic Spring** at *St. Louis*, near the centre of the State, reached by the Saginaw Valley and St. Louis Railroad, is much frequented by invalids from the South. The **Midland Magnetic Springs**, on the line of the Flint and Père Marquette Railway are receiving much attention.

Mackinaw, a favorite resort, is reached by steamers from *Bay City* three times a week, also from Detroit, Cleveland, Buffalo, and Chicago.

Indeed, there are many interesting features being developed on the line of the Flint and Père Marquette Railway, extending from Toledo north-westerly through the State to *Ludington* on Lake Michigan, which may receive attention in the next edition of "Popular Resorts."

PLEASURE ROUTE No. 28.

Richmond (Ind.), Cincinnati, Louisville, and the South, to Grand Rapids, Traverse City, Northern Michigan, and the Straits and Island of Mackinaw.

GRAND RAPIDS AND INDIANA RAILROAD.

This route runs directly north through Indiana, dividing longitudinally the eastern peninsula of Michigan, and leading directly to a large number of popular resorts, at present being developed in that State, many of which, for the pleasures they afford, will soon successfully rival the summer resorts of older sections of the country. Although unprepared to give illustrated descriptions, a few of the more prominent places of interest will be briefly alluded to.

"The triangular part of Northern Michigan, lying between the Straits of Mackinaw on the north, the Grand Rapids and Indiana Railroad on the east, the Manistee River on the south, and Lake Michigan with its indentations on the west, offers a comparatively unknown region, with many charming spots which, once visited, always command a return.

"The streams swarm with the finest fish, and the many beautiful lakes scattered throughout the forests are rare jewels set by the hand of Nature. To the Grand Traverse region we draw particular attention. The healthfulness of this section, like its scenery and fishing, is unsurpassed. Good hotel accommodations are generally obtainable. Visitors should not expect the conveniences and *cuisine* of city houses; but good food, well cooked and presentable, with clean and comfortable beds, can generally be procured.

"**Richmond, Ind.**, is the southern terminus of the Grand Rapids and Indiana Railroad, and the junction of the same with the Pittsburgh, Cincinnati, and St. Louis, and Cincinnati, Hamilton, and Dayton Railroads.

"In the depot, which is occupied by all of the above roads, the traveller will find a good dining room and lunch counter.

"The Grand Rapids and Indiana Railroad runs its through coaches over the Cincinnati, Hamilton, and Dayton Railroad to and from Cincinnati, thus offering a through railroad line from the Ohio River to Petoskey, on Little Traverse Bay, and Traverse City, on Grand Traverse Bay."

The following are among the places of interest reached: —

Fort Wayne, Ind. (twenty minutes for dinner), Rome City, Ind., Kalamazoo, Mich., Grand Rapids, Grand Haven, with its **Mineral Springs**, Howard City, Reed City, where the line crosses the Flint and Pere Marquette Railroad, Tustin Station, near which is *Pine River* (good Grayling fishing), *Diamond Lake, Hewitt Lake, Rose Lake,*—all well stocked with fish.

Clam Lake is twelve miles from Tustin, with a village of twelve hundred inhabitants, fine fishing, and comfortable hotels. At Mayfield Station, near which is *Boardman River,* is the trout stream of Northern Michigan; Traverse City, with two thousand inhabitants, located on Grand Traverse Bay, Lake Michigan, with good fishing, is the next place of importance. Near by are *Cedar Lake, Bass Lake, Betsey Lake, Long Lake,* and *Traverse Lake,* each noted for their fine fish. Brook trout are very abundant in the vicinity. *Fife Lake* is four miles north of Walton Junction.

Kalkaska, county seat of Kalkaska County, is literally surrounded by small lakes, plentifully supplied with fish. Four miles west from Nelsonville, through an unbroken forest, is *Intermediate Lake,* narrow, but twenty-four miles long, fed by many streams filled with trout. Guides are required. The fishing is superb.

Returning to the railroad, Petoskey is the northern terminus of the main line, 494 miles from Cincinnati. A steam yacht runs to Little Traverse, near which are *Round* and *Crooked Lakes;* and a swift steamer connects with *Mackinaw Island,* five hours distant. "This lovely island is one of the finest summer resorts in Northern Michigan. Its locality, scenery, bathing, boating, and fishing, present irresistible attractions to the tourist, invalid, and sportsman. *For hay fever and asthmatic affections its air offers speedy relief.*' By a recent act of Congress, the island of Mackinaw has been declared a *National Park.* The points of interest are *Fort Mackinaw,* on a rocky eminence just above the town, which was built by the English about ninety years ago, and *Fort Holmes,* the crowning point of the island, also built by the English, and originally called Fort George.

"There are three hotels on the island, with accommodations for about five hundred guests, and numerous comfortable boarding-houses are located at various points.

"From Petoskey the tourist may proceed to *Charlevoix,* sixteen miles distant, by steamer. This town is charmingly located on a peninsula formed between Pine Lake and a small body of water called Round Lake, at the mouth of the former. The surrounding scenery is romantic. From Charlevoix the visitor may proceed to Traverse City. Fine fishing in the bay, lake, or rivers, with a steamboat trip to *Torch Lake,* and a visit to the large blast furnaces located here, serve to while away the hours. Then boarding the steamer he passes up Elk River a mile, and enters *Elk Lake,* a beautiful expanse of water ten miles in length and from two to three miles in breadth; entering through "the narrows,' to *Round Lake,* so called, whose length is about four miles with a breadth of two; thence running up *Torch River,* a stream three miles long, always interesting with its curious windings and tortuous channels, its thickly-wooded banks and wild scenery. From the head of this stream he enters the lower end of Torch Lake, one of the loveliest in the world, rivalling the famous Lake George, eighteen miles long, and from two to three miles broad. Reaching Torch Lake village, near the head, he finds good hotels, splendid scenery, and excellent fishing. *Look out for lying guides at this point!*

After wearying of this locality, the visitor may take the little steamer, and be deposited with his boat at the mouth of *Clam Lake*, flowing into the east side of Torch Lake. Passing up Clam Lake, he will presently come to a "narrows," leading into *Grass Lake*, which is joined to Intermediate Lake by Grass River, a very swift stream, not navigable, but affording fine fishing.

"Having gone the rounds with the visitor, we now leave him to his own inclinations, and continue our travels farther west."

Chicago is also the most important railway centre in the Western States, and its location serves to make it the most prominent business emporium. Although but seventy-five years old, it numbers over four hundred and fifty thousand souls. Its history has been varied and interesting, and singularly fraught with mishaps. It was settled in 1804 by John Kinzie; on the 7th of April, and 12th of August, 1812, the garrison was massacred by the Indians. In 1830 it had increased to but twelve or fifteen houses, and about one hundred inhabitants, all told. It became an incorporated town three years later, and was incorporated a city in 1837, when the population had increased to 4,170. Its growth has been steady and rapid. Its location, on the shore of Lake Michigan, at the mouth of the Chicago River, gives it peculiar business facilities. It is generally level, having a slight incline to the lake. No city in the country has been so unfortunate as this. In the great fire of 1871, 13,800 buildings were destroyed, by which 74,450 persons were made homeless. Yet with an energy and perseverance unprecedented the city has been rebuilt, and is to-day more beautiful than before. Chicago, being the great intermediate railway centre of the West, from which radiates a network of lines so conveniently arranged and so adroitly managed that every section is opened to commerce and to travel, must always maintain its prominence in business affairs. It is noted for its elegant buildings, — improved since the fire, — fine bridges, and extensive tunnels, while its parks and boulevards are attractive and beautiful. Although not a popular resort, it contains much to interest the tourist.

BEYOND CHICAGO.

HITHERTO we have said but little about the railways or the country beyond Chicago. Many of our readers have "worn out," by constant visits, much of the scenery and tourists' resorts of the East, and are looking for "forests new and pastures green" elsewhere. We would invite them to look to the great North-west. Here they can find lakes and hills, mountains and valleys, woods and streams, new, and untrod by the Eastern tourist. To get to Chicago, you have various and good routes.

Beyond Chicago, we would name the great Chicago and North-Western Railway, as a route having more terminal points, more connecting lines, and reaching more points of interest, than any other; and it may be taken, should you be on business or pleasure bent, or should you desire to reach Denver, and the mountains of Colorado, the wilds of Idaho and Montana, or the Pacific slope, or north-westwardly the woods and streams and lakes of Wisconsin, Minnesota, or the Lake Superior country of Northern Michigan.

HISTORY OF THE LINE.

In 1849 there was not a mile of completed railroad in the vicinity of Chicago. In that year W. B. Ogden was endeavoring to interest the people in the northern portion of Illinois to subscribe money to build a *tramway* along the route of what afterwards became the Galena and Chicago Union Railroad. From that small beginning great results have grown. In time the Chicago and North-Western Railway Company was formed, and took unto themselves this first effort of Mr. Ogden, and also other lines projected by him and others. This great corporation now owns and operates over two thousand miles of road, that, radiating from Chicago like the fingers of a man's hand, extend in all directions, and cover about all of the country north, north-west, and west of Chicago. With one branch it reaches Racine, Kenosha, Milwaukee, and the country north thereof; with another line it pushes through Janesville, Watertown, Oshkosh, Fond du Lac, Green Bay, Escanaba, to Negaunee and Marquette; with another line it passes through Madison, Elroy, and for St. Paul and Minneapolis; branching westward from Elroy, it runs to and through Winona, Owatonna, St. Peter, Mankato, New Ulm, and stops not until Lake Kampeska (Dakota) is reached; another line starts from Chicago, and continues through Elgin and Rockford to Freeport, and connects for Warren, Galena, and Dubuque, and the country beyond. Still another route runs almost due westward, and passes through Dixon, Clinton (Io.), Cedar Rapids, Marshalltown, Missouri Valley Junction, to Council Bluffs and Omaha. This last-named is the "Great Trans-Continental Route," and the *pioneer overland line* for Nebraska, Colorado, Utah, Idaho, Montana, Nevada, California, and the Pacific Coast. It runs through the garden of Illinois and Iowa, and is the shortest and quickest route for Omaha, Lincoln, and other points in Nebraska, and for Cheyenne, Denver, Salt Lake City, Sacramento, San Francisco, and all other points west of the Missouri River.

POINTS REACHED BY THIS LINE.

See, then, what one company can do for you. If you want to go to Milwaukee, Fond du Lac, Sheboygan, Janesville, Watertown, Oshkosh, Green Bay, Ripon, Madison, Baraboo, Eau Claire, Hudson, Stillwater, St. Paul, Minneapolis, Duluth, Breckenridge, Fort Garry, Winona, Owatonna, New Ulm, Freeport, Warren, Galena, Dubuque, Waterloo, Fort Dodge, Sioux City, Yankton, Council Bluffs, Omaha, Lincoln, Denver, Salt Lake City, Sacramento, San Francisco, or a hundred other northern, north-western, or western points, this great line affords the amplest accommodations. The track is of the best steel rail, is well ballasted, and all the appointments are first-class in every respect. The trains that run over this route are made up of elegant new Pullman palace drawing-room and sleeping coaches *built expressly for this line;* luxurious, well-lighted, and well-ventilated day-coaches, and pleasant lounging and smoking cars, all built by this company in their own shops. The cars are all equipped with the celebrated Miller safety platform, and patent buffers and couplings, Westinghouse safety air brakes, and every other appliance that has been devised for the safety of passenger-trains. All trains are run by telegraph. It is acknowledged by the travelling public to be the popular line for all points in Northern Illinois, Wisconsin, Minnesota, Northern Michigan, Dakota, Western Iowa, Nebraska, Wyoming, Colorado, Utah, Montana, Idaho, Nevada, California, and the Pacific slope.

SUMMER RESORTS.

The Devil's Lake country around Baraboo (171 miles from Chicago), the beautiful lake country around Madison (140 miles from Chicago), Sparta (a celebrated magnetic-spring resort), Escanaba, Fond du Lac, Green Lake, Elkhart Lake, Marquette, St. Paul, Minneapolis, the Falls of Minnehaha, Lakes Calhoun, Harriett, and Minnetonka, White Bear Lake, Lake Como, and many other points on this route, are well worth visiting. Good hotels, and pleasant boarding and farm houses, abound ; near which you can find fishing, sailing, boating, and bathing to your heart's content. Lake Geneva, the gem of Western lakes, is also on this line; and at its head is the town of that name, appropriately named "The Saratoga of the West." The Chicago and North-Western Railway each year places on sale excursion-tickets at cheap rates to all of its summer resorts in the North-west, and to Denver and other points in Colorado.

PLEASURE ROUTE No. 29.

The Chicago and North-Western Railroad: its Branches and Connections.

THE GREAT NORTH-WEST.

To the majority of the citizens of the West, the Chicago and North-Western Railway is well known; not so, however, to those in many parts of our extended country.

Operating over two thousand miles of the best road in the world, and running through five great States, with numerous branches, it is not to be expected that the traveller from the South or the East could, unaided, understand the vast system of lines owned and operated by this company.

Summering.[*] — "Not more regularly does the warm season recur, than does the desire, begotten of it, to get away from home; to throw every thing aside, if only for a day or two, and go away somewhere. And what a blessed comfort it is to shake off the dust of the city, to leave behind and forget its hot pavements and dusty walls, and hurry off to some cool, leafy nook nestling far away among the mountains; or on the shore of some one of the many inland lakes which, like jewels, bestud our northern landscape! With the warm breezes wafted by the opening summer's days there comes an inward longing to be off, a desire for a change of scene, — a yearning for that abandon which can only be found in nature's more secluded haunts."

As we have seen, the great North-West is rich in lovely retreats.

WISCONSIN. — "The 'Badger State' is yearly becoming more widely and more favorably known to the summer tourist, and to the seeker after rural pleasures. While comparatively a new State, it is yet old in many respects. As long ago as the middle of the seventeenth century, it was visited by French missionaries and traders, who took home with them glowing accounts of the fertility of its soil, the splendor of its scenery, the freshness of its odorous pine-clad hills, its flashing, rapid running streams, full of fish, its clear, deep, cold, pure, and beautiful lakes, of which the State has many hundreds, and its delightful, balmy, and invigorating summer climate. The stories of these advantages were not lost on the beauty-loving French, and soon colonies were formed for the settlement of this charming 'Neekoospara,' as they had learned from the

---

[*] The following matter has, by permission, been collated largely from several small guide-books of the North-West. It is intended to give special attention to this region before the issue of 1876.

Indians to call the country we have since named Wisconsin. It may rightly then, be inferred that the French were the first whites to make homes along the bays, lakes, and rivers of this well favored land.

"Distributed, if we may so speak, all over the State, can be found objects of interest to the lover of the picturesque and to the antiquary. Scattered over her undulating plains are found earthworks modelled after the forms of men and animals: these are evidently the work of a race different from those who possessed the country at the period of the arrival of the French. At *Aztalan*, in Jefferson County, is an ancient fortification, seventeen hundred feet long and nine hundred feet wide, with walls five to six feet high and more than twenty feet thick; this, with another near the *Blue Mounds*, near Madison, resembles a man in a recumbent position. Another, near Madison, in Dane County, resembles a turtle; one at the south end of the *Devil's Lake*, in Sauk County, closely resembles an eagle; and one near Cassville, in Grant County, on the Mississippi River, resembles the extinct mastodon. The Blue Mounds, in Dane County, rise to two thousand feet above the surrounding country, and are prominent landmarks in that prairie region. This State shares with Minnesota the beautiful *Lake Pepin*, an expansion of the Mississippi River, mostly walled in by precipitous shores which rise in places to five hundred feet. Connected with almost every cliff or promontory along the shores of this beautiful lake, are legends of the Indians who formerly had homes here. Along the rivers of this State are found many waterfalls, rivalling in beauty those of older States. In the St. Louis River are the **Dalles**, which have a descent of three hundred and twenty feet. The *Dalles* of the St. Croix are also well known. **Quinnessec Falls**, in Menomonee River, have a perpendicular pitch of over fifty feet, and a general descent of one hundred and fifty feet in a mile and a half, besides many other rapids, where the river tosses and dashes through narrow and tortuous defiles. The *Chippewa Falls* and the *Big Bull Falls* might also be noted. Along the Wisconsin River are many grand and picturesque views; in Richmond County the banks of the river rise to a height of two hundred to two hundred and fifty feet, and in Sauk County it passes through narrow gorges where the banks rise to five hundred to six hundred feet elevation. *Grandfather Bull Falls*, the greatest rapids of the Wisconsin River, are in 45° north latitude, and are a series of wild cascades dashing through a gorge a mile and a half long and one hundred and fifty feet perpendicular height; on the same river, near latitude 40°, is *Petenwell Peak*, an oval mass of rock, nine hundred feet long by three hundred wide and two hundred high, and from which commanding views can be obtained. About seventy feet of the upper portion of this rock is cut and split into fantastic shapes, many of the fragments resembling castles, towers, and turrets.

A few miles distant is *Fortification Rock*, which rises perpendicularly several hundred feet. At the Dalles this river is compressed for five or six miles between red sandstone cliffs, averaging over one hundred feet in height; while in places the river is compressed to a width of fifty-five feet.

"The principal lakes are *Lake Winnebago*, — in the south-eastern portion of the State. This lake is about thirty miles long and ten miles wide, and communicates with Green Bay (an arm of Lake Michigan) through Fox or Neenah River, — *Horicon Lake, Devil's Lake, Lake Koshkonong, Lake Geneva, Lake Zurich*, and the four lakes around Madison; these are the larger of this lake-studded State. Along all the rivers, and at their "heads," *hundreds* of little lakelets are found, like gems glittering in the sunshine.

"**Waukegan, Ill.**, is the county seat of the county of Lake, so called because of the *fifty-six* beautiful lakes within its boundaries. The city has between seven and eight thousand inhabitants: is situated on a bluff on the western shore of Lake Michigan, which it overlooks. The town proper stands about one hundred feet above the lake, and in point of attractiveness as a summer resort stands unrivalled in all the Western States. Its acknowledged beauty, fine drives, society, schools, picturesque scenery, ravines, brooks, and general loveliness, fashioned by Nature's own hands, aided by cultivated taste, combine to make it a place which will be sought for by the thousands of private families who yearly, more and more, seek to avoid the heat, dust, and noise of a busy metropolis, for the health-giving quiet and retirement of the country.

"The **Glen Flora Mineral Springs.** — The waters which have been for perhaps untold ages gurgling from their cool, rocky depths, and flowing in miniature rivulets into Lake Michigan, have this year been proven beyond question, by scientific analysis, to be of the most valuable medicinal character.

"The Glen Flora Springs are easy of access, being located on the line of the Milwaukee Division of the Chicago and North-Western Railway. About sixteen trains pass and repass between the cities of Chicago and Milwaukee daily. The railroad station, 'Glen Flora,' is only about one-quarter of a mile from the Springs, which are reached by a newly graded road, leading up the bluff in close proximity to them. The location, for picturesque beauty and romantic surroundings, is unsurpassed in this country. They are nestled in a beautiful ravine or glen, originally named 'Floral Glen,' because of the remarkable profusion of wild flowers which grow and thrive spontaneously from end to end of its labyrinthian traceries.

"**Milwaukee.** — Cities, says an eminent writer, have always been the fire-places of civilization, whence light and heat radiate out into the dark, cold world; and the union of men in large masses is indispensable to their development and growth.

"Fifty years ago, and all there then was of the now prosperous and beautiful city of Milwaukee — lowland, shore, and forest — was in the undisputed possession of the Indian.

"The first white man to invade this beautiful retreat was Solomon Juneau. He came here in the autumn of 1818, and built him a log cabin, which gradually assumed the distinctive features of a store, in which he kept a few goods suitable for barter. For seventeen years he was not only the only merchant in the place, but the only white settler. A few Indian traders occasionally came, but none made a permanent location.

"Milwaukee is now the commercial emporium of the State of Wisconsin and one of the most important cities, in many respects, in the North-West. It has a population of about a hundred thousand; and is built largely of the famous cream-colored Milwaukee brick, which are produced here in large quantities. The situation of the city on Lake Michigan shore, at the mouth of Milwaukee River, is very pleasant, healthful, and attractive. The river furnishes one of the best and most commodious bays and harbors on the lakes."

Tourists will find at the *Newhall House* and *Plankinton House*, first-class accommodations.

Lake Dells is a popular resort in the suburbs, with a fine water view, which is receiving considerable attention.

"During a few years past numbers of residences and cottages have been built upon the summit of the bluff, or on the plateau beneath, and many more are planned for the coming season.

"During the late summer and early fall months the temperature of the water is delightful for bathing, averaging in the shallow bay off Lake Dells, by actual test and record, sixty-eight to seventy degrees.

"**Sheboygan, Wis.**, the county seat of Sheboygan County, is a thriving manufacturing city of about seven thousand inhabitants. It is the eastern terminus of the Sheboygan and Fond du Lac Railway, and is the most important station on the line of the Milwaukee, Lake Shore, and Western Railway north of Milwaukee. The former of these roads makes close connections with the Wisconsin Division of the Chicago and North-Western Railway at Fond du Lac, and the latter is practically an extension of the Milwaukee Division of the Chicago and North-Western Railway.

"Sheboygan is one of the most delightful summer resorts in the West. It has a commanding location on a bluff overlooking Lake Michigan. The river affords unsurpassed opportunities for rowing, while the lake is a favorite resort for those who enjoy sailing. Pleasure boats of all kinds may be had here. The fishing is good; the fisheries off Sheboygan are among the most important on Lake Michigan, and a summer day can hardly be more pleasantly spent than in visiting them in one of the steam fishing-smacks. The drives in the vicinity of the city are fine. This is especially true of the excursion up the river five miles, to the charming village of Sheboygan Falls. But, after all, the chief attractions which Sheboygan holds out to the summer tourist are the healthfulness and coolness of its climate. Lying, as it does, ten miles out in the lake, it is fanned by deliciously cool and invigorating lake breezes from the north, east and south, and consequently, the intensely hot weather that generally prevails in the interior during the summer months is unknown here. The *Beekman House* will furnish a good home to visitors.

" From Sheboygan, the popular summer resorts along the line of the Sheboygan and Fond du Lac Railway may be easily reached. It is only an hour's ride from Sheboygan to beautiful *Elkhart Lake*, a three hours' ride to the famous Mineral Springs at Fond du Lac ; and a five hours' ride to that gem of Wisconsin scenery, *Green Lake*.

" Wisconsin is famous for her beautiful lakes, the annual resort of thousands of people in quest of health and pleasure. Among the loveliest of these is **Lake Elkhart**, acknowledged to be one of the most healthful places in the West. It is fifty-seven miles north of Milwaukee, and can be reached by the Chicago and North-Western Railway and its immediate connection, the Sheboygan and Fond du Lac Railway. The station, *Glenbeulah*, is three miles from the Lake. Omnibuses run to the hotels from all the trains arriving at this station.

" Cincinnati, Louisville, St Louis, Indianapolis, Chicago, and Milwaukee are largely represented at Elkhart. Instead of going to some 'fashionable' watering-place, where dress, flirtation, and the usual consequences of folly, prevail, our Western people wisely prefer these secluded spots, where Nature is so profuse of her charms.

" Elkhart Lake covers about eight hundred acres, and is pleasingly diversified by bays, which coquettishly wind around jutting bluffs, beneath whose shades the crystal water slumbers, so pure and clear that the white sand and gravel of the lake bottom can be plainly seen at a depth of twenty to twenty-five feet. As viewed from the elevated veranda of *Marsh's Swiss Cottage*, the scene is beautiful beyond description. The surrounding hills are verdant with pine, spruce, maple, basswood, and red and white cedar ; while wheat fields, seen through forest vistas, afford to the eye, as it feasts upon the varying charms, a most pleasing variety.

" A steamboat and barge, sail and row boats, will run in connection with the trains of the Sheboygan and Fond du Lac Railroad, and convey passengers to any part of the lake desired.

"**Fond du Lac.** — Passengers desirous of reaching Fond du Lac, or any point north thereof, can go *via* Milwaukee, or *via* Janesville and the Wisconsin Division.

"This is called the 'second city of the State.' It lies in one of the richest agricultural districts in the West, seventy miles from Madison, sixty-three miles from Milwaukee, forty miles from Sheboygan, sixty miles from Green Bay, and one hundred and forty-eight miles from Chicago, — directly connected with all of the above-named cities by railroads; in short, the railroad facilities for coming to Fond du Lac are equal to those of any other interior city in the Union, as they radiate from the city at nearly all points of the compass. It is rich in manufacturing resources, and in wealth and population is second only to Milwaukee among all the cities of Wisconsin. Fond du Lac is blessed with enterprising, liberal-minded manufacturers and business men.

"**Fountain Mineral Spring,** in the neighborhood, has gained great celebrity for the many cures its waters have performed.

"Several yacht clubs navigate *Lake Winnebago* in elegant rakish craft, for prizes in sportive contests. Steamboats, with pleasure parties, often make excursions around the lake, which is thirty miles long by ten broad — whose borders furnish beautiful landscape views, and the most gorgeous scenery. Besides, Fond du Lac is surrounded with other pleasant places of resort. **Lake de Neveu**, a beautiful sheet of water, is romantically situated, about three miles south-east of the city. Eastward is *Elkhart Lake*, already famous for its natural beauties; and westward lies *Green Lake*, a noted summer resort. On all these lakes are pleasure boats propelled by steam, wind, and man power. Their waters furnish a plentiful and various supply of fresh-water fish, where piscatorially inclined ladies and gentlemen can enjoy plenty of amusement in that line.

Parties visiting Fond du Lac on pleasure bound, or for the purpose of being restored to health, will find hotel accommodations of every variety and grade, — some elegantly fitted and furnished, rivalling the best houses in the country. There are also a number of private boarding-houses, some in the immediate vicinity of the *Magnetic Springs*, where board may be obtained at reasonable rates.

Although we do not usually recommend cities as popular resorts, still there will be found at Fond du Lac much to interest the tourist, both in the town and its vicinity. The drives are interesting, and the roads good, which, combined with the other attractions named, entitle this resort to the favorable reputation it has won.

"**Green Lake** (Dartford P.O.), Wis., is the name of a village situated on the banks of Green Lake, Wisconsin; is a station on the Sheboygan and Fond du Lac Railroad, and is accessible by the Chicago and North-Western and the Sheboygan and Fond du Lac Railways. The public houses are first-class, and pleasant boarding places are readily obtained at the farm-houses in the vicinity.

"The natural scenery around Dartford is unrivalled in variety and beauty. Groves of primeval grandeur, far-stretching prairies, and extensive lake views, greet the eye from every point. The grounds around the lake have been terraced, furnished with swings, promenades, and otherwise ornamented, to render them pleasant and attractive. The lake averages a length of fifteen miles, and a width of three miles. Its banks vary from beautiful grassy slopes to high rocky cliffs, bordered with evergreens, presenting the greatest diversity of physical character, and affording unlimited natural advantages for pleasing and romantic rambles. Its waters are very pure, and so transparent that their pebbly bed may be seen at a depth of from twenty to thirty-five feet. A great variety and abundance of the finny tribe inhabit this beautiful sheet of water; and good fishing-boats and tackle for lovers of sport, and excellent sail-boats for seekers of pleasure, are furnished for the accommodation of visitors.

"*The Oakwood* is the name of a hotel situated on the banks of Green Lake. The location of this much-sought summer resort and delightful watering-place is only one mile from Green Lake Station and depot. Omnibus and carriages await the arrival of all trains. The *Sherwood Forest House*, and *Walker House* also furnish first class accommodations.

"**McHenry, Ill.**, is situated on the Fox River, only sixty miles from Chicago, *via* the Chicago and North-Western Railway. It has about fifteen hundred inhabitants. About six miles north, by a pleasure steamboat, we come to a chain of four lakes, unsurpassed anywhere in the West for their beautiful scenery. They are skirted on all sides by miniature mountains, little islands dot the surface from one end to the other, varying in size from two to sixty acres each. These lakes are all deep, and abound in fish.

"**Lake Zurich** is named after one of the most splendid lakes in Switzerland; and when once seen, and its scenery and loveliness enjoyed, no one would for a moment think that he who named it was guilty of any presumption in the christening. This place is situated on the edge of Lake County, four miles north by east from *Barrington Station*, on the Wisconsin Division of the Chicago and North-Western Railway. An omnibus is in waiting on the arrival of trains, to convey passengers to the *Lake Zurich House*. The road runs through a splendid farming country, teeming with all the exuberant richness of a bounteous soil,

GREEN LAKE, WIS.

Oakwood House.

alternating with woodland and prairie, hills and valleys, fields of waving grain, and farmhouses embowered in shrubbery,—making one of the finest landscapes in this region.

"**Oshkosh** as a summer resort and watering-place possessed a rare combination of natural features for a delightful summer residence before the late fire, by which it suffered fearfully. The climate is not surpassed in healthfulness; the air is pure and dry; and the invigorating breezes from the lake temper the heats of summer. The scenery is lovely; the lake a most magnificent sheet of water with beautiful shores and good harbors that are accessible in every direction, thus affording the best of yachting facilities. The surrounding country is beautiful, with excellent roads, affording delightful drives, and picturesque views of lake and river scenery. Wild game is abundant in the vicinity, blue and green winged teal, mallard and wood duck, snipe, woodcock, quail, and prairie chickens. The waters abound in black and white bass and other fish, and brook trout are plentiful in streams within a day's travel."

Lake Winnebago is really the special attraction. *Fox River* is a beautiful stream, both before it enters Winnebago at Oshkosh, and after it leaves it at *Menasha*, from which place to Green Bay it is a series of rapids falling a hundred and seventy-five feet, which are overcome by *locks* and *dams*.

The *Seymour House* of this place has long been noted for its homelike accommodations, and is much frequented by visitors.

"**Menasha**, being situated at the foot of Lake Winnebago and Lake Buttes des Morts, and embracing part of Doty's Island, furnishes picturesque and entertaining scenery, unsurpassed by any Western town.

"The climate is salubrious, and no place on the continent is freer from epidemics. The air is mild and bracing, and yields a vigor and endurance to the system that is above all price.

"*Lake Winnebago* provides boating, sailing, and bathing facilities. Steamboats ply on its limpid waters, and sailing vessels can always be had, furnishing ample means for pleasure excursions. Rowboats are kept in connection with the *National Hotel*, always at the service of guests, furnishing a most agreeable and healthful exercise, on the ever placid waters of Fox River and adjacent bays of the lake. In connection with other sports, fish and game are plenty, and the sportsman need have no lack of pleasurable novelties.

"The city of **Neenah** is located in the county of Winnebago, and most romantically and beautifully situated on the *Fox River*, at the outlet of Lake Winnebago, and on the line of the Chicago and North-Western Railway. The present population is over four thousand.

"**Pewaukee, Wis.** — Among the many points of attraction within the borders of our already famous and justly celebrated country, Pewankee holds a fair share of possessions, and in some considerations is superior to any other. Located nineteen miles west of Milwaukee, and situated at the foot of Pewaukee Lake, which everybody concedes to be the gem of all others.

"**Oakton Springs** are situated in a beautiful grove on the south side of Pewaukee Lake, which is five miles long from east to west, and from half a mile to one and a half in width, with an average depth of fifty feet, and two hundred and sixty-three feet above the level of Lake Michigan. *Oakton Springs Hotel* is in every way attractive, and affords every desired convenience.

"**Geneva Lake, Wis.**, is reached from Chicago *via* the Chicago and North-Western Railway, by the Wisconsin Division and Crystal Lake, and *via* the Freeport and Fox River lines *via* Elgin. Through trains run by both these routes.

"Geneva has but recently been reached by the railroad, and has sprung at once into notice as a delightful place for summer resort. The village is situated at the foot of the lake. It has three hotels, all well kept.

"Hills, retaining a portion of their native forest, surround the village on three sides; and from its centre spreads out towards the south a charming bay nearly a mile in width, with high and wooded banks. Beyond the points which bound it, are seen the main lake and the southern shore about three miles away. Bending to the right, it extends south-westwardly some eight or nine miles, its width varying from half a mile to two miles at different points. Its banks are high and wooded, especially near its head, where they rise to bold bluffs. The water is principally derived from springs, and is pure and clear as crystal. The scenery of this lake can scarcely be surpassed for picturesqueness and beauty.

"The lake abounds in fish, such as pickerel, black and rock bass, perch, sun-fish, &c. *Como Lake* and other smaller lakes, also full of fish, may be easily reached from the village. Small boats in great numbers are kept here for hire, and yachts and other sailboats constantly utilize the summer breezes. A nicely finished and well-conducted steamboat, capable of accommodating three or four hundred passengers, was put upon the lake in 1873, and added much to its attractions.

"**Mineral Springs of Palmyra, Wis.** — There are two groups of springs at this place, differing somewhat in their chemical properties, and both claiming their special devotees. Several are chalybeate, while others are strongly impregnated with the different chlorides, bicarbonates, and sulphates.

"**Submerged Spring.** — Here is found one of the most notable of all objects of interest yet discovered in this region, which excites unceasing wonder. Imagine an expanse of water covering an area of one-fourth of an acre, and from one to two feet in depth. Rowing out a few yards, you find yourself suddenly looking down into a cave twenty-five feet in length and twelve across, the sides perpendicular, and covered with verdant foliage. The color of the water next attracts your attention, which appears of the most delicate opaline tints. At the bottom of the cave, fifteen feet below the surface, the fine white sand boils and bubbles unceasingly, like a seething mammoth caldron; and here is the source of the spring. Its finny inhabitants seem suspended in air.

"**Green Bay** is on the east bank of *Fox River*, one mile from where it empties into the bay from which the town of Green Bay takes its name. A fine bridge connects it with Fort Howard. The bay and river afford a perfectly secure harbor, and make this the principal shipping point for Northern Wisconsin. The climate is mild and exhilarating in summer. The cool nights cause mosquitoes to give the place a wide berth. There is excellent fishing in the bay and river, and every facility for boating. At this point the river is a quarter of a mile broad. The bay is dotted with little islands covered with forests, and admirably adapted for picnic and pleasure parties. The adjacent woods are filled with partridge, woodcock, pigeon, and deer in their season, while the bay and river abound in the water-fowl peculiar to this northern latitude. Neighboring streams are stocked with speckled trout, and the tributaries of the bay furnish excellent bass fishing. The place is the most notable in the State for the gayety of its people and select parties. Small steamers, yachts, and boats of all kinds, are to be had for trips on the river and bay. The place is a favorite summer resort for parties from Chicago, St. Louis, the South, and East. The best hotels are the *First National Hotel* and *Beaumont*.

"**Pensaukee** is situated on the west shore of Green Bay, twenty-five miles north of Green Bay City, and five miles south of Oconto City, on the line of the Chicago and North-Western Railway. As a resort for pleasure-seekers and sportsmen it cannot be excelled by any point in Northern Wisconsin. In June and July the streams abound with trout, and Green Bay affords better bass fishing than any point on the entire chain of lakes. In September, October, and November, there is an abundance of deer, duck, and snipe, in the immediate vicinity of the *Pensaukee House*.

"**Escanaba** is one of the pleasantest summer resorts there is in the West. The water of the bay, clear as crystal, washes the streets of the city on two sides, while the Escanaba River forms the third, and the aromatic 'Piney Woods' close well down on the other side. The *Tilden House* offers quiet and comfortable quarters for tourists who may wish to spend days or weeks here fishing, boating, or bathing. White Fish Bay in this vicinity offers rare sport for the fishermen; and every little stream, and they are numerous, is almost alive with the ever beautiful spotted brook trout. From Escanaba excursions are fitted out in various directions. Those not caring for fishing can take to the 'woods,' and find bear and deer in abundance, to say nothing of ducks, geese, brants, partridges, and smaller feathered game.

"**Marquette**, situated on the south shore of Lake Superior, at the eastern terminus of the Marquette, Houghton, and Ontonagon Railroad, is the centre of the great iron region of Lake Superior. The principal business interest is that of mining. It is well supplied with excellent hotels and large summer boarding-houses.

"On the bay you have unequalled facilities for boating, and its waters are filled with *gamey* fish which seem eager to reward the angler, as they are caught in great abundance with but little labor. A few miles out in the bay are several large islands covered with virgin forests. These islands are favorite resorts for picnic parties, who reach them by sail-boats, by steamer, or by small row-boats, of which any number may be hired in Marquette at any time.

"For the invalid, or for the resident of our Southern States, Marquette offers many inducements as a summer resort. Its air is pure and clear, its days not hot, its nights pleasantly cool and yet not too cold, and its healthfulness unquestioned.

"From Marquette you can take steamer for Hancock and Houghton on the Kewenaw peninsula, or for Sault St. Mary, for Isle Royal, St. Ignace Island, Fort William, or any point on the north shore of Lake Superior. On that shore you will find Nature in all her wildness. The white man's arts and ways have not yet penetrated its wilds, and the Indian with his peculiarities can be found without seeking far. This Indian is not the savage of the plains or mountains, but he who has been tamed by the kindly teachings of the patient Catholic missionary, who has been a dweller in the tents of the uncultured child of the forest for generations, and who has lived there really and truly for the Indian's good, and not for the white man's aggrandizement, as is too often the case with the so-called friend of the Indian. No finer trout fishing is to be had anywhere on the broad earth than can be found on the north shore of this great inland ocean. *Speckled trout, weighing from five to twelve pounds,* are often caught by the few adventurous spirits who have for several years sought these favored shores. The rivers *Nipigon* and *Michapacoton* are the best known of the trout streams of the north shore. Guides to these streams can be easily hired at Marquette, and fishing parties can be fitted out with little expense or labor. And here we might drop a hint that may be useful to the stranger: take an Indian for your guide if you go to the north shore to fish; see that you get one who does not love "fire-water," and one that is not afraid of work. Plenty of lazy white men will tender their services, and boast of their knowledge and skill, but touch them not. They are utterly worthless, either in your boat or out of it.

MARQUETTE TO DULUTH, VIA THE NEW LINE OF SIDEWHEEL STEAMERS.

"These elegant boats leave Marquette in the morning, and pass by the Huron Islands, Manitou Island, around Kewenaw Point, past Fort William, Eagle Harbor, Eagle River, Ontonagon, the Pewabic Copper Mines. Copper Harbor, Ashland, Bayfield, and so up to Fond du Lac and Duluth. We have an attractive trip on magnificent boats over the largest lake in the world. Lake Superior is noted for its clear, cold water (it being so clear that from the deck of the steamer you can plainly see the great lake trout playing in the water forty feet below the surface); you pass within sight of the shores of the lake, which are in many places mountainous, and clothed in the verdure of the pine, hemlock, spruce, fir, and other evergreen trees. A more delightful trip for the hot days of summer cannot be had within the bounds of the American Continent. The steamers are large, stanch, finely equipped, and commanded by officers whose superiors in courtesy and kindness cannot be found anywhere.

CHICAGO AND ST. PAUL LINE.

"This is a through line between Chicago and St. Paul of 409 miles, and Chicago and Minneapolis and St. Anthony of 420 miles. *One management controls the route, and trains run through to St. Paul without change of cars.*

"Leaving the Canal and Kinzie Streets Depot of the Chicago and North-Western Railway, you pass through many pleasant villages, such as Irving Park, Desplaines, Barrington, Crystal Lake, and Woodstock, and reach **Harvard**, sixty-two miles from Chicago, where you cross the track of the Kenosha and Rockford Division of this company's lines. At **Caledonia**, seventy-seven miles out, you cross another branch of this road. At **Beloit**, ninety-three miles from Chicago, you cross the Western Union Railroad, and enter the State of Wisconsin. At **Hanover**, one hundred and six miles from Chicago, you cross the Monroe Branch of the Milwaukee and St. Paul Road, running from this point to Monroe, twenty-seven miles distant.

"At **Madison** you are 140 miles from Chicago, and at the capital of Wisconsin, a city of over fifteen thousand people.

"The city is pleasantly situated on an isthmus about three-fourths of a mile wide, between Lakes Mendota and Monona, in the centre of a broad valley, surrounded by heights from which it can be seen at a distance of several miles. *Lake Mendota* lies north-west of the town, is six miles long and four miles wide, with clean gravelly shores, and a depth sufficient for navigation by steamboats. Lake Monona is somewhat smaller.

"At this point you reach a branch of the Milwaukee and St. Paul Railroad, which runs northward to Portage City.

"**The City of Madison.** — A great many efforts have been made to depict the beauties of the location, but no words can convey an adequate idea of what is, indeed, indescribable. The reason of this is, that every new point of observation creates a shifting panorama,—that no two exhibit the same scenery. From any considerable elevation, a circuit of nearly thirty miles in every direction is visible. Four lakes lie embosomed like gems, shining in the midst of groves of forest trees; while the gentle swells of the prairies, dotted over by fields and farms, lend a charm to the view which words cannot depict. On the west, the lofty peak of the *West Blue Mound*, twenty-five miles away, towers up against the sky, like a grim sentinel guarding the gateway to the setting sun, while the intermediate setting is filled in with swelling hills, majestic slopes, levels, and valleys of rivers and rivulets. Madison is the centre of a circle whose natural beauties compass all that is charming to the eye, grateful to the senses, pleasing to the imagination, and which, from its variety and perfection, never grows tedious or tiresome to the spectator. With good taste the citizens have preserved the native forest trees, so that its dwellings are embowered in green and buried in foliage in the proper season, to such an extent that the whole city cannot be seen from any point of view. It is in itself unique, like its surroundings, and the transient traveller gains no conception of the place by barely passing through it.

"**Devil's Lake** is prominent among the summer resorts of the Northwest, thirty-six miles north of Madison, and one hundred and seventy-six miles from Chicago, on the Madison Division of the Chicago and North-Western Railway.

"The bluffs of the Wisconsin, at the point where the *Baraboo River* embouches into the valley, are six hundred feet in height. In the midst of this enormous rocky stratum is a deep fissure, or gorge, depressed over four hundred feet from the surface, hemmed in by mighty precipices, which constitute the basin of a body of water about a mile and a half in length by a half a mile in breadth, known as the Devil's Lake. The level of the waters is one hundred and ninety feet above the Wisconsin river, and it is supposed that the bottom reaches below that of the river. A two-hours' ride on the cars from Madison will land the visitor directly upon the shore.

"The lake is one of the most wonderful and romantic spots in existence; and it has an abiding attraction for tourists, who return to it again and again to admire and enjoy it, to wonder at it, and to puzzle over it. Here, ages ago, probably some terrible internal convulsion rent the earth's surface, and piled various strata of rock, of immense size, from three hundred to six hundred feet high, and disposed it in every conceivable fantastic form. Within the basin thus made lies nestled a beautiful, placid lake of clear, pure water, which reflects on its mirror-like surface the rugged and awe-inspiring barriers which environ it. It has no visible inlet or outlet, but abounds in fish. This marvellous place has attracted the attention of geologists and scientific men for many years; and various theories exist regarding its formation, but not one which has met with general acceptance, or which appears to satisfactorily explain it. Increasing numbers of tourists include it in their round now that it has become so easily accessible by rail.

"On every side of the lake you see 'rock piled on rock' in every conceivable form, and in immense columns, pillars, piles, and masses of very great magnitude and height. The railroad runs along the shore of the lake on a bed that was literally blasted out of the sides of the mountain. From the car windows all the beauties of this wonderful and weirdly mysterious region can readily be seen.

"MINNESOTA.— This picturesque State lies between 43° and 49° north latitude and 89° and 97° west longitude, and is about 380 miles long by 300 miles wide. The surface of the country is undulating. It has no mountains, yet it is the most elevated tract of country between Hudson's Bay on the north and New Orleans on the south. The soil varies very much; in the valleys it is excellent, and especially so in the valley of the St. Peter. Above the Falls of St. Anthony the soil is of 'drift' formation.

"Minnesota has so long been celebrated for its dry, healthful, and invigorating climate, and has been so well brought to the notice of health and pleasure seekers, that it is useless for us to take up your time with any lengthy description of the State, its cities, or in fact, any thing relating thereto: we feel sure that all you need to know is where to go, and *how to get there*, and your own good judgment or choice will supply the rest.

"**St. Paul** is the largest city in the State, and is *the place* to go to *first ;* from that point you can reach all other summer resorts in the State. This city is well supplied with good hotels and first-class boarding houses, in which the summer visitor can find all the comforts and accommodations he may wish. The charges for board are reasonable.

"From St. Paul you can, in one hour or less, reach **Minneapolis** and **St. Anthony,** by the St. Paul and Pacific Railroad. Several trains are run each day between these points and St. Paul. Around St. Paul, and within easy carriage-driving distance, you will find the Falls of St. Anthony, Minnehaha Falls, Carver's Cave, Fountain Cave, White Bear Lake, the Bridal Veil Falls, Lake Como, Fort Snelling, the Fawn's Leap, and so on. A short distance farther off, but within easy access from the city are the beautiful Lakes Harriet and Calhoun, Lake Minnetonka, Cedar Lake, the Lake of the Isles, Twin Lakes, Crystal Lake, and many others 'too numerous to mention.' Each has beauties peculiarly its own. All are full of fine fish; and boats can be had for fishing or sailing at any time during the season. On the shores of most of these lakes are hotels specially fitted up for summer visitors; and around all the lakes are fine farms, in the homes of which you are always welcome, and where you can board for the summer, or for a few days or a few weeks, at nearly nominal rates.

"For those who leave home for a summer of dissipation, of balls and parties, and late hours and fashion's show, with all of its accompanying frivolity and worthlessness, this is *not* the place to find them; and, seek them as you may, you cannot find them here. It *is* the place, however, for those who want to enjoy themselves, to rest, to gain health and strength and relaxation, and to fit themselves for their busy lives at their homes during the balance of the year."

The **Sparta Mineral Well** is one of the attractions of this vicinity, located near the village of Sparta, Minn.

"**Wabash** is situated on the Mississippi River, at the foot of Lake Pepin, nearly opposite the mouth of the Chippewa River.

As a point where the invalid, and the wearied and worn man of business, can rest free from the bustle and cares of life, and recuperate their depleted bodies, no better point can be found.

"To the tourist, the man of pleasure, and the sportsman, rare enjoyments are presented, and no place in the North-West offers as great a variety.

"To the angler it is his Eldorado. Hundreds of spring brooks, tributaries of the Chippewa, O'Buf, and Tombia Rivers, which empty into the Mississippi near the city of Wabasha, are well stored with that sure sporting fish, the speckled trout.

"For larger and more exciting prey, the angler must resort to the Mississippi, where will be found the black, white, and striped bass, weighing from four to ten pounds, which congregate in innumerable numbers at the foot of Lake Pepin, just where the waters of that lake fall into the river proper, making it the most superb trolling grounds in the world for this noble fish. Pike and pickerel abound in the Mississippi in large numbers, and are taken, weighing from five to forty pounds. This exciting sport can be enjoyed by ladies and children, within the city, in perfect safety.

"For the sportsman who follows his dog, and carries his breech-loader, his rapacity here can be glutted; for this is the home of the prairie chicken, the ruffled and pinnated grouse, the woodcock, the quail, and the innumerable family of duck and goose tribe, together with that noble game, the elk, the deer, and the bear; all of which are easy of access by water or land. The visitors at all times can find guides among the remnants of the Indians and their relations, to conduct them to the most certain spots for good sport.

"Ascending the bluffs back of the city, a loveliness and grandeur of scene is spread out before the eye, stretching far away over Lake Pepin, the Horicon of the West, and far up the great valley of the Chippewa River, giving you an isometric view rivalling the far-famed Yo Semite Valley, which is awaiting the pencil of the artist to give it a world-wide fame.

"**Frontenac** has a national reputation for the beauty of its scenery, its perfection of the Minnesota climate, the varied interest of its drives, and its facilities for boating, bathing, trout-fishing, and grouse-shooting.

"Yachts and rowboats of the finest model invite the guests to the pleasures of aquatic sports. A small steamer makes daily trips to the various lake ports. The steamboat landing is on the Lake side Hotel grounds.

"Frontenac is the centre of the finest region in the North-West for both trout-fishing and grouse-shooting. It is one of the very few places where these two sports can be had together; and there is no place where they can be had together in such excellence and with such ease of access to the sportsman. A row or sail of half an hour from the hotel, across the lake, and a pleasant walk of half a mile, introduces the angler to an inexhaustible and never disappointing supply of trout in Pine Creek; and a drive of ten miles will make him acquainted with the three and four pound trout of Rush River, and with the dark forests, the overhanging rocks, and the sparkling waters of that finest of all the trout streams. The hotel accommodation is excellent.

"**Fond du Lac** presents many attractions to the invalid, the sick, and the afflicted, as well as to the robust and healthy. It is surrounded with pleasant places of resort. Lake de Neveu, a beautiful sheet of water, is romantically situated about three miles south-east of the city. Eastward is Elkhart Lake, already famous for its natural beauties, and westward lies Green Lake, a noted summer resort. On all these lakes are pleasure-boats propelled by steam, wind, and man power. The waters of all these lakes furnish a plentiful and various supply of fresh-water fish, where piscatorially inclined ladies and gentlemen can enjoy ample amusement in that line.

"Fond du Lac also claims superiority over other Wisconsin summer resorts, for the season that within her boundaries she has *Magnetic Mineral Springs* that are claimed to be equal, or even superior, to those of Northern Michigan.

"'**Maiden Rock.**'— On the eastern shore of Lake Pepin, about twelve miles from its mouth, there stands a bluff which attracts attention by its boldness. It is about four hundred and fifty feet in height, the last hundred of which is a bald, precipitous crag. It is the Maiden's Rock of the Dakota, and a thrilling Indian tradition attends its history.

PLEASURE ROUTE No. 30.
Chicago to Omaha and the Far West.

GIBBON the historian dwells glowingly on the highway from end to end of the Roman world; that is, from Glasgow to Jerusalem, a distance of 3,709 miles. But this stupendous work was only one-twelfth the length of the present railroad system of the United States, and was not much longer than the miles of track already laid down in Iowa.

Railroads were in the outset far less perfect, while more complicated and costly, than they have since become. The law of progress has governed them, as it indeed governs every thing. Some Western roads have profited by Eastern experience, and from their beginning have introduced those modern appliances for reducing the *risks of travel to a minimum*, and the *comfort to a maximum*.

Specimens of such improvements are the Miller Platform and Coupler, Westinghouse Safety Air Brake (which would have prevented many of the fearful railway catastrophes which sicken the heart), the Pullman Sleeping and Dining-room Cars, and Passenger Coaches, running through from cities widely distant. Travellers from Europe and the far East wonder at seeing these things beyond the banks of the Mississippi.

The **Chicago, Burlington, and Quincy Railroad**, which extends from Chicago to several terminal points on the Missouri River, is a good illustration of a complete highway. This road crosses the Mississippi at Burlington, on one of the finest iron bridges in the world.

No expense has been spared in its construction. Its length is 2,237 feet, resting on piers of solid masonry 18x155 feet at the bottom, and 9x23 at the top, and rising twelve feet above the highest water-mark known. From Burlington the road extends westward to Leavenworth, Atchison, Kansas City, St. Joseph, Plattsmouth, Nebraska City, and Omaha; at the latter point making direct and close connection with the "Union Pacific" for San Francisco and the Territories. The Hotel Car, most convenient of all modern inducements to travellers, is constantly in use on this line; also the celebrated Pullman Sleeping-Car, wherein the traveller can sleep as comfortably while travelling at the rate of thirty miles an hour as in his own bed at home. It is these adjuncts of travel that make this the favorite route across the continent, a trip which should be taken by all who desire to know more of the customs and scenery of the great and growing West.

Assuming that the traveller is familiar with the route as far as Chicago, or at least that he knows how to get there, let him on any fine morning take the 10.15 train on the Chicago, Burlington, and Quincy Railroad, *than which there is no better in the country, en route* for San Francisco. For the first few miles he will be surprised and delighted with the large number of suburban towns, and the completeness of their construction.

NIGHT EXPRESS.
"Blessings on the man who invented the Pullman sleeping-car."

The "Illinois Central Railroad" crosses our route at Mendota, which place we reach after three hours' ride. Here dinner awaits us, the quality and abundance of which are among the noticeable features; and, what is better than all, we have a plenty of time in which to discuss it. Again we are *en route;* and the train is whirled along an immense prairie region, through fields of corn, studded with enterprising towns and thriving farms. At 6, P.M., we reach Galesburg; yet so evenly ballasted is this road that we have not yet thought of fatigue. Galesburg seems like a New England town, magically transplanted to a Western State. The society also is said to be unexceptionable; and few places East or West exhibit more taste or refinement.

PLEASANT VALLEY,
Pacific R.R.

Here a Hotel Car is attached to the train. The safety, pleasure, and comforts of railroad travel have been wonderfully improved during the last few years, but this, one of the latest, will unquestionably be pronounced the *best* of the comfort-seeking inventions yet produced. For the extremely low sum of seventy-five cents the wants of "the inner man" are supplied, — broiled steak and quail, and cakes smoking hot, and no cry of "All aboard!" from the conductor. In a word, while moving without exertion through the air like a bird of passage, we eat, drink and sleep, surrounded by the luxuriant ease and comforts of home; and passengers thus sumptuously regaled are lost to distance. With the setting sun we find ourselves approaching the Mississippi, which we cross to Burlington; the next morning we reach Council Bluffs and at 11, A.M., connect with the Union Pacific Railroad at Omaha.

PLEASURE ROUTE No. 31.

Omaha and the East to Colorado, Wyoming, Montana, Idaho, Utah, and the Pacific Coast.

THE UNION AND CENTRAL PACIFIC RAILROADS.

To publish a book on the popular resorts of the United States without allusion to the many points of interest reached by the Union and Central Pacific Railroads, would, indeed, be a repetition of the old adage of the play of Hamlet with the character of Hamlet left out. Numerous and beautiful as are the resorts and pleasure routes of the East, popular and growing as are those of the North-West, and interesting as are those of the South, the natural attractions of the Far-West eclipse them all. Here Nature's works are displayed in her grandest moods. The mountains of the Atlantic States are but foot-hills to the Sierra Nevadas, and the grandest features of the East would be lost in the awful cañons of the West.

DAKOTA. — The Territory of Dakota is being brought into notice by the recent gold excitement, the *Black Hills* country especially. The opening up of the mining interest will develop the scenic beauties of that region. *Yankton* is reached over the lines just described. *Bismark* is the present terminus of the Northern Pacific Railroad.

The Union Pacific Railroad, receiving Eastern tourists from the several lines converging at Omaha, extends west towards the Pacific, and, with the Central Pacific, forms the main artery of travel. Visitors to Colorado diverge to the left by the Colorado Central Railroad, which leads to one of the most interesting regions of the United States, and opens up some of the grandest scenery in America. Colorado is filled with pleasure resorts, and the facility with which they are reached adds materially to their popularity. To do justice by description (which cannot be done without illustrations) would require a volume of itself.

Denver is the metropolitan centre and business entrepôt of the State. "The traveller who desires to visit the summer resorts of *Colorado Springs*, or the pleasant cities of far Southern Colorado, *Pueblo, Cañon*, or *Trinidad*; who wishes to go into the mountains to Black Hawk or Central, the gold centres ; to *Idaho Springs*, with her wonderful baths and beautiful cañons and peaks ; to Georgetown or Caribou, the silver producing districts ; to Gray's Peak, Long's Peak, Pike's Peak, or Mount Lincoln, the watch-towers of the continent ; to the Parks, with their wealth of mountain scenery, fish, game, mineral springs, and rocky gorges; to *Manitou*, with her medicinal waters ; to the marvellous rocks of the *Garden of the Gods*, or the disintegrated sandstone sentinels of Monument ; whatever section it is desired to reach, — Denver stands to it as the departure point."

MONTANA. — If **the National Park of the Yellowstone** be the objective point, the tourist will continue on the Union Pacific Railroad to Corinne, Utah, at present the nearest approach by rail. From Corinne, the trip is completed partly by stage and by saddle, but should only be undertaken by persons of strong physical endurance, after special preparation.

The region of the Yellowstone, with its stupendous waterfalls, enormous cañons, unrivalled geysers, and its boiling springs, is unquestionably one of the most thrillingly interesting localities on the American continent, and when better known, with improved transportation facilities, will become the "Mecca" for tourists.

It should be understood that a territory about fifty miles square, embracing the head waters of the Yellowstone River, Yellowstone Lake, the Grand Cañon of the Yellowstone, and the Geyser Basins, was, by Act of Congress approved March 1, 1872, set apart "for the benefit and enjoyment of the people," to be known as the "**Yellowstone National Park.**"

It is not proposed to give, in this edition, a detailed description of this interesting section, save a few extracts, by permission, from the official reports of F. V. Hayden, United States Geologist, and N. P. Langford, Superintendent of the Yellowstone National Park.

"The park is at present accessible only by means of saddle and pack trains. . . . It can be visited any time between the last of April and the first of November, but it appears to the best advantage during the months of July, August, and September. . . . Tourists desirous of reaching it by the most picturesque route will proceed by railroad to Corinne, Utah, where they can purchase their outfits cheaper and to better advantage than at any advanced point." * The Yellowstone region can also be reached by the Northern Pacific Railroad to Bismark, Dakota Territory, and continued from there by stage; but tourists familiar with both routes give their preference to the former.

The **Geyser Basins,** Upper and Lower, prove features of great interest to the tourist. The *Lower Basin* is first approached. "The geysers here, though comparatively small, are very wonderful to the eyes of the visitor who first beholds them." Ten miles farther by an interesting route is the *Upper Basin*, in which "there are at least two thousand hot springs, large and small, . . . and of this number probably two hundred are geysers. The whole basin is enveloped in steam, and seen at a distance is like the approach to a cluster of manufactories."

---

* The visitor to any point west of Omaha is advised to correspond first with Thomas L. Kimball, Esq., Omaha, Neb., General Passenger Agent of the Union Pacific Railroad. The frequent changes in routes, excursion-rates, &c., make it desirable to receive the latest information from headquarters.

The geysers project water with terrific force, and in fabulous quantities, and in every conceivable form, to heights varying from twenty to two hundred and fifty feet. These seen in the rays of the midday sun, or in the beams of a full moon, are inexpressibly grand. It is fifteen miles from the basin to Yellowstone Lake." . . . . The lake is nearly eight thousand feet above the ocean. It is twenty-five miles in length, embosomed amid mountains, gemmed with green islands, in form unique, and surrounded on all sides by hot springs in great variety, number, and beauty. Jets of steam may be seen issuing from hot springs, from the islands, even from the bosom of the lake itself. Some of the loftiest and most inaccessible mountain-ridges on the continent lift their snow-clad summits in the immediate vicinity. The scenery is colossal and full of savage grandeur.

GIANT GEYSER,
Yellowstone Park, Montana Territory.

"Following down the river from the foot of the lake nine miles, we reach *Sulphur Mountain*, *Mud Geyser*, *Mud Volcano*, and the *Blowing Cavern*, all objects of separate interest."

CRYSTAL FALLS, ON CASCADE CREEK, 129 FEET.

"Ten miles farther are the two great cataracts, and the **Giant Canon of the Yellowstone**, perhaps the most stupendous elements of scenery in the park. The upper fall is one hundred and fifteen feet in height. Between the two falls, *Cascade Creek* flows into the Yellowstone from the west. A short distance above its mouth is located the picturesque *Crystal Falls*, or cascade, for it is made up of three distinct falls, the aggregate height of which is one hundred and twenty-nine feet. The lower fall of the Yellowstone, which plunges directly into the cañon, is three hundred and fifty feet high, — higher than Bunker Hill Monument, or the spire of Trinity Church, New York; and the cañon itself, varying from one to three thousand feet in depth, is forty miles in length, and for the whole distance presents to the eye the most wonderful chasm in the world. . . . Lieut. Doane, who in 1870 succeeded in reaching the bottom, at a point where the walls are nearly three thousand feet in height, in his official report says. ' It was about three o'clock, P. M., and *stars could be distinctly seen*, so much of the sunlight was cut off from entering the chasm.'

" Evidences of volcanic action are everywhere visible." There is in this Yellowstone Range an unlimited field for the artist, the scientist, and pleasure-seeker.

Before resuming our tour westward on the line of the Central Pacific Railway, which we struck at *Ogden*, the tourist should devote some attention to Salt Lake City and the valley in which it lies. This valley is about ninety by fifty miles in extent, and has an elevation of four thousand feet above the sea, and is noted for the remarkable purity of its atmosphere.

Salt Lake City now numbers about twenty thousand souls; it covers an area of twenty-seven square miles, and is surrounded by rugged snow-capped mountains. The *Salt Lake House*, *Townsend*, and *Revere* are the principal hotels. The city contains much to interest a stranger, and should not be ignored by the tourist.

The **Warm Springs** are only two miles from the city, and will attract attention: but the **Hot Springs** are the most interesting. The water is projected forcibly from an aperture in the rocks, at a temperature too high to admit of bathing, and sufficient, it is claimed, to cook eggs. It is highly impregnated with sulphur.

Continuing our trip westward, if the tourist has still a desire to visit the rush of mighty waters, he should see the **Great Shoshone Falls** of Idaho, located about three hundred miles north-west from Great Salt Lake City, and one hundred and eighty-five miles from Boisé City, and within six miles of the stage road. The *Shoshone River* has here cut a cañon to the depth of one thousand feet, and nearly a half mile in width, widening to one mile at the falls. A few hundred yards above the falls, the stream is divided into six channels by five islands standing in the bed of the river; uniting again, the current rushes on to the final leap, which is made in an unbroken sheet, plunging sheer down to the depth of two hundred feet, where it is broken into clouds of foam and spray.

CALIFORNIA. — The Western tourist who continues his travels to the Pacific coast should allot sufficient time to study the wonderful features of this interesting State. Its many resorts should be illustrated to be appreciated, which it is hoped the next edition of "Popular Resorts" will supply. **Lake Tahoe** and *Donner Lake* are the two principal bodies of fresh water, and have already attained a great popularity with tourists.

Lake Tahoe, one of the most beautiful sheets of water west of the Rocky Mountains, lies in two States and five counties, California and Nevada dividing the honor of its ownership. It is twenty-two miles in length, and ten in width. Its shores are indented with lovely little bays, two of which, *Emerald* and *Cornelian Bays*, have been for several years the favorite resort of tourists. The visitor will leave the cars at *Truckee City*, at which place he will find three well-conducted hotels, the largest of which, the *Truckee House*, is considered the headquarters of tourists. Daily stages leave for *Lake Tahoe* and *Donner Lake*, the first twelve miles south, and the latter three miles to the north-west. Tahoe City, at the foot of the lake, contains excellent accommodations ; a good hotel, livery stable, pleasure boats, &c. A steamboat makes daily pleasure trips about the lake for the entertainment of tourists.

Donner Lake lies within three miles of Truckee, and, though but small compared with the latter, by its romantic surroundings is destined to become famed as a resort. It is three and a half miles long and one mile wide. It has been sounded to the depth of two thousand feet, but no bottom has ever been found. It is supposed to be the filled-up crater of an extinct volcano. Its waters are intensely cold and clear. On all sides of it wooded mountains rise abruptly to a considerable height.

Calistoga is one of the most popular of the summer resorts on the Pacific coast, and is near San Francisco Bay. The medicinal qualities of its springs annually draw large numbers of visitors, and in the heat of the summer it is a California Saratoga. The hunting and fishing in the neighborhood are unsurpassed, while the surrounding scenery is delightful. It is reached from San Francisco by steamer to Vallejo, distance twenty-eight miles, at which point the cars are taken over the Napa Valley Railroad, forty-three miles more. The hotels are as good as can be found in the State, and the conveniences all that could be asked for. Five miles south-east of *Calistoga Hot Springs* is located the **Petrified Forest**, of recent discovery, and great scientific interest. The petrifaction is perfect, and the place well deserves the attention of tourists. Many of the trees are so well preserved that the species of wood and character of growth are easily determined.

Mount St. Helena, an extinct volcano, having an altitude of 4,343 feet, is but ten miles distant from Calistoga, and is attracting a good share of attention.

The Great Geyser Springs are in Sonoma County, and are impregnated with iron, sulphur, and soda; and the famed localities are severally known as the *Iron and Sulphur* spring, temperature 73° Fah.; *Alum and Iron* incrusted, temperature 97°; Medicated Geyser Bath, temperature 88° 8; *Boiling Alum and Sulphur*, temperature 156°; *Black Sulphur; Epsom Salts Spring*, temperature 146°; close by a spring of iron, sulphur, and salt, at boiling point, and *Boiling Black Sulphur;* the *Witches' Caldron*, seven feet in diameter, is in continual ebullition, temperature 195° 5'; near this, *Alum Spring*, temperature 176°; *Intermittent Scalding Spring*, projecting water fifteen feet; Steamboat Geyser, ejecting steam with a great noise. *Scalding Steam Iron Bath*, temperature 183°. There are many other objects of interest in this vicinity.

Yosemite Falls and Big Trees. — This must always form an object point for the visitor to the Pacific slope. Its description is a proper theme to close a work on "Popular Resorts." It is a grand and fitting after-piece for such an interesting subject. The Yosemite Valley is situated in Mariposa County, one hundred and forty miles south by east from San Francisco, though two hundred and fifty by the route of approach. It is drained by *Merced River*. This whole region must be described by superlatives. Its mountains reaching an altitude of 6,450 feet, its stupendous waterfalls to the heights of 600 feet, 700, 940, 2,634, and 3,300, are beyond comparison, and must be seen to be fully appreciated. Its trees rival the waterfalls, and form one of the most remarkable wonders of the world. This valley is but forty-eight hours from San Francisco by the Central Pacific Railroad, and no one should fail to see it.

Within the past few months a new and wonderful feature has been discovered in the Yosemite region. It is in **Tuolumne River Canon**, seventeen miles north of the Yosemite, now so well known. The Tuolumne River which is much wider than the Merced, runs through this cañon.

The *Tuolumne Canon*, with its connections, has an unbroken length of forty miles. For twenty miles of this distance it is shut in by vertical clean-cut walls of granite. Some of these walls *are several hundred feet higher than the very highest in the Yosemite Valley.* The falls of the Merced Yosemite surpass those of the Tuolumne Canon in unbroken volumes of descending water, but in endless variety of cascades and water-shoots the Tuolumne Canon is far superior.

The Big Tree Grove, Calaveras County, known as the "Big Trees of California," can be reached with only forty miles staging. The region is provided with excellent hotel accommodations, with good guides, and telegraph communications.

INDEX.

Abraham, Mount, Me., 94.
Adams, Mount, 30.
Aiken, S C , 280.
Alabama and Chattanooga Railroad, 298.
Alabama Central Railroad, 299, 305.
Albany and Susquehanna R.R., 142, 169.
Albany, N.Y., 194.
Alburgh Springs, Vt., 324.
Alexandria Bay, N.Y., 178, 196.
Allegash, Lake, Me., 95.
Alleghany Mountains, 253, 279.
Alleghany Springs, Va., 302.
Alleghany Station, Va., 279.
Allentown, Penn., 145, 146, 219.
Alpina, Mich., 314
Altoona, Penn., 251.
Alton, N.H , 39.
Alton Bay, N.H., 41, 70.
Alum and Iron Springs, Cal., 346.
Alum Spring, Cal., 346.
Amicalolah Falls, Ga., 297.
Ammonoosuc Falls, N H. (cut), 61.
Ammonoosuc River, N.H., 31, 37, 51, 56.
Ancient Mounds, Ky., 312.
Androscoggin River, 31, 55, 92.
Annanance (Willoughby Mountain), Vt , 63.
Appomattox, Va., 301.
Ardmore Station (cut), 249.
Artesian Letitia Springs, N.Y., 195.
Artist's Falls, N.H , 72.
Ashland, N.H., 45.
Ashland, N Y., 138.
Ashley, Penn., 164.
Ashley Lake, Mass., 123.
Ashley River, S.C., 291.
Ashville, N C., 295.
Atlanta, Ga., 298.
Atlantic City, N.J , 230.
Atlantic Coast Line, 289.
Atlantic and Great Western R.R , 209, 210.
Atlantic and Gulf Railroad, 292.
Augusta Stribling Springs, Va , 275.
Au Sable Chasm, N.Y., 178.
Au Sable Forks, N.Y., 179
Au Sable Lake (lower), N.Y., 179.
Au Sable Lake (upper), N.Y., 179.
Au Sable River, N.Y., 179.
Avalanche Lake, N.Y , 180
Avon Springs, N.Y., 208
Aztalan, Wis., 321
Bailey's Springs, Ala., 306.
Baker's Inn, N.Y , 180
Bald Head Cliff, Me , 70.
Bald Knob, Va., 303.
Bald Mount, N.C., 297.
Ballston, N Y., 195
Baltimore, Md., 256.
Baltimore and Potomac Railroad, 255.
Baltimore Steam Packet Company (Old Bay Line), 241.
Baraboo River, Wis., 353.
Barden House, Me , 94.
Bar Harbor, Me., 78

Bass Lake, Mich., 316.
Bastion, N Y , 140
Bath Alum Springs, Va., 277.
Battle House, Ala , 299.
Bay City, Mich., 314.
Bay of St. Louis, Miss., 299.
Bay View, N.H., 39.
Bear Cliff Falls, N H., 207.
Bear Mount, Vt., 64.
Bear's Den, N Y., 139.
Beaumont Hotel, Wis., 330.
Beaverkill, N.Y., 138.
Bedford Alum Springs, Va , 304.
Bedford Springs, Penn., 254.
Beech Lake, N.Y , 182.
Beecher's Falls, N H , 90.
Beersheba Springs, Ala., 306
Belknap Mountain, 41
Bellows Falls, Vt., 119, 122.
Belmont Glen, Penn., 214.
Belmont, N.H., 39.
Belvidere Railroad, Penn., 145.
Berkshire Hills, Mass., 123.
Berlin Falls, N.H., 92.
Berry Pond, Mass., 123.
Berwick, Me., 70.
Bethel, Me., 92.
Bethlehem, Penn., 146.
Bethlehem, N.H., 53, 56.
Betsey Lake, Mich., 316.
Beyond Chicago, 318.
Big Bone Cave, Tenn., 305.
Big Bull Falls, Wis., 224.
Big Indian, N.Y., 138.
Big Moose Lake, N.Y., 182.
Big Tree Grove, Cal., 346.
Biloxi, Miss., 299.
Binghamton, N Y., 207.
Birdsboro', Penn., 218, 237.
Birmingham Station, Ala., 313.
Bismark, Dakota, 340.
Blackberry Mount, Can., 65.
Black Chasm Falls, N Y , 140.
Black Head, N Y , 141.
Black Hills, Dakota Ter., 340.
Black Mount, N.C , 295.
Black River Rapids, N.Y., 196.
Black Rock, N.Y., 209.
Black Sulphur Spring, Cal., 346.
Black Sulphur Springs, Tenn., 305.
Blood's Inn, N.Y., 180
Bloomsburg, Penn , 229.
Blout's Springs, Ala., 306.
Blowing Cavern, Mon., 343.
Blue Lick Springs, Ky., 309.
Blue Mounds, Wis., 324, 333.
Blue Mountain Lake, N.Y., 182.
Blue Mount, Me , 94
Blue Mount, N.Y., 180.
Blue Ridge Mountain, Va., 274.
Blue Springs, Ala., 313.
Boardman River, Mich., 316.
Boar's Head, N.H., 66.

347

Boiling Alum and Sulphur, Cal., 346.
Boiling Black Sulphur Spring, Cal., 346.
Bolton Falls, Vt., 120.
Bolton Springs, Vt., 64.
Bomaseen Lake, Vt., 125.
Bon Aqua Springs, Tenn., 305.
Bonsack's Station, Va., 302.
Boston, Mass., 25, 33.
Boston and Albany Railroad, 119.
Boston and Maine Railroad, 73.
Boston and Providence Railroad, 126.
Boston, Lowell, and Nashua Railroad (cut 34), 119.
Bowling Green, Ky., 310.
Brandywine Springs, Penn., 237.
Brantingham Lake, N.Y., 196.
Bread-Loaf Inn, Vt., 125.
Bread-Loaf Mount, Vt., 125.
Bridal Veil Falls, Minn., 334.
Bridgeport, Conn., 130.
Bridgeport, Penn., 259.
Bridgton, Me., 84.
Bristol, Va., 304.
Brookside, Penn. (cut), 223.
Bryant's Pond Station, Me., 92.
Bryn Mawr, Penn. (cut), 245.
Buffalo, N.Y., 209.
Burkeville, Va., 301.
Buttermilk Falls, N.Y., 194, 201.
Buttes des Mortes Lake, Wis., 328.
Buzzard's Bay, Mass., 109.
Cæsar's Head Mountain, N.C., 296.
Cairo and Fulton Railroad, 308.
Calaveras Big Tree Grove, Cal., 346.
Caledonia, Wis., 332.
Calhoun Lake, Minn., 334.
California, 24, 345.
Calistoga Hot Springs, Cal., 345.
Camden, Me., 79.
Camden, S C., 290.
Camden and Atlantic Railroad, 230.
Camel's Hump Mount, Vt., 120.
Camp Stove, 17.
Campton, N.H., 49.
Canada, 64.
Canandaigua, N.Y., 267.
Canandaigua Lake, N.Y., 202.
Canaseraga Creek Falls, N.Y., 198.
Canon, Col., 340.
Canon River, Cal., 346.
Canterbury, N.H., 39.
Cape Arundel, Me., 66, 73.
Cape Cod, Mass., 101.
Cape Fear River, N.C., 290.
Cape May, N.J., 252.
Cape Vincent, N.Y., 197.
Caribou, Col., 340.
Carter Mount, N.H., 93.
Carter's Falls, Vt., 125.
Carver's Cave, Minn., 334.
Cascade Bridge, N.Y., 206.
Cascade Creek, Mon., 343.
Casco Bay, Me., 76.
Castine, Me., 78.
Castle Pinckney, S.C., 291.
Catawissa Creek, Penn., 226.
Catawissa, Penn. (cut), 228.
Catawissa Railroad, Penn., 157, 225.
Catoosa Springs, Ga., 298.
Catskill Mountains, N.Y., 136.
Cancomgomosis Lake, Me., 95.
Cancomgomuc Lake, Me., 95.
Caughnawaga, Can., 65.
Causilor's Springs, N.C., 289.

Cauterskill Clove, N.Y., 140.
Cauterskill Falls, N.Y., 140.
Cauterskill Lakes, N.Y., 140.
Cauterskill Station, N.Y., 139.
Cave (Big Bone), Tenn., 305.
Cave City, Ky., 309.
Cave, Diamond, Ky., 312.
Cave Hotel, Ky., 309.
Cave House, Howe's Cave (cut), 172.
Cave, Indian, Ky., 312.
Cave, Mammoth, Ky., 309.
Cave, White's, Ky., 312.
Cayuga Lake, N.Y., 200.
Cazenovia, Canastota, and De Ruyter Railroad, 198.
Cazenovia Lake, N.Y., 198.
Cedar Lake, Mich., 316.
Cedar Lake, Minn., 334.
Centre Harbor, N.H., 41, 70.
Central Pacific Railroad, 340.
Central Railroad of New Jersey, 142, 182, 225.
Central Vermont Railroad, 119, 196.
Chalybeate Spring, Tenn., 305.
Chalybeate Springs, Ga., 298.
Champlain, Lake, 178.
Champlain Springs, Vt., 121.
Chandler Spring, Ala., 299.
Charleston, S.C., 291.
Charleston, West Va. (cut), 287.
Charlevoix, Mich., 316.
Charlottesville, Va., 274.
Charlottetown, P.E.I., 96.
Chatham Four Corners, N.Y., 119.
Chattanooga, Tenn., 306.
Chautauqua Lake, N.Y., 210.
Chelsea, Mass., 66.
Chelsea Beach, Mass., 66.
Cherokee Springs, Ga., 298.
Cherry Valley Branch Railroad, 171.
Chesapeake and Ohio Railroad, 272.
Chesapeake Bay, 238.
Cheshire Railroad, 119.
Chesuncook Lake, Me., 95.
Chicago, 314, 317, 332.
Chicago and North-Western Railroad, 320.
Chicago and St. Paul Line, 332.
Chicago, Milwaukee, and St. Paul Railroad, 332.
Chickahominy Earthworks, Va. (cut), 270.
Chippewa Falls, Wis., 321.
Chippewa River, Minn., 335.
Chimneys, N.C., 296.
Chilhowee Springs, Va., 304.
Christiansburgh, Va., 303.
Chittenango, N.Y., 198.
Chocorua, Mt., N.H., 30, 71, 84.
Cincinnati, Cumberland Gap, and Charleston Railroad, 305.
City Hotel, La., 300.
City of Rocks, Tenn., 306.
Clam Lake, Mich., 316.
Clarendon Springs, Vt., 124.
Clarksville, Ga., 297.
Clay, Mount, N.H., 30.
Clear Lake, Minn., 334.
Cleveland Springs, N.C., 289.
Cleveland, O., 314.
Clifford House, Mass., 99.
Clinton, N.H., 30, 91.
Cloud Point, Penn. (cut), 160.
Coal Breaker (cut), 156.
Coal Transport (cut), 230.
Coatesville Bridge, Penn. (cut), 242.

Cobles Kill Junction, N.Y., 171.
Cohasset, Mass., 99.
Cohasset Narrows, Mass., 102.
Colden Lake, N.Y., 180.
Colden Mt., N.Y., 180.
Colebrook, N.H., 56.
Colorado, 24, 340.
Colorado Springs, 340.
Columbia, S.C., 290.
Columbia Bridge (cut), 214.
Columbia Hotel, N.J., 232.
Columbia Hotel, S.C., 290.
Como Lake, Minn., 334.
Concord, N.H., 37.
Concord Railroad, 33, 119.
Congaree Falls, S.C., 290.
Congress Hall, N.J., 232.
Conneaut Lake, Penn., 210.
Connecticut Lake, N.H., 56, 93.
Connecticut River, 37, 55.
Conway, N.H., 41, 70, 84.
Conway Branch Railroad, 70.
Cooper House, N.Y., 170, 207.
Cooper River, S.C., 291.
Cooperstown, N.Y., 169.
Cove Hill, N.Y., 180.
Covington, Ky., 308.
Cornelian Bay, Cal., 345.
Corning, N.Y., 207.
Cotuit Port, Mass., 101.
Covington, Va., 279.
Cow-Pasture River, Va., 277.
Coyner's Spring, Va., 302.
Crawford House, Boston, 118.
Crawford House, N.H., 61, 72, 90.
Cresson, Penn., 233.
Cristfield, Del. (cut), 235.
Crystal Cascade, N.H., 72, 93.
Crystal Falls, Mon., 343.
Crystal Lake, Minn., 334.
Crystal Lake, Wis., 332.
Croton Lake, N.Y., 193.
Croton Point, N.Y., 135, 193.
Crooked Lake, Mich., 316.
Crooked, or Keuka Lake, N.Y., 202.
Crosby Mt., N.H., 119.
Crosby Side, N.Y., 178.
Crowders Mt., N.C., 296.
Crown Point, N.Y., 178.
Cumberland Gap, Tenn., 304.
Cumberland Valley Railroad, 254.
Cumberland Mt., Tenn., 304.
Dakota Ter., 340.
Dalles of the St. Croix, 321.
Dalles of the St. Louis, 321.
Dalton, N.H., 53.
Damariscotta Lake, Me., 95.
Danville, Penn., 229.
Dartford, Wis., 326.
Day Boats (Hudson River), 136.
Decatur, Ala., 306.
Delaware and Hudson Railroad, 163, 207.
Delaware Bay, N.J., 234.
Delaware, Lackawanna, and Western Railroad, 145, 186, 207.
Delaware Water Gap, Penn., 145.
Denver, Col., 340.
Detroit, Mich., 314.
Devil's Lake, Wis., 321, 322, 333.
Devil's Oven, N.Y., 208.
Diamond Cave, Ky., 312.
Diamond Lake, Mich., 316.
Dowdy Creek, Va., 284.
Duluth, Wis., 332.

Dial Mt., N.Y., 180.
Diana's Baths, N.H., 72 (cut 87).
Dismal Pool, N.H., 86.
Dix's Peak, N.Y., 183.
Dixville Notch, N.H., 55, 93.
Donner Lake, Cal., 345.
Dover, N.H., 53.
Drennon Black Sulphur Springs, Ky., 309.
Dunkirk, N.Y., 219.
Dunmore Lake, Vt., 125.
Dutchman's Run, Penn. (cut), 261.
Duxbury, Mass., 99.
Eagle Hotel, N.H., 57.
Eagle Lake, N.Y., 142.
Eastatia Falls, Ga., 297.
Eastern Provinces, 96.
Eastern Railroad, 66.
Easton, Penn., 145.
East Pennsylvania Railroad, Penn., 219.
East Tennessee and Virginia Railroad, 305.
Echo Lake, N.H., 49, 72.
Edgartown, Mass., 117.
Eggleston Springs, Va., 303.
Elephantis Mt., Canada, 64.
Elizabeth, Ky., 309.
Elkhart Lake, 336.
Elk Lake, Mich., 316.
Elkmont Springs, Tenn., 313.
Elk River, Mich., 316.
Elkton, Del., 233.
Elliott's Knob, Va., 276.
Elmira, N.Y., 207, 262.
Emerald Bay, Cal., 345.
Empire Falls, N.Y., 261.
Enfield Glen Falls, N.Y., 201.
Enterprise, Ga., 293.
Ephrata Springs, Penn., 219.
Epsom Salt Spring, Cul., 346.
Equinox House Vt., 124.
Equinox Mt., Vt., 124.
Erie Railway, 136, 204, 232.
Escanaba, Wis., 330.
Esopus Valley, N.Y., 140.
European and North American R'way, 96.
Eutaw Springs, S.C., 290.
Fabyan House, 33, 53 (cut 60), 90, 91.
Fairhaven, Mass., 100.
Fairmount Park, Penn., 214, 246.
Fairy Springs, N.Y., 139.
Falling Spring Falls, Va., 278.
Fall River, Mass., 33.
Falmouth Heights, Mass., 101.
Falmouth House, Me. (cut), 75.
Falls:—
  Alhambra, N.Y., 195.
  Amicalolah, Ga., 297.
  Amoskeag, N.H., 37.
  Artist's, N.H., 72.
  Bastion, N.Y., 140.
  Bear Cliff, N.Y., 207.
  Beecher's, N.H., 90.
  Bellows, Vt., 119, 122.
  Berlin, N.H., 93.
  Big Bull, 321.
  Black Chasm, N.Y., 140.
  Black River, Mass., 122.
  Black River (Rapids), N.Y., 196.
  Bolton, Vt., 120.
  Bridal Veil, Minn., 334.
  Buttermilk, N.Y., 194.
  Canaseraga Creek, N.Y., 198.
  Carter's, Vt., 125.
  Cauterskill, N.Y., 140.
  Chippewa, Wis., 321.

Falls *continued:* —
Clyde River, Vt., 64.
Congaree. S.C., 290.
Crystal Cascade. N.H., 72, 93.
Crystal, Mon. (cut). 343.
Eastatia, Ga., 297.
Elk, Mich., 316.
Empire, N.Y., 263.
Enfield Glen, N.Y., 201.
Falling Spring Falls, Va., 278.
Falls of the Yellowstone (upper), Mon., 344.
Falls of the Yellowstone (lower), Mor., 344.
Fawn's Leap Falls, Minn., 334.
Fawn's Leap Falls, N.Y., 140.
Flume Falls, N Y., 201.
Foaming Falls. N Y., 201.
Forrest Falls, N.Y., 201.
Ganoga Falls, Penn., 186.
Genesee Falls, N.Y., 202.
Gibbs Falls, N.H., 91.
Glen Ellis Falls, N.H., 72, 93.
Goodrich Falls, N.H., 72.
Grandfather Bull Falls, 321.
Great Falls of the Catawba, N.C., 296.
Great Shoshone Falls, Idaho, 344.
Hain's, N.Y., 140.
Hector, N.Y. (cut), 265.
High, N.Y., 193, 205.
Hooksett, N.H., 36.
Horse Shoe, N Y., 208.
Ithaca, N.Y., 201.
Jones, Md., 257.
Kauterskill, N.Y., 137.
Lawrence, Mass., 37.
Linnville, N.C., 295.
Little Stony, Va., 303.
Lowell, Mass., 37.
Lyons, N.Y., 196.
Middle, N.Y., 208.
Mildam, N.Y., 193
Minnehaha, Minn., 334
Nayaug, Penn. (cut), 165.
Niagara, N.Y., 203, 267.
Ossipee, N.H., 71.
Passaic, N.J., 204.
Portage, N.Y., 208.
Pulpit, N.Y., 201.
Puncheon Run, Va., 302.
Rainbow, N.Y. (cut), 264.
Ramapo, N.Y., 204.
Richmond, Va. (cut), 283.
Rocky, N.Y., 201.
Rocky Heart, N.Y., 195.
Rumford, Me., 92.
Sawkill, N.Y., 206.
Shawanagan, Can., 65.
Sherman, N.Y., 195.
Slicking, N.C., 297.
St. Anthony, Minn., 24, 334.
Steep, Me., 82.
Sugar, N.Y., 196.
Tallulah, Ga., 297.
Taughannock, N.Y., 201.
Thompson's, N.H., 72.
Toccoa, Ga., 297.
Towalaga, Ga., 297.
Trenton, N.Y., 195.
Triphammer, N.Y., 201.
White Water Cataracts, N.C., 297.
Yo Semite, Cal., 346.
Falls of St. Anthony, Minn., 24.
Falls Village Bridge, Penn. (cut), 212.
Falls Village, Penn., 216.
Farmington, Me., 94.
Farnville, Va., 301.
Fawn's Leap Falls, Minn., 334
Fawn's Leap Falls, N.Y., 140.
Fernandina. Fla., 283.
Fife Lake, Mich., 316.
First National Hotel, Wis., 339.
Fisher's View, Va., 302.
Fishkill N.Y., 194.
Fitchburg Railroad, 119, 121.
Flint and Père Marquette Railroad, 314.
Florida, 24, 282.
Flume (Dixville Notch), N.H., 55.
Flume (Franconia Mountains), N.H., 49.
Flume Falls, N.Y., 201.
Foaming Falls, N.Y., 201.
Fond du Lac, Wis., 332, 336.
Forest House, Wis., 326.
Forrest Falls, N Y., 201.
Fort Holmes, Mich., 316.
Fortification Rock, Wis., 322.
Fort Lee. N.Y., 135.
Fort Mackinaw, Mich., 316.
Fort Moultrie, S.C., 291.
Fortress Monroe, Va., 239.
Fort Point, Me., 79.
Fort Pownal, Me., 79.
Fort Pulaski, Ga., 292.
Fort Ripley, S.C., 291.
Fort Snelling, Minn., 334.
Fort Trumbull Conn., 130.
Fort Tryon, N.Y., 135.
Fort Washington, N.Y., 135.
Fort William Henry Hotel, N.Y., 177.
Fountain Cave, Minn., 334.
Fox River, Wis., 328.
Franconia Mountains, N.H., 35, 49, 50.
Franconia Notch, N H., 48.
Frankinstein Cliff, N.H., 88.
Franklin Mt., N.H., 30, 32.
Franklin, Penn., 210.
Fredericton, New Brunswick, 96.
Frontenac, Minn., 336.
Frost's Point, N.H., 69.
Fryeburg, Me., 84.
Fulton Lake, N.Y., 182.
Gainesville, Ga., 298.
Ganoga Falls, Penn., 186.
Garden of the Atlantic Coast, Del., 233.
Garden of the Gods, Col., 340.
Gas Spring, N Y., 210.
Genesee Canal, N.Y., 208.
Genesee Falls, N.Y., 202.
Geneva Lake, Wis., 322.
Georgia Railway, 258.
Georgetown, Col., 340.
Gettysburg, Penn., 250, 257.
Geyser Basins, Mon., 341.
Giant Cañon of the Yellowstone, Mon., 343.
Giant Geyser. Mon., 341.
Gibbs Falls, N.H., 91.
Gibraltar Mount, Penn., 218.
Gilmanton, N.H., 39.
Ginger Cake Rock, N.C., 296.
Glade Spring Station, Va., 304.
Glen Ellis Falls, N.H., 72, 93.
Glen Excelsior, N.Y., 263.
Glen Flora Mineral Springs, Wis., 322.
Glen House, N.H., 72, 87, 93.
Glen Mountain House, N.Y. (cut 106).
Glen Onoko, Penn., 154.
Gloucester, Mass., 66, 67.
Glen Station, N.H., 87.

INDEX.   351

Glen Thomas, Penn. (cut), 161.
Godbald's Mineral Wells, Miss., 307.
Goodrich Falls, N H., 72.
Gordonsville, Va., 274.
Gorham, N.H., 92.
Goshen, N.Y., 205.
Grand Central Depot, N.Y. (cut), 129, 193, 197.
Grandfather Bull Falls, Wis., 321.
Grand Rapids and Indiana Railroad, 315.
Grand Traverse Bay, Mich., 316.
Grand Traverse Region, Mich., 315.
Grand Trunk Railway, 92, 119.
Grass Lake, Mich., 317.
Grass River, Mich., 317.
Gray's Peak, Col., 340.
Great Falls of the Catawba, N.C., 296.
Great Falls, N.H., 70.
Great Geyser Springs, Cal., 346.
Great Shoshone Falls, Idaho, 344.
Great Southern Mail Route, 289.
Green Bay, Wis., 330.
Greenbrier River, Va., 281 (cut 282, Kanawha).
Greenbrier White Sulphur Springs, Va. (cut), 280.
Green Cove Springs, Fla., 292.
Green Lake, Minn., 336.
Green Lake, Wis., 326.
Green Lakes, N.Y., 198.
Green Mountains, Vt., 171.
Green Ridge, Penn., 163.
Greensboro', N.C., 294.
Green Springs, Fla., 293.
Greenville, N.C., 296.
Greenwood Lake, N.Y., 204.
Greenwood Springs, Miss., 307.
Gretna, La., 300.
Greylock Mt., Mass., 121, 124.
Griffith's Knob, Va., 277.
Gulf Shell Road, Ala., 299.
Hain's Falls, N.Y., 140.
Halifax, E.P., 96.
Hamilton and Dayton Railroad, 315.
Hampton Junction, N.J., 145.
Hampton, N.H., 66, 68.
Hampton, Va., 239.
Hancock, Mich., 331.
Hanging Rock, N.C., 296.
Hannah's Hill, N Y., 170.
Hanover Junction, Penn., 257.
Hanover, Wis., 332.
Harbor and Coastwise Excursions, 21.
Harlem Railroad, 119.
Harriet Lake, Minn., 334.
Harrisburg, Penn., 145, 251 (cut 259).
Harrison, Me., 82.
Harrodsburg Springs, Ky., 308.
Hart's Ledges, N.H., 72.
Harvard, Wis., 332.
Harvey's Lake, Penn., 182.
Harwick Lake, N.Y., 207.
Havana Glen, N.Y., 263.
Haverhill, Mass., 41.
Haverhill, N.H., 51.
Haverstraw, N.Y., 135.
Havre de Grace, Md., 238.
Hawk's Bill, N.C., 296.
Hawk's Nest Mount, Va., 271.
Hawley, Penn., 166, 206.
Haystack Mount, Vt., 125.
Healing Spring, S.C., 290.
Healing Springs, Va., 277, 279.
Hensonville, N.Y., 138.

Herdic House, Penn., 260.
Hermitage, Tenn., 313.
Herndon, Penn. (cut), 224.
Hewitt Lake, Mich., 316.
Hickory Nut Gap, N.C., 295.
High Falls, N.Y., 205.
Highgate Springs, Vt., 121.
Highland Lake, Penn., 17, 186.
High Peak, N.Y., 141.
Hillsboro', N.C., 294.
Hingham, Mass., 98.
Hints to Tourists, 23.
Holston Springs, Va., 304.
Hooksett Mount, 35.
Hoosac Mount, Mass., 125.
Hoosac Tunnel, Mass., 123.
Horicon Lake, Wis., 322.
Hor Mount, Vt., 64.
Horse Shoe Falls, N.Y., 208.
Hot Springs, Ark., 307.
Hot Springs, Utah, 344.
Hot Springs, Va., 277, 279.
Houghton, Mich., 331.
Howard's Creek, Va., 281.
Howe's Cave, N.Y., 171.
Hudson Highlands, N.Y., 194.
Hudson Railroad, 119.
Hudson River, 134, 179.
Hunter, N.Y., 137, 140.
Hunter's Mount, N.Y., 140.
Hunter's Glen, N.Y., 140.
Huntington, W. Va. (cut 288).
Huntsville, Ala., 306.
Huron Islands, Lake Superior, 332.
Hyannis, Mass., 100.
Hygeia Hotel, Va., 240.
Idaho Springs, Dakota Ter., 340.
Indian Cave, Ky., 312.
Indian Rock, Ky., 3 2.
Indian Springs, Ga., 298.
Inside Line, 21.
Intermediate Lake, Mich., 316.
Intermittent Scalding Spring, Cal., 346.
International and Great Northern R.R., 308.
International Bridge, N.Y., 209.
Iowa, 23.
Iron and Sulphur Springs, Cal., 346.
Irvington, N Y., 193.
Irvington Park, Wis., 332.
Isle Royal, Lake Superior, 331.
Isles of Shoals, N.H., 66, 69.
Ithaca, N.Y., 207.
Ithaca Falls, N.Y., 201.
Iuka, Ala., 306.
Jackson's River, Va., 278.
Jacksonville, Fla., 292.
James River, Va., 270, 278.
Jamestown, N.Y., 210.
Jay Peak, Vt., 64.
Jefferson Mount, Penn., 152.
Jefferson, N.H., 55.
Jerry's Run, Va., 299.
Johnson's Wells, Ala., 306.
Jones Falls, Md., 237.
Jones Lake, Penn. (cut), 167.
Juniata River, 251.
Kalkaska, Mich., 316.
Katahdin Mount, Me., 95.
Katama Bay, Mass., 108.
Kauterskill Falls, N.Y., 137.
Kearsarge, N.H., 119.
Keeseville, N.Y., 179.
Kennebec River, Me., 27.
Kentucky Central Railway, 309.

Kewenaw Point, Mich., 331.
Kiarsarge (Pequawket), N.H., (cut 86), 119.
Killington Peak, Vt., 124.
Kimball House, Atlanta, Ga., 299.
Kineo Mount, Me., 95.
King's Mount, N C., 296.
Kingston, N.Y., 137, 205.
Kingston, R I., 128.
Kitchen Creek, Penn., 186.
Kittery, Me., 66, 70.
Knoxville and Charleston Railroad, 305.
Knoxville, Tenn., 305.
Koshonong Lake, Wis., 322.
Lachine Rapids, Can., 65.
Lackawanna and Bloomsburg Railroad, Penn., 163, 164, 186, 229.
Lackawanna Valley House, Penn., 164.
Lackawaxen, Penn., 206.
Lack's Springs, N.C., 289.
Laconia, N.H., 39 (cut 40).
Lady of the Lake, steamer (cut), 44.
Lafayette Mountain, N.H., 32, 49.
Lake Dells, Wis., 323
Lake de Neveu, Minn., 336.
Lake George, Fla., 292.
Lake George, N.Y 177.
Lake Michigan, 316.
Lake of the Isles, Minn., 334.
Lake Pontchartrain, La., 299.
Lake Shore and Michigan Southern Railroad, 210.
Lake Tahoe, Cal., 345.
Lake Village, N.H., 41.
Lake Zurich House, Wis., 323.
Lakes:—
  Allegash, Me., 95.
  Ashley, Mass., 123.
  Bass, Mich., 316.
  Beech, N.Y., 182.
  Berry (Pond), Mass., 123.
  Betsey, Mich., 316.
  Big Moose, N.Y., 182.
  Blue Mountain, N.Y., 182.
  Bomaseen, Vt., 125.
  Buttes des Mortes, Wis., 328.
  Calhoun, Minn, 334.
  Canandaigua, N.Y., 202.
  Caucomgomosis, Me , 95.
  Caucomgomuc, Me., 95.
  Cauterskill, N.Y., 139.
  Cayuga, N.Y., 200.
  Cazenovia, N Y., 198.
  Cedar Lake, Minn., 316, 334.
  Champlain, 178.
  Chautauqua, N.Y., 210.
  Chesuncook, Me., 95.
  Clam, Mich., 316.
  Colden, N.Y., 180.
  Como, Minn., 334.
  Conneaut, Penn., 210.
  Connecticut, N H , 56, 95.
  Contoocook, N.H , 122.
  Crooked, or Keuka, N.Y., 202.
  Crooked, Mich , 316.
  Croton, N.Y., 193.
  Crystal, Minn., 334.
  Damariscotta, Me., 95.
  Delaware, N J., 234.
  De Neveu, Minn , 336.
  Devil's, Wis., 322.
  Diamond, Mich., 316.
  Donner, Cal , 345.
  Dunmore, Vt , 125.
  Eagle, N.Y., 182.

  Echo Lake, N.H., 49.
  Elkhart, Minn., 336.
  Fife, Mich., 316.
  Fulton, N.Y., 182.
  Geneva, N.Y., 322.
  George, Fla., 293.
  George, N.Y., 177.
  Grass, Mich., 317.
  Green, Minn., 336.
  Green, N.Y., 198.
  Greenwood, N.Y., 205.
  Harriet, Minn., 334.
  Hartwick, N.Y., 207.
  Harvey, N.Y., 182.
  Hewitt, Mich., 316.
  Highland, Penn., 186.
  Horicon, Wis., 322.
  Intermediate, Mich., 316.
  Jones, Penn. (cut), 167.
  Koshkonong, Wis , 322.
  Lake of the Isles, Minn., 334.
  Lawson's, N.Y., 174.
  Little Tupper, N Y., 181.
  Long, Mich , 316.
  Long, N.Y., 181.
  Long (Pond), N.H , 35, 45.
  Macopin, N Y., 205.
  Mahopac, N.Y , 193.
  Melville, Mass , 123.
  Memphremagog, Vt., 51, 64
  Michigan, 322.
  Minnetonka, Minn., 334.
  Mohonk, N.Y , 205.
  Moosehead, Me , 73, 95.
  Newfound, N.H., 119.
  North Branch, N Y., 182.
  Old Man's Wash-Bowl, N H., 49.
  Oneida, N.Y., 198.
  Onondaga, N Y., 200.
  Onota, Mass., 123.
  Ossipee Lake, N.H , 71.
  Otisco, N.Y., 200.
  Otsego, N.Y., 170, 207.
  Owasco, N.Y., 200.
  Paradox Pond, N.Y., 181.
  Pepin, Minn., 333.
  Pepin, Wis , 324.
  Pewaukee, Wis., 329.
  Pinnacle Lake, Can., 65.
  Placid, N.Y., 181.
  Pontchartrain, La , 299.
  Pontoosuc, Mass., 123.
  Rangeley, Me., 73, 94.
  Raquette, N.Y., 181.
  Rockland, N.Y., 193.
  Round, Mich., 316.
  Round, N.Y., 181.
  Rose, Mich., 316.
  San-cha-can-tack-et, Mass., 118.
  Saranac (upper), N.Y., 181.
  Saranac (lower), N.Y., 180.
  Schroon, N.Y., 181.
  Schuyler's, N.Y., 170.
  Sebago, Me., 80.
  Seclusion, Tenn., 206.
  Seminary, N.Y., 207.
  Seneca, N.Y., 201.
  Silver, Mass , 123.
  Skaneateles, N.Y., 200.
  Sodom, N Y., 198.
  Squam, N.H., 45.
  St. Catherine, Vt., 125.
  Superior, 331.
  Sylvan, Mass., 123.

Lakes, continued: —
  Tahoe, Cal., 345.
  Thunder, Mich., 314.
  Torch, Mich., 316.
  Traverse, Mich., 316.
  Tupper (Big), N.Y., 181.
  Twin Lakes, Minn., 334.
  Umbagog, Me., 72.
  Utowana. N.Y., 182.
  Wachusett, Mass., 121.
  Waukawan, N H., 35.
  Wawayandah, N.Y., 205.
  White Bear, Minn., 334.
  Willoughby, Vt., 63.
  Winnebago, Wis., 322.
  Winnepesaukee, N.H., 35.
  Winnesquam, N.H., 35.
  Zurich, Wis., 322.
Lancaster House, N.H. (cut), 53.
Lancaster, N.H. (cut), 54.
Lancaster, Penn., 219.
Lauderdale Springs, Miss., 307.
Lawrence, Mass., 34, 37, 43.
Lawson's Lake, N.Y., 174.
Leather-Stocking Cave, N.Y., 170.
Lebanon Springs, Mass., 124.
Lebanon Valley Railroad, 219.
Lehigh and Susquehanna Railroad, 142.
Lehigh Valley (cuts), 144, 157.
Lehigh Valley Railroad, 142, 262.
Levana Springs, N.Y., 201.
Lewis Mineral Spring, N.C., 289.
Lexington, N.Y., 137.
Lexington, Va., 276, 302.
Liberty, Va., 302.
Linnville Falls, N.C., 295.
Lisbon, N H., 51.
Little Boar's Head, N.H., 66, 68.
Little Falls, N.Y., 195.
Little Schuylkill River, Penn., 219.
Little Stony Falls, Va., 303.
Littleton, N.H., 49, 51 (cut 52).
Little Traverse, Mich., 316.
Littonian Springs, Ky., 308.
Litz Spring, Penn., 219.
Livermore Falls, N.H. (cut), 48.
Lock Haven, Penn., 229.
Lockport, N.Y., 203.
Logan House, Penn. (cut), 252.
Long Branch, N.J., 182, 231.
Long Island Sound, 131.
Long Lake, Mich., 316.
Long Lake, N.Y., 181.
Long's Peak, Col., 340.
Long Pond. Me., 82.
Lookout Mt., Tenn., 298, 305.
Loretto, Penn., 234.
Louisville and Great Southern Route, 308.
Louisville, Cincinnati, and Lexington Railroad, 3 9.
Louisville, Ky., 308.
Lowell Island, Mass., 66.
Lowell, Mass., 37.
Lower Saranac Lake, N.Y., 180.
Lynchburg, Va., 301
Lyons Falls, N.Y., 196.
Mackinaw. Mich., 24, 314, 316.
Macon and Western Railway, 298.
Macopin Lake, N.Y., 205.
Madison, Wis., 332.
Madison Springs, Ga., 298.
Magnetic Spring, Mich., 314.
Magnetic Spring, Minn., 336.
Magog, Can., 64.

Mahanoy Plane, Penn. (cut), 222.
Mahopac Lake, N.Y., 193.
Maiden Rock, Minn., 336.
Maine Central Railroad, 73, 94
Mainville Water Gap, Penn. (cut), 227.
Manatawny Creek, Penn., 218.
Manchester, N.H., 33, 37.
Manchester, Vt., 124.
Manitou Island, Lake Superior, 332.
Manitou Springs, Col., 340.
Mansfield Mt., Vt., 120.
Mansion House, Mauch Chunk, Penn., 149.
Mansion House, Mount Carbon, Penn., 221.
Maplewood Hotel (cut), 57.
Marblehead, Mass., 66.
Marcy (Tahawus), N.Y., 180.
Margaretsville N.Y., 138.
Marquette, Houghton, and Ontonagon Railroad, 331.
Marquette, Wis., 331.
Marshall House, Ga., 292.
Marshfield, Mass., 99.
Massaumkeag Hotel, Me., 79.
Mattakeset Lodge, Mass. (cut), 114.
Mattawamkeag, Me., 95.
Mauch Chunk (cut), 148.
Mayville, N.Y., 210.
McIntire Mt., N.Y., 180.
McMartin Mt., N.Y., 180.
Meadville, Penn., 210.
Medicated Geyser Bath Springs, Cal., 346.
Megunticook Peaks, Me., 79.
Melville Lake, Mass., 123.
Memphis and Charleston Railroad, 306.
Memphis and Little Rock Railroad, 308.
Memphis, Tenn., 307.
Memphremagog Lake, Vt., 51, 64.
Menasha, Wis., 328
Mendota Lake, Wis., 332.
Merced River, Cal., 346.
Meredith, N H., 39, 45.
Merrimack River, 31.
Metaire, La., 300.
Michigan, 23, 314.
Michigan, Lake, 324.
Middle Falls, N.Y., 208.
Middle States, 133.
Middletown, Del., 234.
Middletown, N.Y., 205.
Middletown Springs, Vt., 125.
Midland Magnetic Springs. Mich., 314.
Millbrook, N.Y., 138.
Miller's Falls, Mass., 119.
Miller's Ferry, Va. (cut), 286.
Mills House, S.C., 291.
Milton, Penn., 229.
Milwaukee Lake Shore, and Western Railroad, 323.
Milwaukee, Wis., 323.
Mineral Springs, Ala., 306.
Mineral Springs, Miss., 307.
Mineral Springs, Wis., 329.
Minnehaha Falls, Minn., 334.
Minnequa Springs, Penn., 261.
Minnesota, 23, 334.
Minnetonka Lake, Minn., 334.
Minnow Island, Can., 64.
Missisquoi Springs, Vt., 121.
Mitchell's Peak, N.C., 296.
Mobile, Ala., 299.
Mobile and Montgomery Railroad, 299.
Mobile and Ohio Railroad, 306.
Mohawk River, N.Y., 193.
Mohawk Valley, N.Y., 174.

Mohegan Glen, N.Y., 207.
Mohonk Lake, N.Y., 205.
Monadnock, N.H., 122.
Monckton, E.P., 16.
Monocacy Creek, Penn., 218.
Monona Lake, Wis., 332.
Montana Ter., 341.
Montgomery, Ala., 299.
Montgomery White Sulphur Springs, Va., 302.
Montreal, 63.
Montvale Spring. Tenn., 305.
Moor Mount, N Y, 180.
Moosehead Lake, Me., 73. 95.
Moosehorn Mt., Vt., 125
Moosic Mt Highlands, Penn. (cut), 166, 168.
Moosilauke, N.H. (cut), 50.
Moresville, N Y., 138
Morganton. N.C, 295.
Moriah, N.H., 93.
Morris and Essex Railroad, 145.
Morristown, Penn , 216.
Moss Run, Va , 279.
Mountain House, Canada, 64.
Mountain House, Penn., 253.
Mount Carbon, Penn. (cut), 220.
Mount Desert, Me , 66, 73, 78.
Mount Hayes, N.H., 92.
Mount Kineo, Me., 95.
Mount Kineo House, Me., 95.
Mount Katahdin, Me , 95.
Mount Lincoln, Col., 340.
Mount Morrill, Vt., 64.
Mount Olive, N.C., 289.
Mount Pleasant, Me., 84.
Mount St. Helena, Cal., 345.
Mount Tom, Mass , 123.
Mount Vision, N.Y , 170.
Mount Washington, 61.
Mount Washington Railway (cut 62), 90.
Mount Washington River, N H., 88.
Mount Washington Summit House, 62.
Mount Washington Turnpike, 63, 90.
Mount Whiteface, N H., 72.
Mount Whiteface, N.Y., 180.
Mount Whiteside, N.C., 296.
Moyer's Rock, Penn., 155.
Mountain Valley Springs, Ark., 307.
Mountains:—
 Abraham, Me., 94.
 Adams, N H , 30.
 Adirondack, N Y., 171.
 Alleghany, Penn., 255.
 Annanance (Willoughby), Vt., 63.
 Bald, N.C , 297.
 Bear, Vt , 64
 Bear (cut Mauch Chunk), 150
 Belknap, N H., 41.
 Berkshire (Hills), Mass., 123.
 Blackberry, Can., 65.
 Black Head. N Y., 141.
 Black, N.C., 295.
 Blue, Me., 94
 Blue, N Y , 180.
 Blue Ridge, Va., 274.
 Bread-Loaf, Vt., 125.
 Cæsar's Head, N.C., 296.
 Camel's Hump. Vt., 120.
 Carter, N.H , 93.
 Catskill, N.Y , 136.
 Chocorua, N.H., 30, 71.
 Clay, N.H., 30.
 Clinton, N.H., 30.
 Colden, N.Y , 180.
 Cove (Hill), N.Y., 180
 Crosby, N.H., 119.
 Crowders Knob, N.C., 296.
 Cumberland, Tenn , 3. 4.
 Dial, N.Y., 180.
 Dix's Peak, N.Y., 180.
 Elephantis, Can., 64.
 Elliot's Knob, Va., 276.
 Equinox, Vt., 124.
 Franconia, N.H., 35, 30.
 Franklin Mt., N H , 30, 32.
 Gibraltar, Penn , 218.
 Ginger Cake Rock, N.C., 296.
 Greylock, Mass , 121, 124.
 Griffith Knob, Va., 277.
 Gray's Peak, Col., 340.
 Hayes, N.H., 92.
 Haystack, Vt., 125.
 Hawk's Bill, N C , 296.
 Hawk's Nest, Va , 271.
 High Peak, N.Y., 141.
 Holyoke, Mass , 123.
 Hoosac, Mass., 123.
 Hor, Vt , 63.
 Hunter, N Y., 140.
 Jackson, N H , 30.
 Jefferson, N H., 30.
 Jefferson, Penn , 152.
 Jay (Peak), Vt , 64.
 Katahdin Mt , Me., 95
 Kearsarge, N H , 119, 30.
 Kiarsarge (Pequawket), N H. (cut 86), 119.
 Killington (Peak), Vt., 124.
 Kineo, Me , 95.
 King's Mount, N.C., 296.
 Lafayette, N.H., 32, 49
 Lincoln, Col., 340.
 Long's Peak, Col., 340.
 Lookout, Tenn., 298, 305.
 Madison, N.H , 30.
 Mammoth Cave, Ky., 309.
 Mansfield, Vt., 120.
 Marcy (Tahamus), N.Y , 180.
 McIntire, N Y., 180.
 McMartin, N.Y., 180.
 Meguntioook (Peaks), Me., 79.
 Mitchell's Peak, N.C., 296.
 Monadnock, N.H., 122.
 Monroe, N.H , 30.
 Moor, N.Y., 180.
 Moosehorn, Vt., 125.
 Moosic (Highlands), Penn., 166, 168
 Moosilauke, N.H., 30 (cut 50).
 Moriah, N.H., 93.
 Morrill, Vt , 64
 Lyon, N.H., 55.
 Neversink, Penn , 218.
 Nippleton, N Y., 180.
 North, N.Y., 139.
 North, Penn., 17.
 North, Va., 276.
 Oquago, N.Y , 207.
 Orford, Vt , 64.
 Ossipee, N.H , 71.
 Overlook, N Y., 137.
 Owl's Head, N.H. (cut), 50.
 Owl's Head. Canada, 64.
 Patenwell Peak Wis., 321.
 Pleasant Mt , Me., 84.
 Pleasant, N H., 30.
 Peaks of Otter, Va , 271, 302.
 Pequawket (Kiarsarge), N.H., 72, 84, 119.
 Penn, Penn., 218.

INDEX. 355

Mountains, continued: —
  Percy (Stratford) Peaks, N.H., 55, 93.
  Pike's Peak. 340.
  Pilot, N.C., 294.
  Pilot Range. N H., 55.
  Pine Grove, N.Y., 139.
  Pinnacle, Can., 65.
  Pinnacle Rock, N.C., 297.
  Pisgah, Penn., 148, 152.
  Profile, N H., 30
  Prospect (Hill), Vt., 64.
  Prospect. N.H., 48.
  Pulaski Mount, Va., 63.
  Ragged, N.H., 45.
  Red Hill, N H., 41.
  Rock, Ga., 297.
  Round Top. N Y., 111.
  Saddleback, Mo., 64.
  Sager-Warner, N.Y., 174.
  Saluda, N.C., 296.
  Sandwich, N.H., 71.
  Seward, N.Y., 180.
  Shawangunk, N.Y., 235.
  Sky Top (Peak) N.Y., 205.
  South. N.Y., 130.
  Starr King, N.H., 53.
  St. Catherine, Vt., 125.
  Sterling, Vt., 120.
  St. Helena. Cal, 345.
  Stool. N.C., 297.
  Stratford (Percy) Peaks, N.H., 55, 93.
  Sugar-Loaf, N.H., 119.
  Sulphur, Mon., 343.
  Surprise, N.H., 93.
  Table, N.C., 295.
  Table Rock, N.C. 295.
  Tom, Mass., 123.
  Wachusett, Mass., 121.
  Washington. Mass., 123.
  Washington. N.H., 50.
  Wawayandah, N.Y., 205.
  White Mountains. N.H., 33.
  Webster, N.H., 50, 83.
  Willard, N.H., 68.
  Willey. N.H., 68.
Mud Geyser, Mon., 343.
Mud Volcano, Mon., 343.
Mumfordsville, Ky., 309.
Nahant, Mass., 66.
Nancy's Brook, N.H., 88.
Nantasket Beach, Mass., 98.
Nanticoke Branch Railroad, Penn., 163.
Nanticote, Penn., 103.
Nantucket, Mass., 102.
Narragansett Bay, R.I., 128.
Narragansett Pier, R.I., 128.
Nashville, Chattanooga, and St. Louis Railroad, 305.
Nashville, Ky., 313.
National Hotel (First), Wis., 330.
National Hotel, Wis., 228.
National Military House, Va., 239.
Natural Bridge. Va., 271, 302.
Natural Tunnel, Va., 301.
Nayaug Falls. Penn. (cut), 165.
Neenah, Wis., 328.
Nescopee Branch Railroad, 158.
Nescopec Junction. Penn., 158.
Nesquehoning Valley Branch Railroad, 157.
New Bedford, Mass., 33, 107.
New Brunswick, E.P., 96.
Newbury, Vt., 51.
Newburyport. Mass., 64.
Neversink Mt., Penn., 218.

New Castle, N.H., 69.
New England, 26.
Newfound Lake, N.H., 119.
Newfound River, N H., 119.
Newhall House, Wis., 323.
New Hampshire, 27.
New Haven, Conn., 130.
New London, Conn., 130.
New London Northern Railroad, 119.
New Orleans, La., 299.
Newport, R I., 33, 104.
Newport, Vt., 51, 64, 64.
New River. Va. (cut), 283.
New River Rapids, Va (cut), 285
Newbury Sulphur Springs, Vt., 63.
New York and Canada Railroad, 142, 177.
New York Central and Hudson River Railroad, 130, 134, 174, 193.
New York City, 130.
New York, Kingston, and Syracuse Railroad, 176, 137.
Niagara Falls, N.Y., 103, 237.
Nickerson House, S C., 290.
Nineveh Branch Railroad, 169.
Nippleton Mount, N. Y., 180.
Northampton, Mass., 123.
North Bridgton, Mo., 82.
North Branch Lake, N Y., 182.
North Conway, N H., 70, 72 (cut), 85.
Northern Central Railroad, 201, 207, 229, 230 255.
Northern Railroad, 119.
North Mountain, N.Y., 130, 229.
North Mountain, Penn., 181.
North Mountain, Va., 276.
North Mountain House, Penn., 163, 182, (cut), 183.
North Mountain View (cut), 190.
North Pennsylvania Railroad, 142.
Nova Scotia, E.P., 96.
Northumberland, N.H., 55.
North-West, 23.
Oak Bluffs, Mass., 103.
Oak Orchard Acid Springs, N.Y., 202.
Oakton Springs, Wis., 329.
Oakwood House, Wis. (cut), 326.
Ocean House, Me., 78.
Ogden, Utah, 344.
Ogdensburg, N.Y., 119, 178.
Ohio River (cut 2-8).
Oil City, Penn., 210.
Old Dominion Steamship Company, 241, 271.
Old Colony Railroad, 97.
Old Colony Steamboat Company, 105.
Old Man of the Mountain, N.H., 49.
Old Man's Wash-Bowl, N.H., 49.
Old Orchard, Me., 66, 70, 73.
Old Point Comfort, Va., 239.
Old Sweet Springs, Va., 280, 302.
Omaha, 340.
Oneida Lake, N.Y., 198.
Onondaga Lake, N.Y., 200.
Onota Lake, Mass., 123.
Oquago Mount, N.Y., 207.
Orange and Alexandria Railroad, 274.
Orangeburg, S.C., 295.
Orford, Mount, Vt., 64.
Oshkosh, Wis., 328.
Ossipee Falls, N H., 71.
Ossipee Lake, N.H., 71.
Ossipee Mount, N.H., 71.
Oswego Midland Railroad, 198.
Otisco Lake, N.Y., 200.
Otsego Lake, N.Y., 170, 207.

Otter Ponds, Me., 82.
Overlook Mount, N.Y., 137.
Owasco Lake, N.Y., 200.
Owego Lake, N.Y., 207.
Owl's Head Mount, Can., 64.
Owl's Head, N H. (cut), 150.
Painted Rock, N.C., 296.
Painted Rock, N.Y., 207.
Panacea Springs, N.C., 289.
Panther Gap, Va., 276.
Panther's Leap, N.Y., 170.
Paradox Pond, N.Y., 181.
Parrot House, N.Y., 174.
Pascagoula, Miss., 299.
Passaconaway, N.H., 71.
Passaic Falls, N.J., 204.
Pass Christian, Miss., 299
Passumpsic Railroad, Vt., 33, 51, 63.
Paterson, N.J., 204.
Pavilion Hotel, N.Y., 171.
Peabody River, N.H., 93.
Peach Gathering, Del. (cut), 234.
Peaks of Otter, Va., 271, 302.
Peekskill, N.Y., 135.
Pemigewasset House, 47.
Pemigewasset River, N.H., 35, 46.
Penn Haven, Penn., 158.
Penn Mount, Penn., 218.
Pensaukee House, Wis., 330.
Pensaukee, Wis., 330.
Pennsylvania Coal Company's Gravity Railroad, 14.
Pennsylvania Railroad, 242.
Penobscot River, Me., 79.
Pepin Lake, Minn., 345.
Pequot, Conn., 130.
Pequawket Mount (Kiarsarge), N.H., 72, 84, 86, 119.
Pequannock River, N.Y., 204.
Perkiomen Creek, Penn., 217.
Petenwell Peak, Wis., 321.
Petersburg and Weldon Railroad, 289.
Petrea Island, N.Y., 194.
Petrified Forest, Cal., 345.
Pewaukee Lake, Wis., 329.
Pewaukee, Wis., 328.
Philadelphia, 142, 210.
Philadelphia and Erie Railroad, 229.
Philadelphia and Reading Railroad, 214.
Philadelphia, Wilmington, and Baltimore Railroad, 237.
Phillips, Me., 94.
Phillipsburg, N.J., 145.
Phœnicia, N.Y., 137.
Phœnixville, Penn., 217.
Picton, E.P., 96.
Pictured Rocks, 24.
Piedmont Air Line Railroad, 289, 294.
Piedmont Springs, N.C., 205.
Piermont, N.Y., 145.
Pigeon Cove, Mass., 66, 68.
Pike's Peak, Col., 340.
Pilot Mount, N.C., 294.
Pilot Range, 55.
Pilot Rock, Ky., 312.
Pine Hill, N.Y., 138.
Pine Orchard Mount, N.Y., **139**.
Pine River, Mich., 315.
Pinkham Notch, N.H., 72, 87, 93.
Pinnacle Lake, Can., 65.
Pinnacle Mount, Can., 65.
Pinnacle Mount, N.C., 297.
Pisgah Mount, Penn., 148, 152.
Pittston, Penn., 163.

Pittsfield, Mass., 123.
Pittsfield, N.H., 37.
Pittsburg, Cincinnati, and St. Louis Railroad, 315.
Pittsburg, Penn., 254.
Placid Lake, N.Y., 181.
Plains of Chalmette, La., 300.
Plankinton House, Wis., 328.
Plattekill Clove, N.Y., 140
Plattekill Ravine, N.Y., 137.
Plattsburg, N.Y., 178.
Pleasant Mount, N.H., 30, 91.
Pleasure Route No. 1, 33.
Pleasure Route No. 2, 63.
Pleasure Route No. 3, 66.
Pleasure Route No. 4, 73.
Pleasure Route No. 5, 80.
Pleasure Route No. 6, 92.
Pleasure Route No. 7, 94
Pleasure Route No. 8, 97.
Pleasure Route No. 9, 108.
Pleasure Route No. 10, 109
Pleasure Route No. 11, 126.
Pleasure Route No. 12, 131.
Pleasure Route No. 13, 134.
Pleasure Route No. 14, 142.
Pleasure Route No. 15, 193.
Pleasure Route No. 16, 204.
Pleasure Route No. 17, 214.
Pleasure Route No. 18, 232.
Pleasure Route No. 19, 233.
Pleasure Route No. 20, 239.
Pleasure Route No. 21, 242.
Pleasure Route No. 22, 255.
Pleasure Route No. 23, 272.
Pleasure Route No. 24, 289.
Pleasure Route No. 25, 294.
Pleasure Route No. 26, 301.
Pleasure Route No. 27, 308.
Pleasure Route No. 28, 315.
Pleasure Route No. 29, 320.
Pleasure Route No. 30, 337.
Pleasure Route No. 31, 340.
Pleasure Travel in the Olden Time, 22.
Pleasant Mount, S.C., 291.
Plumb Island, Mass., 66, 68.
Plymouth, Mass., 99.
Plymouth, N.H. (cut 46), 47.
Point Look-off, Penn., 189.
Pulaski Alum Springs, Va., 304.
Pompey's Pillar, Va., 303.
Pontoosuc, Mass., 123.
Pool (Franconia Mountains), N.H., 49.
Port Clinton, Penn., 219.
Port Deposit, Md., 238.
Port Kent, N.Y., 179.
Portage City, Wis., 332.
Portage Falls, N.Y., 208.
Portland and Ogdensburg R.R., 72, 80, 121.
Portland, Me. (cut), 74.
Portland to Mount Desert, 21.
Portsmouth, N.H., 69.
Poteskey, Mich., 316.
Pottstown, Penn., 218.
Pottsville, Penn, 220.
Poughkeepsie, N.Y., 194.
Powder Springs, Ga., 298.
Pratt's Rocks, N.Y., 140.
Prattsville, N.Y., 138.
Preble House, Me., 75.
Presumpscot River, Me., 80.
Presumpscot Valley, Me., 31.
Princeton, Mass., 121.
Profile, N.H., 49.

Profile House, N.H., 49, 54.
Prospect Mount, N.H., 48.
Prospect House, Mass., 124.
Prospect Hill, Vt., 64.
Prospect Rock (cut), 159, Penn.
Providence, R.I., 33, 128.
Provincetown, Mass., 100.
Puncheon Run Falls, Va., 302.
Pulaski Mount, Vt., 63.
Pulaski, Vt., 63.
Pulaski House, Ga., 292.
Pulpit Falls, N.Y., 201.
Quaker Street, N.Y., 174.
Quebec, 60.
Quinnesec Falls, Wis., 221.
Ragged Mountain, N.H., 45.
Railroads and Steamboats: —
 Alabama and Chattanooga, 298, 306.
 Alabama Central, 299, 306.
 Albany and Susquehanna, 142, 169, 207.
 Atlantic and Great Western, 209, 210.
 Atlantic and Gulf, 292.
 Atlantic Coast Line, 289.
 Baltimore and Potomac, 255.
 Baltimore Steam-packet Company (Old Bay Line), 241.
 Belvidere, 145.
 Boston and Albany, 119.
 Boston and Maine, 33, 34, 70, 73, 87.
 Boston and Providence, 126.
 Boston, Concord, and Montreal, 33, 70, 90.
 Boston, Lowell, and Nashua, 33, 34, 119.
 Cairo and Fulton, 308.
 Camden and Atlantic, 239.
 Catawissa, 157, 225.
 Cazenovia, Canastota, De Ruyter, 198.
 Central (New York) and Hudson River, 136.
 Central Railroad of New Jersey, 142, 182, 225.
 Central Vermont, 119, 196.
 Chauncey Vibbard (steamer), 136, 139.
 Cherry Valley Branch, 171.
 Chesapeake and Ohio, 272.
 Cheshire, 119.
 Chicago and North-Western, 320.
 Chicago and St. Paul Line, 332.
 Cincinnati, Cumberland Gap, and Charleston, 305.
 Concord, 33, 34, 119.
 Connecticut River, 33.
 Conway Branch, 70.
 Cumberland Valley, 254.
 Daniel Drew (steamer), 136, 139.
 Delaware and Hudson, 163.
 Delaware, Lackawanna, and Western, 145, 164, 171, 207.
 Eastern, 166.
 East Pennsylvania, 219.
 East Tennessee and Virginia, 305.
 Erie, 136, 204, 262.
 European and North American, 96.
 Fitchburg, 119, 121.
 Flint and Père Marquette, 314.
 Framingham and Lowell, 33.
 Georgia, 296.
 Grand Rapids and Indiana, 315.
 Grand Trunk, 55, 92, 119, 197.
 Great Southern Mail Route, 289, 301.
 Hamilton and Dayton, 315.
 Harlem, 119.
 Hudson, 119.
 International and Great Northern, 308.

 James W. Baldwin (steamer), 136.
 Kentucky Central, 309.
 Knox and Lincoln, 78.
 Knoxville and Charleston, 305.
 Lackawanna and Bloomsburg, 63, 164, 180, 229.
 Lady of the Lake (steamer), 41.
 Lake Shore and Michigan Southern, 210.
 Lebanon Valley, 219.
 Lehigh and Susquehanna, 142, 145.
 Lehigh Valley, 143, 145.
 Louisville and Great Southern Route, 208.
 Louisville, Cincinnati, and Lexington, 308.
 Macon and Western, 298.
 Maine Central, 73, 94.
 Manchester and Lawrence, 33.
 Memphis and Charleston, 306.
 Memphis and Little Rock, 308.
 Mary Powell (steamer), 136.
 Marquette, Houghton, and Ontonagon, 331.
 Milwaukee and St. Paul, 332.
 Milwaukee, Lake Shore, and Western, 323.
 Mobile and Montgomery, 299.
 Mobile and Ohio, 307.
 Morris and Essex, 145.
 Mount Washington, 33.
 Mount Washington (steamer), 41, 70.
 Nauticote, Penn., 164.
 Nashua and Rochester, 33.
 Nashville, Chattanooga, and St. Louis, 305, 306.
 Nescopec Branch, Penn., 158.
 Nesquehoning Valley Branch, 157.
 New London Northern, 119.
 New York and Canada, 142, 177.
 New York Central and Hudson River, 130, 174, 193, 207.
 New York, Kingston, and Syracuse, 135, 137, 205.
 Nineveh Branch, 169.
 Northern, 119.
 Northern Central, 201, 207, 249, 250, 255.
 North Pennsylvania, 142.
 Old Colony, 97.
 Old Colony Steamboat Company, 105.
 Old Dominion Steamship Co., 241, 271.
 Orange and Alexandria, 274.
 Oswego Midland, 198.
 Passumpsic, 33, 51, 63.
 Pennsylvania, 212.
 Pennsylvania Coal Co.'s Gravity, 164.
 Petersburg and Weldon, 289.
 Philadelphia and Erie, 229.
 Philadelphia and Reading, 214, 260.
 Philadelphia, Wilmington, and Baltimore, 253.
 Piedmont Air Line, 289, 294.
 Pittsburg, Cincinnati, St. Louis, 315.
 Portland and Ogdensburg, 72, 89, 121.
 Portland, Bangor, and Machias, Me., 78.
 Portsmouth and Concord, 33.
 Reading and Columbia, 219, 250.
 Rensselaer and Saratoga, 124, 142.
 Rome, Watertown, and Ogdensburg, 196.
 Sackett's Harbor Branch, 196.
 Saginaw Valley and St. Louis, 314.
 Sanford's Independent Line (steamers), 119.
 Selma, Rome, and Dalton, 299.
 Sheboygan and Fond du Lac, 323.

Railroads and Steamboats, continued:—
  Shore Line (cut of Providence Depot), 126.
  Sodus Point and Southern, 202.
  St. Paul and Pacific, 334.
  Stonington Steamboat Line, 33, 131.
  Suncook Valley, 37.
  Switch-back (Mauch Chunk, Penn.), 152.
  Switch-back (Scranton, Penn.), 166.
  Thomas Cornell (steamer), 136.
  Union and Central Pacific, 340.
  Utica and Black River, 196.
  Weston and North Carolina, 295.
  Western and Atlantic, 306.
  West Jersey, 572.
  Wilmington and Reading, 237.
  Wilmington and Weldon, 283.
  Wilmington,Charlotte,and Raleigh,289.
  Worcester and Washington, 34.
Rainbow Falls, N.Y. (cut), 264.
Raleigh, N.C., 294.
Ramapo River, N.Y., 204.
Ramapo Falls, N.Y., 204.
Rangeley, Me., 73, 94.
Raquette Lake, N.Y., 181.
Raquette River, N.Y., 181.
Rattlesnake Range, N.H., 84.
Reading, Penn., 145, 218.
Reading and Columbia Railroad, 219, 250.
Recluse Island, N.Y., 178.
Red Sulphur Springs, Va., 304.
Red Sweet Springs, Va., 281, 502.
Red Sulphur Springs, Ga., 298.
Renova Springs, Penn., Penn., 260.
Rensselaer and Saratoga Railroad, 124, 142.
Revere, Mass., 66.
Revere House, Utah, 344.
Richmond, Va., 276.
Richmond, Ind., 315.
Richfield Springs, N.Y., 170.
Richmond Falls, Va. (cut), 298.
Ridley Station (cut), 256.
Ringtown, Penn., 226.
Rivers:—
  Androscoggin, 55.
  Ammonoosuc, N.H., 37, 51, 56.
  Ashley, S.C., 291.
  Au Sable, N.Y., 179.
  Baker's, N.H., 37.
  Cape Fear, N.C., 290.
  Charles, Mass., 35.
  Chippewa, Minn., 334
  Cooper, S.C., 291.
  Cow-Pasture, Va., 277
  Fox, Wis., 328.
  Grass, Mich., 317.
  Greenbrier (Kanawha), Va. (cut), 282.
  Hudson, 124, 179.
  Israel's, N.H., 37.
  James, Va., 276, 278.
  Juniata, Penn., 251.
  Kennebec, Me., 27.
  Merced, Cal., 346.
  Merrimac, 35.
  Mohawk, N.Y., 195.
  Mount Washington, N.H., 88.
  Mystic, Mass., 35.
  Newfound, N.H., 119.
  New, Va. (cut), 283.
  O'Buf. Minn., 335.
  Ohio (cut), 288.
  Peabody, N.H., 93
  Pemigewasset, N.H., 35, 119.

Penobscot, Me., 79.
Pequannock, N.Y., 204.
Pine, Mich., 315.
Presumpscot, Me., 80.
Ramapo, N.Y., 204.
Raquette, N.Y., 181.
Saco, Me., 80.
Schoharie, N.Y., 174.
Schuylkill River, Penn., 214, 219.
Shoshone, Idaho, 344.
Songo, Me., 82.
Squam, N.H., 45.
St. John's River, Fla., 292.
St. Maurice River, Can., 65.
Suncook, N.H., 35.
Susquehanna, Penn. (cut 224), 238, 251
Thames, Conn., 130.
Tombia, Minn., 335.
Tuolumne, Cal., 346.
Venango, Penn., 210.
Warsaw, Ga., 292.
Wells, N.H., 37.
Winnepesauke, N.H., 35.
Roaring Brook, Penn., 164.
Rockport, Mass., 66, 68.
Rock City, N.Y., 209.
Rochester, N.Y., 292, 208, 267.
Rocky Falls, N.Y., 201.
Rocky Point, R.I., 128
Rockbridge Alum Springs, Va. (cut), 276.
Rockbridge Baths, Va., 276.
Rock, Ga., 2,97.
Rodick House, Me., 78.
Rome, N.Y., 196.
Rome, Watertown, and Ogdensburg Railroad, 196.
Rose Lake, Mich., 316.
Round Island, N.Y., 181.
Round Lake, N.Y., 181.
Round Lake, Mich. 316
Round Top, N.Y., 141.
Round Island, Can., 64.
Rouse Point, N.Y., 178.
Rowland's Mineral Spring, Ga., 298.
Roxbury, N.Y., 198.
Rupert, Penn., 229.
Rumford Falls, Me., 92.
Rustico, P.E.I., 96.
Rye, N.H., 66, 69.
Saco, Me., 80.
Sackett's Harbor Branch Railroad, 196.
Saddleback Mt., Me., 94.
Sager-Warner Mt., N.Y., 174.
Saginaw Valley and St. Louis Railroad, 314.
Saint Anthony Falls, 334.
Saint Paul, Minn., 334.
Salamanca, N.Y., 209.
Salem, Mass., 33.
Salem, N.C., 294.
Salisbury Beach, Mass., 68.
Salisbury, N.C., 295.
Salt Lake City, Utah, 344.
Salt Lake House, Utah, 344.
Salt Pond, Va., 303.
Salt Sulphur Springs, Va., 304.
Saluda, N.C., 296.
Samoset House, Mass., 99.
San-cha-cau-tack-et Lake, Mass., 118.
Sandwich Mt., N.H., 71.
Sanford's Independent Line (steamers), 21.
Sans Souci Springs, N.Y., 195.
Saranac Lake (upper), N.Y., 181.
Saranac Lake (lower), N.Y., 181.
Saratoga Springs, N.Y., 175.

INDEX.

Savannah, Ga., 2?2.
Sawkill Falls, N.Y., 245.
Schenectady, N.Y., 174, 195.
Schoharie, N.Y., 174.
Schoharie River, N.Y., 174.
Schroon Lake, N.Y., 181.
Schuyler's Lake, N.Y., 170.
Schuylkill Haven, Penn., 220.
Schuylkill, Penn., 214, 219.
Scituate, Mass., 100.
Scranton Gorge, Penn. (cut). 165
Scranton, Penn., 164. 229.
Scriven House, Ga , 2?2.
Sea-View Boulevard, Mass., 117.
Sea-View House, Mass (cut), 108, 111.
Sebago Lake, Me., 80 (cut 81).
Seclusion Lake, Tenn., 306
Selma, Rome, and Dalton Railroad, 299.
Seminary Lake, N.Y., 207.
Seneca Lake. N.Y. (cut 167), 241.
Seven Mine Spring. Va., 304.
Seven Springs Mountain House, N.Y., 205.
Seward Mt., N.Y., 140.
Shandaken Valley, N.Y., 137.
Sharon Springs, N.Y., 174, 195.
Sharon Alum Springs, Va , 304.
Shawanegan Falls, Can., 65.
Shawangunk Mt., N.Y., 205.
Sheboygan and Fond du Lac Railway, 322.
Sheboygan, Wis , 3?3.
Shelby Springs, Ala., 299.
Sheldon Springs, Vt., 121.
Shelter Island Park N.Y., 130.
Sherwood House, 326.
Shickshinny, Penn., 185, 229.
Shoshone Falls, Idaho, 344.
Shoshone River, Idaho 344.
Silver Cascade, N H. (cut), 89.
Silver Cascade, Vt., 63.
Silver Lake, Mass., 123.
Silver Springs, Fla., 293.
Sinking Springs, Penn., 219.
Sing Sing, N.Y., 135, 193.
Skaneateles Lake, N.Y., 200.
Skinner's Island, Can., 64.
Sky Top Peak, N.Y., 205.
Sleepy Hollow, N Y., 139.
Slicking Falls, N.C., 297.
Smuggler's Notch, Vt., 120.
Sodom Lake, N.Y., 198.
Sodus Point, N.Y., 202.
Sodus Point Southern Railway, N.Y., 202.
Solomon's Gap, Penn. (cut), 162.
South Mt., N.Y., 139.
Southern States, 268.
South-west Harbor, Me., 78.
Sparkling Catawba Springs. N.C., 289.
Sparta Mineral Well, Minn., 334.
Springs ; —
  Alburgh, Vt., 321.
  Alleghany, Va., 302.
  Alum and Iron, Cal., 346
  Alum, Cal , 346.
  Artesian Lech 1a, N.Y., 195
  Augusta (Stribling), Va., 275.
  Avon, N.Y., 238.
  Bath Alum, Va., 277.
  Bailey, Ala., 306.
  Bedford Alum, Va., 301.
  Bedford, Penn., 254.
  Beersheba, Ala., 306.
  Black Sulphur, Cal., 346.
  Black Sulphur, Tenn., 303.
  Bladen, Ala , 299.

Blount's, Ala., 306, 313.
Blue Lick, Ky., 309.
Blue. Ala., 313.
Boiling Alum and Sulphur. Cal., 346.
Boiling Black Sulphur, Cal., 346.
Bolton, Vt., 64
Bon Aqua, Tenn., 305.
Brandywine, Penn., 237.
Calistoga Hot, Cal., 345.
Catoosa, Ga., 298.
Causilor's. N.C., 289.
Chalybeate, Tenn., 305.
Chalybeate, Ga.. 298
Champlain. Vt., 121.
Chandler, Ala., 299.
Cherokee, Ga , 298.
Chilhowee, Tenn , 305.
Chilhowee, Va., 304.
Clarendon, Vt., 124.
Cleveland Mineral, N.C., 289.
Colorado, Cal., 340.
Coyner's, Va , 302
Drennon Black Sulphur, Ky., 309.
Eggleston, Va., 303.
Elkmont, Tenn., 313.
Ephrata, Penn., 219.
Epsom Salt, Cal., 346.
Eutaw, S.C., 299.
Falling Spring Falls, Va., 278.
Fairy, N.Y., 139.
Gas, N.Y., 210.
Glen Flora. Mineral. Wis., 322.
Godbald's Mineral Wells, Miss., 30.
Great Geyser, Cal , 346.
Green Cove, Fla., 292.
Green, Fla., 292.
Greenwood, Miss., 307.
Green Brier White Sulphur, Va. (cut, 261.)
Harrodsburg, Ky , 308.
Healing, Va., 277.
Healing, S.C., 290.
Highgate, Vt., 121.
Holston, Va., 304.
Hot, Ark., 307.
Hot, Va., 277.
Hot, Utah, 344.
Idaho, 340.
Indian Ga., 298.
Intermittent Scalding, Cal., 346.
Iron and Sulphur, Cal., 346.
Johnson's Wells, Ala., 306.
Lacks. N.C ; 289.
Lauderdale, Miss., 307.
Levana, N Y., 2 11.
Lewis Mineral N.C., 289.
Litz, Penn. 219.
Littonian, Kt., 308.
Madison, Ga., 298.
Magnetic Mineral. Minn., 336.
Magnetic, Mich., 314.
Manitou, Col , 340.
Medicated Geyser Baths, Cal., 346.
Midland Magnetic, Mich., 314.
Middletown, Vt., 125.
Mineral, Miss., 307.
Mineral, Ala., 306.
Mineral of Palmyra, Wis., 329.
Minnequa. Penn., 261.
Missisquoi, Vt.. 121.
Montgomery White Sulphur, Va., 302
Mont Vale, Tenn., 305.
Mountain Valley, Ark., 307.
Newbury Sulphur, Vt., 63.

New Lebanon, Mass., 124.
Oak Orchard Acid, N.Y., 202.
Oakton, Wis., 329.
Old Sweet, Va., 280, 302.
Panacea, N.C., 289.
Piedmont, N.C., 295.
Powder, Ga., 298.
Pulaski Alum, 304.
Red Sweet. Va., 281, 302.
Red Sulphur, Ga., 298.
Red Sulphur, Va., 304.
Renova, Penn., 260.
Richfield. N Y., 170.
Roaring Creek, Penn., 261.
Rockbridge Alum, Va. (cut), 276.
Rockbridge Bath, Va., 276.
Rowland's Mineral, Ga., 298.
Salt Sulphur. Va., 304.
Sans Souci, N.Y., 195.
Saratoga. N.Y., 175.
Scalding Steam Iron Bath, Cal., 346.
Seven Mine, Va., 304.
Sharon Alum, Va., 304.
Sharon, N.Y., 171, 195.
Shelby. Ala., 299.
Sheldon, Vt , 121.
Silver, Fla., 293.
Sinking. Penn., 149.
Sparkling Catawba, N.C., 289.
Sparta Mineral Well, Minn., 334.
Spring Church, N.C., 289.
Steamboat Geyser, Cal., 346.
Submerged, Wis., 329.
Sulphur, Ala., 299.
Sulphur, Ark., 307.
Sulphur, Ga., 298.
Sulphur, N C., 289.
Sulphur, Tenn., 305.
Suwannee, Fla., 292.
Sweet Chalybeate, Va , 280.
Talladega Ala., 299.
Tar, Ky., 309.
Tate, Tenn., 305.
Union, N.Y., 201.
Valhermosa, Ala., 306, 313.
Vallonia, N.Y., 207.
Vermont, 121.
Verona, N.Y., 198.
Virginia, 271.
Warm, Ga., 298.
Warm, N.C., 286.
Warm, Tenn., 305.
Warm, Utah, 344.
Warm, Va., 277.
White Cliff, Tenn., 305.
White Sulphur, Ky., 309.
White Sulphur, Miss., 307.
White Sulphur, N.Y., 198.
White Sulphur, Tenn., 305.
White Sulphur, Va., 271, 281 (cut 280), 302, 304.
Wilson's, N C., 295.
Witch's Culdron, Cal , 346.
Yates, N Y., 198.
Yellow Sulphur, Va., 303.
Shore Line Railroad, 126.
Squam Lake, N.H., 45.
Squam River, N.H., 45.
Stamford, N.Y., 138.
St. Anthony Falls, 334.
Star of the East (steamboat), 21.
Starr King Mountain, N.H., 53.
Staruccca Viaduct, N.Y., 206.
St. Augustine, Fla., 293.

Staunton, Va., 274.
St. Catherine's Hotel, 125.
St. Catherine's Lake, 125.
St. Catherine's Mountain, 125.
St. Charles Hotel, La., 300.
Steep Falls, Me , 82.
Sterling Mt., Vt., 120.
St. James Hotel, 300.
St. John, Can., 119.
St. John, E.P., 96.
St. John's, Fla., 292.
St. Johnsbury, Vt., 63.
St. Lawrence River, 196.
St. Louis Hotel, La., 300.
St. Louis, Mich, 314.
Stockton House, N.J., 232.
Stonington, R.I., 128.
Stonington Steamboat Line, 131.
Stony Clove, N.Y., 137.
Stool Mt , N.C , 297.
Stowe, Vt , 120.
St. Paul and Pacific Railroad. 334.
St. Paul, Minn., 334.
Stratford (Percy) Peaks, N.H., 55, 93.
Strawberry Culture (cut), 233.
Stribling (Augusta) Springs, Va., 275
Submerged Spring. Wis., 328.
Sugar Falls, N.Y., 195.
Sugar-Loaf Mt., N.H., 119
Sugar-Loaf, N.Y., 208.
Sullivan's Island, S.C., 291
Sulphur, Ark., 307.
Sulphur Mount, Mon , 343.
Sulphur Springs, Ala., 299.
Sulphur Springs, Ga., 298.
Sulphur Springs, N.C., 289.
Sulphur, Tenn , 305.
Summerside, P.E.I., 96.
Summit Hill, Penn., 152.
Summit House, N H., 50.
Summit House, Vt , 120.
Sunbury, Penn , 259.
Sunnyside, N.Y., 135, 195.
Superior Lake, 311.
Surprise Mount, N.H., 93.
Susquehanna River (cut, 224), 248, 251.
Susquehanna, N.Y., 206.
Suwannee Springs, Fla., 292.
Swampscott, Mass., 66, 67.
Swananda Gap, N.C., 295.
Sweet Chalybeate Springs, Va., 280.
Switch-back (Scranton), Penn., 166.
Switch-back Railroad (Mauch Chunk), Penn., 148, 152.
Sylvan Glade, N.H., 90.
Sylvan Lake. Mass., 123.
Syracuse, N.Y., 200.
Table Rock, N.C., 296.
Table, N.C., 296.
Taconic House, Vt., 124.
Tahoe Lake, Cal., 345.
Tallulah Falls, Ga., 397.
Talladega Springs, Ala., 299.
Tamanend, Penn., 225.
Tamaqua, Penn., 221.
Tar Springs, Ky., 309.
Tarrytown, N.Y., 135.
Tate Spring, Tenn., 305.
Taughannock Falls, N.Y., 201.
Thames River, Conn., 130.
Thayer's Hotel, Littleton, 51.
Southern Mail Route Railroad, 189, 301.
Thousand Islands House, N.Y. (cut), 199.
Thousand Islands, N.Y., 196.

Thomas Cornell (Steamer), 136.
Three Rivers, Can., 65.
Thunderbolt, Ga., 292.
Thunder Bay, 314.
Ticonderoga, N.Y., 177.
Tilden House, Wis., 330.
Tilton, N.H. (cut), 38.
Toccoa Falls, Ga., 297.
Toledo, Mich., 314.
Thompson's Falls, N.H., 72.
Tombia River, Minn., 335.
Torch Lake, Mich., 316.
Torch Lake Village, Mich., 317.
Townsend House, Utah, 344.
Towaliga Falls, Ga., 297.
Traverse City, Mich., 316.
Traverse Lake, Mich., 316.
Trenton Falls, N.Y., 195.
Trip-hammer Falls, N.Y., 201.
Trinidad Cave, 340.
Troy, N.Y., 194.
Truckee, Cal., 345.
Truckee House, Cal., 345.
Tuckerman's Ravine, N.H., 91.
Tumbling Shoals, Tenn., 306.
Tupper Lake (Big), N.Y., 181.
Tupper Lake (Little), N.Y., 181.
Tuscumbia, Ala., 306.
Twin Lakes, Minn., 334.
Twin Mountain House, N.H (cut), 58.
Umbagog Lake, Me., 92.
Union and Central Pacific Railroad, 340.
Union Springs, N.Y., 201.
University of Virginia (cut), 269.
United States Hotel (cut), 76.
Union Bridge, N H., 39.
Upper Bartlett, N.H., 87.
Up the Hudson, 130.
Utica and Black River Railroad, 196.
Utowana Lake, N.Y., 182.
Utica, N.Y., 195.
Valley Forge, Penn., (cut), 216.
Vallonia Springs, N.Y., 297.
Valhermosa Spring, Ala., 306.
Vermont, 120.
Verona Springs, N.Y., 198.
Vermont Springs, Vt., 124.
Venango River, Penn., 210.
Virginia, 268.
Virginia Springs, 271.
Vineyard Haven, Mass., 110.
Wachusett Mount, Mass., 121.
Wachusett Lake, Mass., 121.
Wagon Riding, 17.
Walden's Ridge, Tenn., 306.
Walker House, Wis., 323.
Wambec House, N.H., 55.
Warm Spring, Utah, 344.
Warm Springs, N.C., 296.
Warm Spring, Tenn., 305
Warm Springs, Va., 277, 270.
Warsaw River, Ga., 292.
Warren, N.H., 59.
Waukegan, Ill., 322.
Washington, D.C. (cut), 235.
Watch Hill, R.I., 128.
Wabash, Minn., 335.
Watkins Glen, N.Y., 201, 264.
Wawayandah Lake, N.Y., 205.
Wawayandah Mount, N.Y., 205.
Webster, N.H., 88.
Wells, Me., 66, 70, 73.
Wells River Junction, Vt., 51, 63.
West Burke, Vt., 63.

West Canada Creek, N.Y., 195.
West Jersey Railroad, 232.
West Hurley, N.Y., 137.
Western and Atlantic Railroad, 306.
West Point, N.Y., 135, 194.
Westkill, N.Y., 137.
West Ossipee, N.H., 70.
Western States, 314.
Western North Carolina Railroad, 295.
Weyer's Cave, Va., 274, 275.
White Bear Lake, Minn., 334.
White's Cave, Ky., 312.
Whitcomb'r Bowlder, Va, (cut), 286.
White Face Mount, N.H., 72.
White Face Mount, N.Y., 189.
Whitefield, N.H., 53.
White Head Cliff, Me. (cut), 77.
White Haven, Penn., 158.
White Mountains, N.H., 35.
White Mountain Notch, N H. (cut 90), 88.
White River Junction, Vt., 119.
White Cliff Spring, Tenn., 305.
White Sulphur Spring, N.Y., 198.
White Sulphur, Miss., 307.
White Sulphur, Ky., 309.
White Sulphur Spring, Tenn., 305.
White Sulphur Springs, Va., 271, 302.
Whiteside, N.C., 296.
White Water Cataracts, N.C., 297.
Weir's, N.H. (cut 42) 41.
Wilkes Barre, Penn., 185, 165, 229.
Willard Mt., N.H., 87.
Willey Mountain, N.H., 88.
Willey Notch, N.H., 88.
Williamsport, Penn., 145, 229, 260.
Willoughby Lake, Vt., 64.
Willoughby Mountain, Vt., 63.
Wilmington and Reading Railroad, 237.
Wilmington and Weldon Railroad, 289.
Wilmington, Charlotte, and Raleigh Railroad, 289.
Wilmington, Del., 237.
Wilmington, N C., 290.
Wilmington Notch, N.Y., 179.
Wilson's Springs, N.C., 295.
Windham, N.Y., 138.
Wing Road, N.H., 55, 56.
Winnebago Lake, Wis., 322.
Winnepesaukee Lake, N.H., 70.
Winslow House, N.H., 119.
Wisconsin, 25, 320.
Wissahickon River, Penn., 246.
Wolfboro' Branch Railroad, 70.
Wolfboro', N.H., 44, 70.
Wollaston, Mass., 98.
Woods Hole, Mass., 102, 109.
Wood's Hotel, N.Y., 174.
Woodstock, N.H., 49.
Woodsville, N.H., 51.
Wrightsville (Branch), 250, 257.
Wyoming House, 168.
Wyoming Ter., 340
Wyoming Valley Hotel, Penn., 163.
Wytheville, Va., 304.
Yankton, Dakota, 340
Yarmouth, E.P., 96.
Yates, N.Y., 198.
Yellowstone National Park, Mon., 341.
Yellow Sulphur Springs, Va., 303.
Yonkers, N.Y., 135.
York, Me., 66, 70.
York, Penn., 237.
Yo Semite Falls, Cal., 346.
Zurich Lake, Wis., 322, 326.

www.ingramcontent.com/pod-product-compliance
Lightning Source LLC
Chambersburg PA
CBHW020223240426
43672CB00006B/403